Zukunftstechnologien für den multifunktionalen Leichtbau

Series Editor
Open Hybrid LabFactory e.V.
Wolfsburg, Niedersachsen, Germany

Ziel der Buchreihe ist es, zentrale Zukunftsthemen und aktuelle Arbeiten aus dem Umfeld des Forschungscampus Open Hybrid LabFactory einer breiten Öffentlichkeit zugänglich zu machen. Es werden neue Denkansätze und Ergebnisse aus der Forschung zu Methoden und Technologien zur Auslegung und großserienfähigen Fertigung hybrider und multifunktionaler Strukturen vorgestellt. Insbesondere gehören neue Produktions- und Simulationsverfahren, aber auch Aspekte der Bauteilfunktionalisierung und Betrachtungen des integrierten Life-Cycle-Engineerings zu den Forschungsschwerpunkten des Forschungscampus und zum inhaltlichen Fokus dieser Buchreihe.

Die Buchreihe umfasst Publikationen aus den Bereichen des Engineerings, der Auslegung, Produktion und Prüfung materialhybrider Strukturen. Die Skalierbarkeit und zukünftige industrielle Großserienfähigkeit der Technologien und Methoden stehen im Vordergrund der Beiträge und sichern langfristige Fortschritte in der Fahrzeugentwicklung. Ebenfalls werden Ergebnisse und Berichte von Forschungsprojekten im Rahmen des durch das Bundesministerium für Bildung und Forschung geförderten Forschungscampus veröffentlicht und Proceedings von Fachtagungen und Konferenzen im Kontext der Open Hybrid LabFactory publiziert.

Die Bände dieser Reihe richten sich an Wissenschaftler aus der Material-, Produktions- und Mobilitätsforschung. Sie spricht Fachexperten der Branchen Technik, Anlangen- und Maschinenbau, Automobil & Fahrzeugbau sowie Werkstoffe & Werkstoffverarbeitung an. Der Leser profitiert von einem konsolidierten Angebot wissenschaftlicher Beiträge zur aktuellen Forschung zu hybriden und multifunktionalen Strukturen.

This book series presents key future topics and current work from the Open Hybrid LabFactory research campus funded by the Federal Ministry of Education and Research (BMBF) to a broad public. Discussing recent approaches and research findings based on methods and technologies for the design and large-scale production of hybrid and multifunctional structures, it highlights new production and simulation processes, as well as aspects of component functionalization and integrated life-cycle engineering.

The book series comprises publications from the fields of engineering, design, production and testing of material hybrid structures. The contributions focus on the scalability and future industrial mass production capability of the technologies and methods to ensure long-term advances in vehicle development. Furthermore, the series publishes reports on and the findings of research projects within the research campus, scientific papers as well as the proceedings of conferences in the context of the Open Hybrid LabFactory.

Intended for scientists and experts from the fields of materials, production and mobility research; technology, plant and mechanical engineering; automotive & vehicle construction; and materials & materials processing, the series showcases current research on hybrid and multifunctional structures.

More information about this series at http://www.springer.com/series/16103

Klaus Dröder · Thomas Vietor
Editors

Technologies for economic and functional lightweight design

Conference proceedings 2020

Springer Vieweg

Editors
Klaus Dröder
Institute of Machine Tools and Production Technology
Technische Universität Braunschweig
Braunschweig, Germany

Thomas Vietor
Institute for Engineering Design
Technische Universität Braunschweig
Braunschweig, Germany

ISSN 2524-4787 ISSN 2524-4795 (electronic)
Zukunftstechnologien für den multifunktionalen Leichtbau
ISBN 978-3-662-62923-9 ISBN 978-3-662-62924-6 (eBook)
https://doi.org/10.1007/978-3-662-62924-6

Responsible Editor: Eric Blaschke
This Springer Vieweg imprint is published by the registered company Springer-Verlag GmbH, DE part of Springer Nature.
The registered company address is: Heidelberger Platz 3, 14197 Berlin, Germany

Contents

Innovative and Smart Production

Life Cycle Assessment of Thermoplastic Hybrid Structures with Hollow Profiles

Alexander Liebsch(✉), Michael Müller-Pabel, Robert Kupfer, and Maik Gude

Institut Für Leichtbau Und Kunststofftechnik, Technische Universität Dresden, Holbeinstraße 2, 1307 Dresden, Germany
alexander.liebsch@tu-dresden.de

Abstract. The combination of innovative materials, process chins and intelligent material-adapted design concepts enables an efficient part production for future mobility applications. Hybrid structures made of thermoplastic pre-impregnated composite (TPC) sheets and injection moulding bulk material have already crossed the threshold to series production. Thanks to a newly developed production technology, hollow profiles with continuous fibres can now also be integrated into thermoplastic composite hybrid structures in the sense of a modular design system. However, the resource consumption of such hybrid structures has not yet been investigated.

This contribution describes the set-up and life cycle analysis (LCA) of a highly automated manufacturing process for the production of complex crash-loaded vehicle structures in thermoplastic composite design. The concept bases on the production of a hollow profile made of hybrid yarn, which is subsequently overmoulded and combined with a TPC sheet in an injection moulding process. For the hollow profile manufacturing, a novel automated preforming technology is used. The textile preform is then consolidated in a variothermal consolidation station, consisting of a temperature control system and an additively manufactured consolidation tool. For the subsequent overmoulding of the hollow profile, methods for stabilising the hollow profile were studied and implemented in an injection moulding complex. A plasma system for activating the surface of the hollow profile was used to create a permanent joint between the hollow profile and TPC sheet. At the example of a backrest with an integrated seat belt, the technology and its potential application in crash-relevant structures was proven.

In addition to the process set-up, manufacturing studies were carried out with the aim of evaluating the ecological potential of the process chain. For this purpose, a cradle-to-gate approach was chosen, in which the process-related energy consumption as well as the material consumption were measured and used for LCA. Thus, the most relevant process steps can be identified and possibilities for increasing efficiency can be derived.

Keywords: Thermoplastic composites · Hollow profiles · LCA · Hybrid structures

K. Dröder and T. Vietor (Eds.): *Technologies for economic and functional lightweight design*, Zukunftstechnologien für den multifunktionalen Leichtbau, pp. 3–16, 2021. https://doi.org/10.1007/978-3-662-62924-6_1

1 Introduction

In order to achieve a high degree of mass reduction and resource efficiency, the development and technological implementation of new design and manufacturing concepts using modern material systems is indispensable. Highly integrative multi-component designs with a combination of torsion and bending-resistant hollow profiles and flat components offer particularly promising possibilities for weight reduction. This effect can be even further increased by using fibre-reinforced composites. Due to typical cycle times of less than one minute, composites with a thermoplastic matrix are predestined for high-volume applications, e.g. in the automotive industry [1, 2]. While the combination of thermoplastic pre-impregnated composites (TPC) sheets and injection moulding compounds has already crossed the threshold to series production [3], an additional integration of TPC hollow profiles with continuous fibre course has not been realised so far.

A well-known approach for the production of complex shaped TPC hollow profiles uses braided hybrid yarns that combine thermoplastic and reinforcing fibres [4, 5]. Multiple studies focused on the use of braiding technology for the production of textile preforms, taking into account both thermoplastic [6] and thermoset matrices [7]. For near-net-shape preforms, direct braiding is used, in which the fibres are placed directly on a three-dimensional core. It allows a high variety of the fibre orientation, the cross section and the high mechanical properties [8]. However, manual and time-consuming process steps like changing the bobbins [9] or the fixation of new layers on top of the previous one lead to long cycle times. Alternatively, preforms can be assembled from layers of continuous hoses with constant cross-sections that are braided fully automated with high productivity [5, 10].

There are several technologies to consolidate TPC hollow profile made of hybrid yarn. The pultrusion process is a high productive method to manufacture TPC profiles [11]. Despite the high productivity, the process is limited to profiles with constant cross sections. Changing the cross sections is associated with a change of the pultrusion tool. Alternatively, the blow moulding process in combination with a variothermal mould can be used to consolidate TPC preforms with a high geometric complexity during short cycle times [5].

During the overmoulding of a TPC hollow profile, a high injection pressure acts on the profile, causing it to deform significantly or even collapse. Therefore, strategies to support the hollow profile are necessary. In [12] and [13] fluids are used to build up an internal pressure that stabilises the hollow profile during overmoulding. This approach sets high requirements on the hollow profiles as well as the injection moulding technology. On the one hand a media-tight hollow profile is needed to enclose the supporting structure. On the other hand special devices are needed to seal the profiles surface against the mould and to build up the internal supporting pressure [13]. Furthermore, only small sub-structures can be moulded on the profiles surface. As the internal pressure presses the hollow profile against the surface of the mould, it could deform inside the mould cavity, which could lead to a clogging of the flow paths or even damage the hollow profile due to excessive deformation [13]. Alternatively, fusible, soluble and mechanically demoldable cores can be used. However, these must be filled into the hollow profiles at great expense and removed afterwards [14–16]. Another proven approach is to use particles as supporting structures. In [14] it was shown, that the particle-based supporting structure behaves like a solid core if it is sufficiently compacted. Also, the requirements on the injection mould are limited as it is has only to be ensured that the particles are encapsulated inside the hollow profile.

Independent of the supporting strategy, the hollow profiles need to stay in a cold and stable condition to withstand the high injection moulding pressures. To ensure a reliable connection between the hollow profile and the bulk material, a pre-treatment is necessary. Flame [18], plasma [19, 20] or corona [21] surface pre-treatments are established physical methods to increase the surface energy of polymers and thus their ability to bond. Alternatively, surface micro-structuring by using laser technology is a promising approach to increase the bonding between cold TPC structures and injection moulding bulk material [22].

A comprehensive evaluation of different lightweight designs also requires the determination of the ecological effects of the production process. The high diversity of the different manufacturing options and the dynamic technology development lead to limited data availability in life cycle assessment (LCA) databases [23]. This applies particularly to upcoming technologies that have not yet been transferred from research to industrial use. For example, currently available LCA data on manufacturing of composite hollow profiles only cover pultrusion and resin transfer moulding processes [24]. Therefore, the calculation of environmental impacts in most cases bases only on few validated data. Furthermore, the various TPC processes show numerous interdependencies between the diverse process parameters, resource consumption and part quality [25]. General recommendations for energy efficient hybrid manufacturing processes are given in [26]. The influence of different process parameters of a tape placement with subsequent forming process is discussed in [27]. Following this, the main influence on the ecological impact lies in the processing speed. Though, the database for LCA of TPC is inadequate leading to limited forecasting accuracy when a specific process strategy is considered.

The present work describes the set-up of a continuous process chain for the production of a complex thermoplastic hybrid structure with hollow profiles. Due to the lack of information on the manufacturing-related resource consumption, a combination of detailed process description and life cycle inventory (LCI) is provided. Using a cradle-to-gate analysis, the energy consumption during the production of the lightweight structure is determined for one production scenario. Summarising, the ecologic potential of the design and manufacturing concept is discussed and necessary future work is derived.

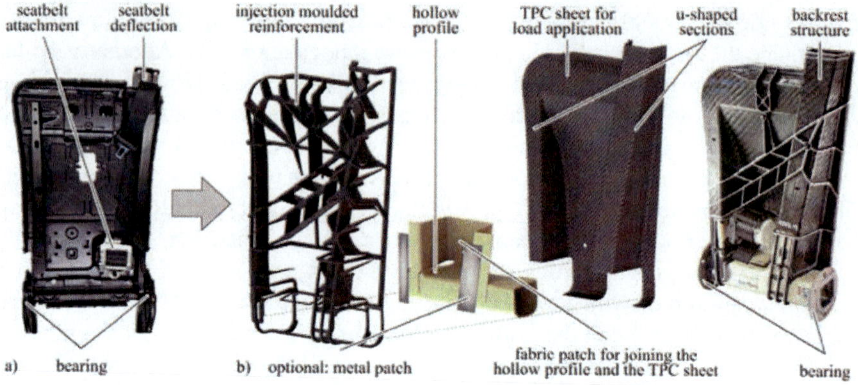

Fig. 1. Backrest structure: a) reference steel structure; b) thermoplastic composite design

2 Functionalised Thermoplastic Hollow Profile

The investigations base on the hybrid structure shown in Fig. 1, representing a car backrest with an integrated seatbelt (Brose). During a front crash, high loads are introduced into the seatbelt deflection. These loads have to be transferred in equal shares to the both bearings, where they are transferred as torsional moments to the peripheral devices. To fulfil this complex structural task, the composite structure is divided into three substructures. The bending load is transmitted by a TPC sheet with two u-shaped sections. The injection moulded rib structure supports the load distribution on the two bearings. The bending moment is introduced into the hollow profile made of glass-fibre (GF) polyamide (PA) hybrid-yarn and transmitted to the bearings as a torsional load. To join the TPC sheet with the hollow profile, a fabric patch is applied to the hollow profile during previous consolidation. The materials of the single components are summarised in Table 1. Polyamide is chosen as basis polymer to withstand the high structural loads also under elevated temperature conditions. GF reinforcement is preferred over the use of carbon fibres (CF) due to the high costs and resource consumption generated during current CF production processes [23, 24].

Table 1 Overview of used materials.

Semi-finished product	Supplier	Specification
TPC sheet	Lanxess	Tepex Dynalite102 RG 600 (2/2 twill weave, 600 g/m^2, 47% fibre volume ratio)
Injection moulding bulk material	EMS	Grivory GV-5H, PA66+PA6I/X (Long fibre reinforced, 50% fibre mass ratio)
Hybrid yarn	PHP	Enka TecTape (1800 tex, 67% glass fibre mass ratio)

3 Process Chain

The production of the backrest structure bases on a two-step process chain as shown in Fig. 2. In the first sub-process the hollow profile is manufactured. As the manufacturing focusses on quantity scenarios in the automotive industry, a cost- and energy efficient preforming of the hollow profiles are required. Therefore, the hollow profile preforms are made by processing of standardised continuous braided hoses that are automatically cut and assembled to the desired layer architecture. As the geometry of the hollow profile is simple, this approach is suitable to reduce the effort for the textile processing. In order achieve low processing times, the consolidation takes place in a variothermal mould. To functionalise the hollow profile and combine it with the TPC-sheet in the second sub-step, an injection moulding process is chosen as it can be fully automated and fits with the desired processing times. To support the hollow profile during the injection moulding process particles are chosen. It is known from previous studies that it is a robust supporting strategy with only modest demands on the injection moulding technology [17].

For a better understanding of the resource consumption of the whole process chain as well as the sub-processes are described in detail and the required materials, energy types and cut-offs are summarised.

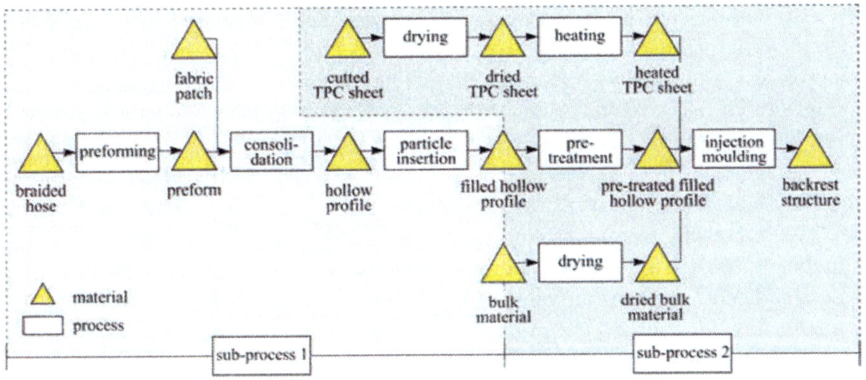

Fig. 2. Schematic process chain for the production of a functionalised thermoplastic hollow profile

3.1 Sub-process 1: Production of the Hollow Profile

The preforming concept uses rolled-up, continuously braided hoses that are cut into individual segments and stacked to the desired layer structure. By varying the length and position of the individual segments, specific local reinforcements in the layer structure can be realised. The central advantage of this preforming concept is the outsourcing of the production of the braided hose, preventing the part manufacturer from installing braiding technology. In contrast to direct braiding, process interruptions caused by changing braiding bobbins or fixing new layers are avoided. In a sequential process, the following steps are repeated for each textile layer (s. Fig. 3).

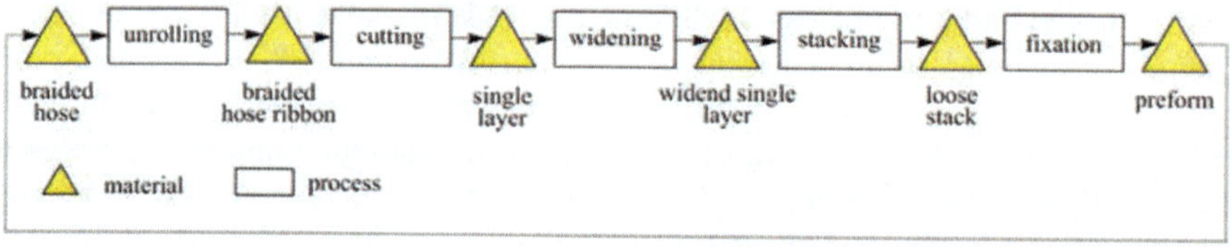

Fig. 3. Process chain of preforming [10]

First, the braided hose (Deutsche Institute für Textil- und Faserforschung Denkendorf) is unwound into a flat belt. Then a segment of a certain length is cut off using electric scissors [10]. This segment is picked up, expanded and pulled over a core structure or previous layer. In order to prevent the individual segments from slipping during the subsequent processing steps, they are fixed using infrared (IR) heaters [10]. To compensate manufacturing uncertainties, the preform is slightly longer than the final profile. This extra amount is cut off after the consolidation. Above that, only electricity is needed to run the preforming process. In addition, fabric patches consisting of the same hybrid yarn as the braided hoses (s. Fig. 1) are cut with a laser cutting unit.

The preforms and the fabric patch are consolidated and joined in a variothermal blow moulding process using a silicon bladder. It is necessary to heat the hybrid yarn to approx. 20 K above the melting point of the polymer in order to achieve a good impregnation of the glass fibres. Since the required thermal energy is introduced via the mould, the cavity must be heated from about 80 °C to 250 °C and cooled down again after impregnation. To enable fast heating-cooling cycles, a 3D-printed variothermal mould is used (Werkzeugbau Siegfried Hofmann) in combination with an oil tempering unit (GWK). By additive manufacturing, the tempering channel grid was placed approx. 2 mm below the cavity contour. Furthermore, the web-like backing structure of the mould significantly reduces the mass of the tool to be tempered [28]. In order to reduce the energy consumption of the process, the required amounts of hot and cool oil are stored in two reservoirs and both circles are separated from each other until the oil enters the mould. Figure 4 gives a schematic overview of the heating and cooling process gives.

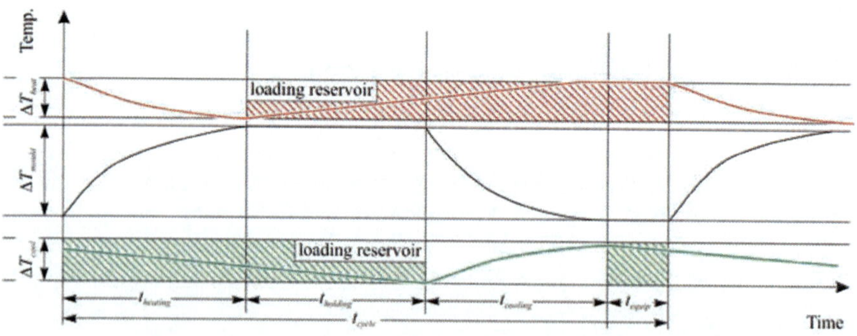

Fig. 4. Schematic heating and cooling cycle. (Source: GWK)

During the mould heating time ($t_{heating}$), the stored amount of heat in the reservoir (ΔT_{heat}) is discharged and the storage reservoir can be slowly recharged during the consolidation ($t_{holding}$), cooling and equipment period ($t_{cooling}$ and t_{equip}). The same functionality is realised during the cooling process using the cooling reservoir. The installed electrical capacity of a heating unit with storage reservoirs can thus be considerably lower as the time required to charge the storage reservoir is longer than the effective heating or cooling times. Required resources are electricity for the heating and cooling of the reservoirs and nitrogen for the inflation of the silicon bladder. The bladder has to be replaced after 15 cycles.

After consolidation and cutting off the edges (cf. preforming), supporting particles for the stabilisation of the hollow profile during injection moulding are filled into the hollow profile under periodic vibration. Since the durability of the particles is unknown it is assumed, that the particles can be used repeatedly. The particle-filled profiles are then stored until further processing. The materials and consumables in sub-process 1 are summarised in Fig. 5 and Table 2.

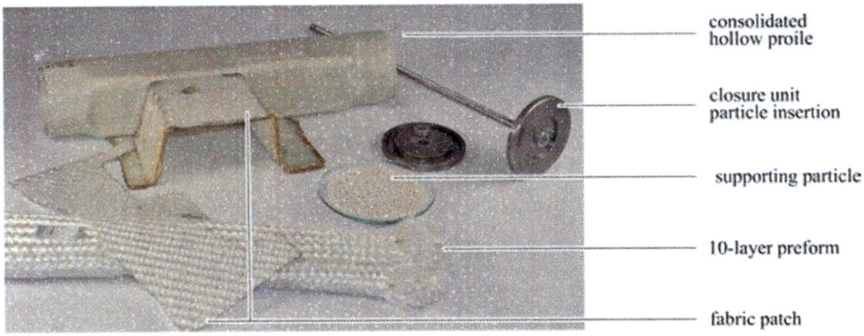

Fig. 5. Materials for sub-process 1

Table 2 Material and energy consumption for sub-process 1.

Process step	Material input	Material waste / consumables	Energy and media
Preforming	Braided hose	–	Electricity
Fabric patch generation	Fabric	Cut-off of fabric	Electricity
Consolidation	Preform, fabric patch	Cut-off of profile Silicon bladder	Electricity, Nitrogen
Particle insertion	–	–	Compressed air

3.2 Sub-process 2: Combination and Functionalisation via Injection Moulding

For overmoulding, the hollow profile needs to stay in a cold and stable condition. Its surface has to be pre-treated to enable a good bonding to the TPC sheet and the injection moulding bulk material. Basing on previous work [22], open-air plasma

(PLASMATREAT) is used as it is capable for inline use. To operate the plasma unit, only compressed air and electricity are needed.

In contrast to the hollow profile, the fabric patch can be heated above the melting temperature of the matrix to ensure an adhesive bonding with the TPC sheet during the overmoulding process. However, it is essential to keep the hollow profile in a cold condition. For this purpose, a combined heating and transfer device was designed (Fig. 6a) and mounted on the top linear robot system of the injection moulding machine. The required heating times are measured by applying temperature sensors on the backside of the patch (Fig. 6b).

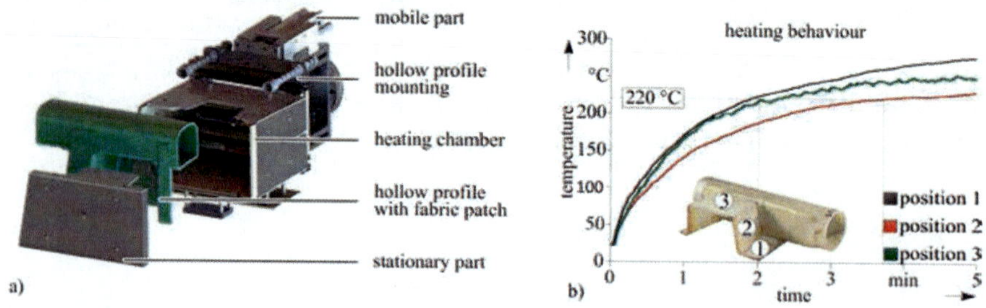

Fig. 6. Heating of the fabric patch: a) device for selective heating and b) measured heating behavior of hollow profile

The TPC sheet itself is trimmed via waterjet cutting according to an optimised cutting plan to produce minimal material. After drying for four hours at 80 °C (SOMOS T100) it is heated in an IR oven (KRELUS). 20 TPC sheets were dried in parallel.

After the heating phase, the hollow profile with the warmed patch is inserted into the injection mould (ELRINGKLINGER) by the linear robot system. Simultaneously, a draping gripper (ARBURG) on a 6-axis-robot transfers the heated TPC sheet into the mould. During transfer, the TPC sheet gets pre-shaped by the gripper to prevent unwanted distortions in the mould. After the TPC sheet is taken over by the mould, the robot moves back, the injection mould is closed and the overmoulding starts. The process times are shown in Fig. 7.

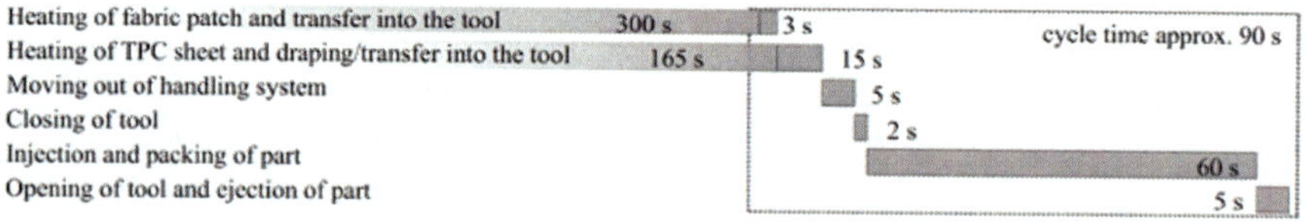

Fig. 7. Process times of overmoulding process

For the heating processes and injection moulding only electricity is needed. The bulk material was dried for four hours at 80 °C (SOMOS T100). During injection moulding no consumables are needed, leading to a waste free process step (hot runner and valve gate). Table 3 summarises the material and energy demand for sub-process 2. The build-up physical process chain is shown in Fig. 8.

Table 3 Material and energy consumption for the sub-process 2.

Process step	Material input	Material waste / consumables	Energy and media
Plasma pre-treatment	–	–	Electricity, compressed air
Heating of fabric patch	–	–	Electricity
Waterjet cutting of TPC sheet	TPC sheet	TPC cut-offs	Electricity, water with cutting compound
Drying TPC sheet	–	–	Electricity
Drying bulk material	–	–	Electricity
Heating of TPC sheet	–	–	Electricity
Injection moulding	Injection moulding bulk material		Electricity

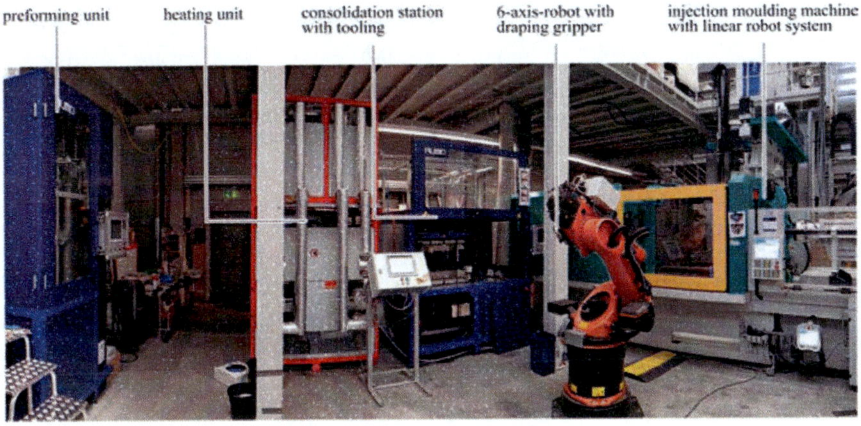

Fig. 8. Built-up process chain

4 Determination of the Ecological Potential

Based on the above described inputs and outputs of each process step, the system boundary for LCA is defined as cradle-to-gate. While the production of raw materials and semi-finished products is included by using values taken from literature, the gate-to-gate process steps given in Fig. 2 are analysed by in-process measurements as no suitable datasets are available. In order to incorporate the contribution of the raw materials and cut-offs, a mass balance is performed and combined with literature values for primary energy demand (PED) and mass specific emissions during production (Table 4). Reinforcement fibres and polymer material are evaluated independently except for the injection moulding bulk material. The textile processes of weaving and braiding are of minor importance according to the data published in [33] and [34]. The production of carbon fibre reinforced TPC was investigated in [34]. For the current investigation, it is assumed that the type of fibre does not affect the required PED.

In-process measurements were performed during manufacturing trials in order to generate a reliable data basis. A PCE-830 power analyser was used in combination with several couplings and adapters to cover the different ranges of amperage. The importance of changing process parameters was previously discussed in [35]. Here, only the results of a reference state are shown.

Table 4 Mass balance and input data for the cradle-to-gate-process [28–34].

Semi-finished product	Production step	Mass fraction	Material input in g	Mass in part in g	Cut-offs /%	PED in MJ/kg	Emissions in CO_2 eq./kg
Braided hose	PA6 prod.	0.33	279	222	20.62	128.80	6.70
	GF prod.	0.67	567	450		20.50	2.03
	Braiding	–	846	672		0.95	0.13
Fabric patch	PA6 prod.	0.33	62	38	39.46	128.80	6.70
	GF prod.	0.67	127	77		20.50	2.03
	Weaving	–	189	115		0.40[a]	0.05
TPC sheet	PA6 prod.	0.33	353	278	21.76	128.80	6.70
	GF prod.	0.67	727	568		20.50	2.03
	TPC prod.	–	1080	846		4.92	0.67
Injection moulding bulk material	PA6 prod.	0.50	545	545	–	118.04	7.20
	GF prod.	0.50	545	545			
	Compounding	–	1090	1090			
	Overall mass in g	–	3204	2723	–	–	–

[a]Convert from the PED data of a specific weaving surface presented in [33]

In Table 5 the part specific PED is given. The resource consumption generated by the use of compressed air and nitrogen is neglected due to the limited impact on the overall results. The thermal oil can be used over years and is thus not considered as a consumable. The silicon bladder has a weight of 100 g and must be replaced after 15 manufacturing cycles. The emissions resulting from silicone production are negligible (0.016 kg CO_2 eq./kg part) [29].

Table 5 PED values of the gate-to-gate processes measured during manufacturing.

Process step	Preforming	Consolidation	Pre-treatment	Drying of TPC and bulk material	IR heating	Robot handling	Injection moulding
PED in MJ/kg part	0.35	23.20	0.04	0.14	4.27	0.16	5.33

In order to calculate the global warming potential (GWP) the current electricity mix of Germany is assumed which corresponds to an emission factor of 0.135 kg CO_2 eq./MJ. The overall results of the calculated GWP are given in Fig. 9. As can be seen, the production of the materials and especially the injection moulding bulk material dominates the consumption of resources which is due to the high mass fraction of PA6 and the PED for compounding.

Regarding the gate-to-gate processes, the consolidation is of major importance for the GWP results. It should be noted that the process chain can be characterised as prototypic which gives reason to expect substantial improvements in case of industrialisation.

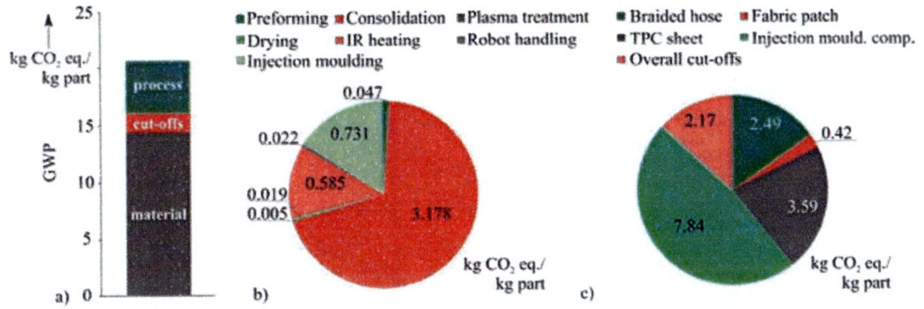

Fig. 9. GWP of backrest-structure: a) weight specific GWP, b) GWP contributions related to the gate-to-gate process steps and c) GWP contributions related to the semi-finished products and cut-offs.

5 Conclusion and Evaluation of the Process Efficiency

In this work, a two-stage manufacturing process for thermoplastic hybrid structures composed of composite hollow profiles, TPC sheets and injection moulded substructures is described and analysed regarding its resource consumption. In the first sub-process, hybrid-yarn based braided hoses are stacked to multi-layered preforms which are consolidated in a variothermal mould. Afterwards, the resulting hollow profile and the TPC sheet are combined in an overmoulding process. For life cycle inventory, each process step is discussed in detail and materials, energy consumption and consumables are given.

Based on the life cycle inventory, the global warming potential (GWP) is estimated for a single hybrid part. It shows clearly, that main influence on the environmental impact is the variothermal consolidation step. A big impact is also found in the usage of polyamide as matrix material having a high energy demand during its production.

As publically available LCA data for composite materials are limited to simple geometries like curved plates and straight profiles [23], a direct benchmark against competitive approaches is not feasible. According to [36], the weight specific GWP for production of hybrid metal-composite structures ranges from 5 to 13 kg CO_2 eq./kg part. Production technologies of further weight optimised structures based on CF material require 25 to 60 kg CO_2 eq./kg part [23]. Given the highly integrative design of the presented structure made of thermoplastic composites with a GWP of 21 kg CO_2 eq./kg part, the developed prototypic process chain represents a promising approach for industrialisation of complex and crash-relevant vehicle structures.

Future work will focus on the reduction of the energy demand during the consolidation process. Especially the energy demand of the variothermal tempering has to be reduced to decrease the environmental impact. A further decrease of the GWP could be achieved by transferring the presented technology to polypropylene as material system. Thereby, not only the needed processing temperatures would decrease but also the environmental impact during the polymerisation would decrease significantly.

Acknowledgements. The research projects are funded by the Federal Ministry of Education and Research (funding ref.: 02P14Z000_02P14Z010).

References

1. Kaufmann, R., Bider, T., Bürkle, E.: Lightweight parts with a thermoplastic matrix. Kunstst. Int. **03**, 65–68 (2011)
2. Kopp, J.W.: BMW I – automotive CFRP-production and potentials for thermoplastic FRP. In: Borgmann, F. (ed.) 2nd International Conference and Exhibition on Thermoplastic Composites ITHEC 2014, p. 11. Messe Bremen, Bremen (2014)
3. Götze, C.: From a Niche product to standard series production – highly resilient lightweight components produced from Organo sheet. Kunstst. Int. **03**, 35–38 (2016)
4. Bernet, M., Bourban, M.: An impregnation model for the consolidation of thermoplastic composite made from commingled yarns. J. Compos. Mater. **08**, 751–772 (1999)
5. Hufenbach, W., Adam, F., Krahl, M., Geller, S.: Ganzheitliche Lösungsstrategien bei der Entwicklung von Faserverbundkomponenten für automobile Leichtbauanwendungen. In: Kawalla, R. (ed.) 2nd International Conference on Advanced Metal Forming Processes in Automotive Industry AutoMetForm 2010, pp. 58–69. Institut für Metallformung, Freiberg (2010)
6. Küppers, S., Milwich, M., Gresser, G.T.: Energyefficient pultrusion process for producing fiber composite components with termoplastic matrix in series applications – Tpult. In: Hillmer, J. (ed.) 9th Aachen-Dresden International Textile Conference, p. 17. DWI – Leibniz-Institut für Interaktive Materialien e. V., Aachen (2015)
7. Gude, M., Lenz, F., Gruhl, A., Witschel, B., Ulbricht, A., Hufenbach, W.: Design and automated manufacturing of profiled composite driveshafts. Sci. Eng. Compos. Mater. **02**, 187–197 (2015)
8. Masselter, T., Haushahn, T., Schwager, H., Milwich, M., Nathanson, R., Gude, M., Cichy, F., Hufenbach, W., Neinhuis, C., Speck, T.: Biomimetic fibre-reinforced composites inspired by branched plant stems. Design Nat. V, 411–420 (2015)
9. Braley, M., Dingeldein, M.: Advancements in braided materials technology. In: 46th International SAMPE-Symposium, pp. 2445–2455 (2005)
10. Liebsch, A., Kupfer, R., Defranceski, A., Rösler, B., Janik, J., Gude, M.: Automated preforming of braided hoses made of thermoplast-glass fiber hybrid yarns. Procedia CIRP **66**, 57–61 (2017)

11. Schäfer, J., Gries, T.: Braiding pultrusion of thermoplastic composites. Advances in braiding technology – specialized techniques and applications. Woodhead Publishing Series in Textiles, pp. 405–428. (2016)
12. Engelmann, U., Layer, M., Albert, A., Reese, E., Landgrebe, D., Kroll, L., Nendel, W.: Towards the functionalized FRP tube – a merger of hydroforming and injection molding technologies is driving lightweight design. Kunstst. Int. **12**, 11–14 (2017)
13. Landgrebe, D., Kräusel, V., Rautenstrauch, A., Awiszus, B., Boll, J., Markov, L.: Energy-efficiency and robustness in a hybrid process of hydroforming and polymer injection moulding. Proceedia Manufact. **08**, 746–753 (2017)
14. Jelínek, P., Adámkov, E.: Lost cores for high-pressure die casting. Arch. Foundry Eng. **02**, 101–104 (2014)
15. Jiang, W., Dong, J., Lou, L., Liu, M., Hu, Z.: Preparation and properties of a novel water soluble core material. J. Mater. Sci. Technol. **03**, 270–275 (2010)
16. Michels, H., Bünck, M., Bührig-Polaczek, A.: Suitability of lost cores in rheocasting process. Trans. Nonferrous Met. Soc. China (English Edition) **20**, 948–953 (2010)
17. Liebsch, A., Andricevic, N., Maass, J., Geuther, M., Adam, F., Hufenbach, W., Gude, M.: Battery mount in a hybrid design – a combination of fiber-reinforced thermoplastic and aluminum replaces a steel part. Kunstst. Int. **09**, 46–49 (2015)
18. Pijpers, A.P., Meier, R.J.: Adhesion behaviour of polypropylenes after flame treatment determined by XPS(ESCA) spectral analysis. J. Electron Spectrosc. Relat. Phenom. **121**, 299–313 (2001)
19. Tusek, L., Nitschke, M., Werner, C., Stana-Kleinschek, K., Ribitsch, V.: Surface characterisation of NH3 plasma treated polyamide 6 foils. Colloids Surf. A **195**, 81–95 (2001)
20. Tahara, M., Cuong, N.K., Nakashima, Y.: Improvement in adhesion of polyethylene by glow-discharge plasma. Surf. Coat. Technol. **173–174**, 826–830 (2003)
21. Oosterom, R., Ahmed, T.J., Poulis, J.A., Bersee, H.E.N.: Adhesion performance of UHMWPE after different surface modification techniques. Med. Eng. Phys. **28**, 323–330 (2006)
22. Liebsch, A., Koshukow, W., Gebauer, J., Kupfer, R., Gude, M.: Overmoulding of consolidated fibre-reinforced thermoplastics – increasing the bonding strength by physical surface pre-treatments. Procedia CIRP **89**, 209–214 (2019)
23. Hohmann, A., Albrecht, S., Lindner, J.P., Voringer, B., Wehner, D., Drechsler, K., Leistner, P.: Resource efficiency and environmental impact of fiber reinforced plastic processing technologies. Prod. Eng. Res. Devel. **12**(3–4), 405–417 (2018)
24. Herrmann, C., Dewulf, W., Hauschild, M., Kaluza, A., Kara, S., Skerlos, S.: Life cycle engineering of lightweight structures. CIRP Ann. **67**(2), 651–672 (2018)
25. Großmann, K., Wiemer, H., Großmann, K.K.: Methods for modelling and analysing process chains for supporting the development of new technologies. Procedia Mater. Sci. **2**, 34–42 (2013)
26. Landgrebe, D., Kräusel, V., Rautenstrauch, A., Albert, A., Wertheim, R.: Energy-efficiency in a hybrid process of sheet metal forming and polymer injection moulding. Procedia CIRP **40**, 109–114 (2016)
27. Brecher, C., Schmitt, R., Lindner, F., Peters, T., Emonts, M., Böckmann, M.G.: Increasing cost and eco efficiency for selective tape placement and forming by adaptive process design. Procedia CIRP **57**, 769–774 (2016)
28. Beck, J., Kupfer, R., Hofmann, S., Liebsch, A., Gude, M.: Multi-Material-Design für die automobile Großserie - Faserverbund-Hohlprofile und Spritzgieß-Knotenstrukturen in einem Prozess. VDI-Bericht 2343: Plastics in Automotive Engineering PIAE EUROPE (2019)

29. Brandt, B., Kletzer, E., Pilz, H., Hadzhiyska, D., Seizov, P.: Silicon-chemistry carbon balance – an assessment of greenhouse gas emissions and reductions. CES – Silicones Eur. **05**, (2012)
30. N.N. Environmental Product Declaration: Polyamide 6 (PA6). PlasticsEurope, Product Group Engineering Polymers **02**, (2014)
31. Boustead, I.: Eco-profiles of the European plastics industry: glass filled polyamide 6. PlasticsEurope, product group engineering polymers **03**, (2005)
32. N.N. Life cycle assessment of CFGF – Continuous Filament Glass Fibre Products. PwC – Sustainable Performance and Strategy, **10**, (2016)
33. Stiller, H.: Material intensity of advanced composite materials – results of a study for the Verbundwerkstofflabor Bremen e. V. Wuppertal Papers Nr. 90 (1999)
34. Hohmann, A., Albrecht, S., Lindner, J.P., Wehner, D., Kugler, M., Prenzel, T., Pitschke, T., Seitz, M., Schüppel, D., Kreibe, T., von Reden, T.: Recommendations for resource efficient and environmentally responsible manufacturing of CFRP products. Carbon Composites e. V., Spitzencluster MAI Carbon (2017)
35. Müller, M., Liebsch, A., Kupfer, R., Stegelmann, M., Gude, M.: Process chain based data capture for a flexible and reliable life cycle inventory of a glass fiber-reinforced thermoplastic lightweight structure. Procedia CIRP **89**, 32–36 (2019)
36. Delogu, M., Zanchi, L., Dattilo, C.A., Pierini, M.: Innovative composites and hybrid materials for electric vehicles lightweight design in a sustainability perspective. Mater. Today Commun. **13**, 192–209 (2017)

Interdisciplinary Research for the Development and Realization of a Structural Component in Multi-Material Design Suitable for Mass Scale Production

Benjamin Bader[1](✉), Werner Berlin[2], and Michael Demes[2]

[1] Technische Universität Braunschweig, Institute for Engineering Design, Hermann-Blenk-Straße 42, 38108 Braunschweig, Germany
b.bader@tu-braunschweig.de

[2] Technische Universität Braunschweig, Institute of Machine Tools and Production Technology, Langer Kamp 19b, 38106 Braunschweig, Germany
{w.berlin,m.demes}@tu-braunschweig.de

Abstract. Lightweight design offers a reduction of local emissions and increase the range and handling dynamic of automobiles while driving. With a 40% share of the total vehicle mass, the body-in-white, consisting of high-strength steel materials, is a significant lever for weight reduction, but it is reaching its limits as the sheet thickness is further reduced, especially in crash-relevant structures such as A- or B-pillars, bumper cross beams e.g. (Friedrich in Leichtbau in der Fahrzeugtechnik, Springer, ATZ/MTZ-Fachbuch. Wiesbaden, 2017; Liu et al. in Int. J. Adv. Manuf. Technol. 69:211–223, 2013). An alternative approach are hybrid structures made of a combination of fibre-reinforced plastics (FRP) and metal sheets. The standard metal components can be improved by the targeted use of FRP in many ways. It enables a production of completely new components with increased performance and higher level on functional integration. This paper describes the transition of a lower A-pillar of a high-volume segment vehicle by showing the changeover from the monolithic steel design to the hybrid plastic-metal design. Therefore, the conceptual design of the component and its production are explained.

Keywords: Multi material lightweight design · Hybrid structures · A-pillar · Manufacturing process · Coated semi-finished sheet metal product

1 Introuction

Hybrid lightweight structural design is characterized in particular by a combination of materials that meets the requirements for fulfilling component specifications while taking the manufacturing processes into account. In addition to the lightweight construction aspect, advantages such as the integration of additional functions and/or the

K. Dröder and T. Vietor (Eds.): *Technologies for economic and functional lightweight design*, Zukunftstechnologien für den multifunktionalen Leichtbau, pp. 17–24, 2021. https://doi.org/10.1007/978-3-662-62924-6_2

reduction of production costs represent high benefits compared to conventional design methods. Also as an alternative to the monolithic use of FRP and light metal alloys from aerospace, hybrid lightweight structural design with FRP-metal hybrids (based on steel) expects justifiable expenditures in established process chains and costs of large-scale automotive production [3–6].

However, there is a need for research into the further development of virtual methods for securing components, joining technology and process technology for the manufacture of components for series application of FRP-metal hybrids. This is addressed by the cooperation project "Organically coated semi-finished sheet metal products for hybrid lightweight design applications" initiated in 2016 using the example of an A-pillar reinforcement. As part of the project, the companies Volkswagen AG, BASF SE, Thyssenkrupp Steel Europe AG and the Technische Universitaet Braunschweig at the Open Hybrid LabFactory (OHLF) research campus in Wolfsburg have jointly developed approaches to solutions, which are presented in the following sections.

2 Development and Production of a Hybrid Prototype Structure

An important basis for the development and implementation of hybrid lightweight structures on a large scale production is the cross-divisional cooperation of the divisions of product development, joining technology as well as manufacturing and process technology. While in the conventional product development process these instances are usually connected sequentially, hybrid lightweight design requires parallel intervention of the fields of competence due to the many potential improvements in the product and manufacturing process. Since optimizations using the design usually influence all instances, a holistic view of the product and process improvements is necessary. A conventional development process with a sequential sequence of instances is not suitable in all cases for this purpose, which means that the potential of the construction method cannot be fully exploited. Therefore, an iterative product development process is established for the successful development and implementation of the hybrid A-pillar. This enables each instance to influence the product status at almost every stage of development, from the duration of the development to the implementation of the component and to avoid conflicts during the development of the hybrid structure. The result of the iterative product development process is presented in the following. The development is based on a new type of semi-finished steel product that offers significant advantages for the implementation of hybrid lightweight structures suitable for large-scale production.

2.1 Coil-based Hybrid Semi-finished as a Basis for Manufacturing of Hybrid Parts

The basic material used is a steel-based semi-finished sheet metal product which is provided with a coupling layer in a coil coating process developed by thyssenkrupp Steel Europe AG to optimize material adhesion to thermoplastics. The application is carried out on a coil coating line suitable for large series production. In experimental investigations, the contact interface with the thermoplastic coupling layer shown in Fig. 1 shows a 40% higher bond strength between the semi-finished sheet metal and the FRP (PA6 matrix) under shear tensile stress compared with conventional powder- and foil-based adhesion promoter systems [7].

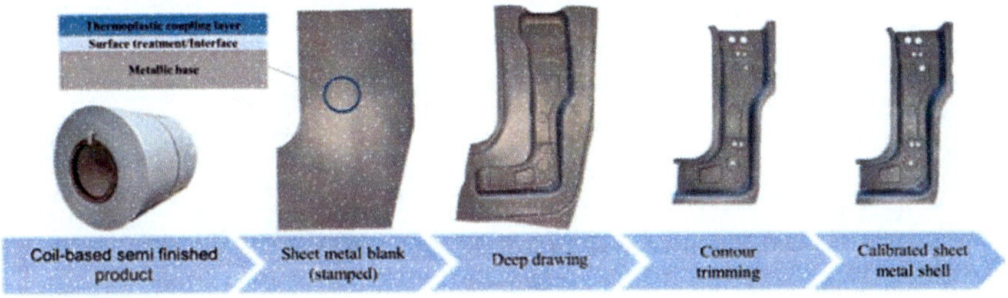

Fig. 1. From coil to calibrated sheet metal shell with organic pre-coating

For the forming steps of the organic pre-coated sheet metal blank shown in Fig. 1, no significant damage to the organic coupling layer can be detected in the process. As with conventionally used steel materials, there is no need for post-processing of the coating. Forming process steps in the processing of the FRP semi-finished products can bind necessary functional elements to the sheet metal component and save joints such as hole anchorages or undercuts.

2.2 Load-Path-Compatible Material Combination and Functional Integration Through Multi Material Design

The selected A-pillar acts as part of a body node and, in addition to the operating loads of the vehicle, is exposed to complex load collectives of the door hinges and overall body in case of a crash. In order to meet the component requirements, the conventional structure (monolithic steel construction) uses seven individual components made of sheet metal in the form of a welded assembly and two additional plastic components (expanded foam parts) to decouple airborne noise and separate the wet and dry areas of the cars body (Fig. 2, left).

The hybrid version of the A-pillar has the same metal sheet in terms of geometry. The steel where originally used is replaced by the organically pre-coated semi-finished metal sheet described above. The internal structure of the assembly has been significantly modified. The deep-drawn and forged parts originally used for structural reinforcement and connection of the door hinges are completely replaced by a newly designed plastic-metal insert. A continuous fiber-reinforced FRP in combination with a load-optimized short glass fiber reinforced ribbed structure made of thermoplastic materials are used to reinforce the metal sheet over its entire surface. In addition, a metallic thread insert is pressed into the semi-finished FRP product to reduce the setting effects when screwing attachments [8]. The design of the complex hybrid structure is based on an iterative optimization strategy consisting of simulation and topology optimization, which is supported by extensive tests on coupon specimen and component level.

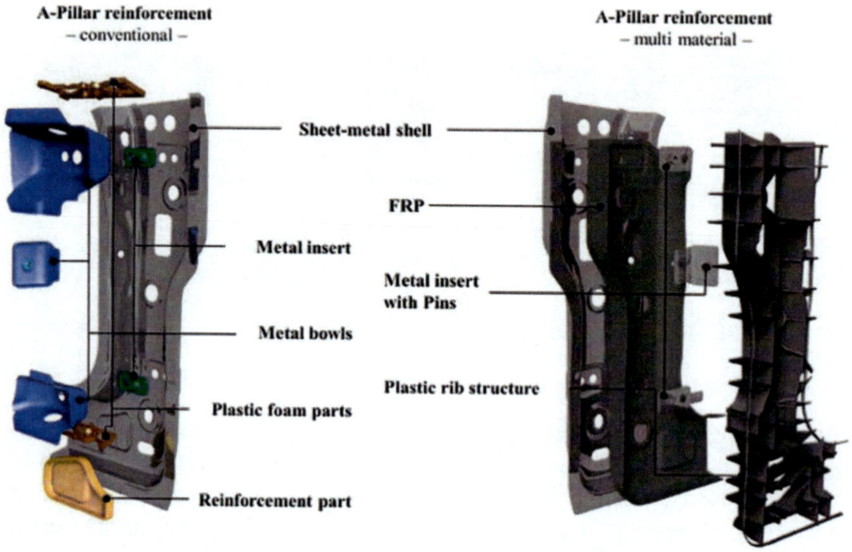

Fig. 2. Left: Structure of the conventional monolithic A-pillar reinforcement made of nine components; Right: Structure of the hybrid A-pillar reinforcement made of five components and injection molded rib application

As a result of the conversion of the assembly architecture from conventional sheet metal to hybrid design, a total weight reduction of the assembly by 20% is achieved with a simultaneous reduction of the individual components. The outer limits of the component are retained, so that it fits into joining sequences of the body shell in the same way as the conventional series part.

2.3 Plant Layout and Tool Structure for Investigation of a Fully Automated Manufacturing Process for Hybrid A-pillar Reinforcement Prototype

The investigation of the production of hybrid A-pillar reinforcements is carried out on the equipment in the OHLF's pilot plant (Fig. 3). The application of the FRP and the plastic ribs is executed using a mould with two functional units on a vertically closing injection moulding machine.

An infrared oven (IR oven) is used to heat the thermoplastic FRP. The required semi-finished products (sheet metal blank, thermoplastic FRP and pin inserts) are provided in defined layers via supply units (Fig. 3 left). A six-axis robot on a movable seventh axis is handling them, using a multifunctional gripper with four functional sides as an end effector. The gripper is thus able to take over all processing steps from inserting the materials into the tool and transferring them between the separate tools until the removal of the component at the end of the process. An infrared radiator (IR radiator) integrated into the gripper prevents the FRP from cooling down during transport from the furnace to the tool. The four-sided and 360° rotatable construction ensures that all tasks of the complex model process can be handled with only one robot and be reproduced with high speed and quality.

Fig. 3. Experimental Manufacturing cell of the hybrid A-pillar reinforcement in the pilot plant of Open Hybrid LabFactory e. V.

Figure 4 shows a schematic diagram of the two-stage mould in open condition. Both the fixed and the moving half of the mold consist of a basic structure with two functional units each. The basic tool structure is isothermally tempered to 40 °C and contains pneumatic as well as hydraulic cylinders for realization of inner mould movements (ejection, movements of slides and leading pins). The functional units are separately tempered isothermally and insulated from the basic structure. The units of the first mould stage show temperatures near the melting temperature of the used thermoplastic during operation. The forming and draping of a heated flat FRP into the previously placed, pre-coated semi-finished sheet metal is realized in this unit. The unit of the second tool stage is tempered for the processing of the plastic used in injection moulding. This forms the stiffening plastic ribs in the injection moulding process.

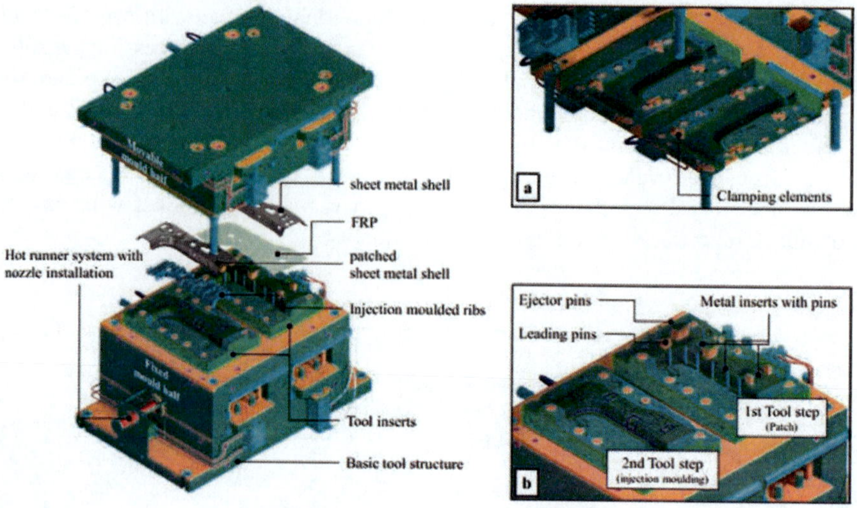

Fig. 4. Mold for the production of hybrid A-pillar reinforcements in the open state and exploded view of the associated products; a: movable mold half with view into the cavity; b: fixed mold half with view into the cavity

Due to the division into two tool stages, the tool can be classified as a conversion tool. In this way, it is possible to react specifically to the temperature requirements of the process in each tool stage. Compared to process control with variothermal and interchangeable tools, the conversion tool offers great potential for optimized production of hybrid structural components due to the use of both tool stages in parallel cycles [9, 10].

2.4 Process Sequence of the Hybrid A-pillar Reinforcement Prototype

The eight process steps for the production of the hybrid A-pillar reinforcement shown in Fig. 5 schematically represent the manufacturing process. In the initial state, the injection molding machine is open and the plasticizing unit is fully dosed. The control and component handling is extensively automated by a robot using a multifunctional gripper. The IR oven heats the FRP, meanwhile the multifunctional gripper is loaded with the pre-coated sheet metal and the pin inserts. After depositing the materials in the respective halves of the 1st tool stage, the semi-finished sheet metal and the pre-applied bonding agent layer heats up as a result of conduction through the tool. The multi-function gripper removes the heated FRP by means of a needle gripper and places it on the extended leaders. An IR radiator integrated in the gripper minimizes a temperature loss during transportation. When the tool is closed, the heated FRP connects to the pre-coated sheet metal. After thermoforming, the resulting material bond cools and the resulting patched semi-finished product (FRP-metal hybrid) is converted to obtain the application of the ribbed structure by injection moulding in the second stage.

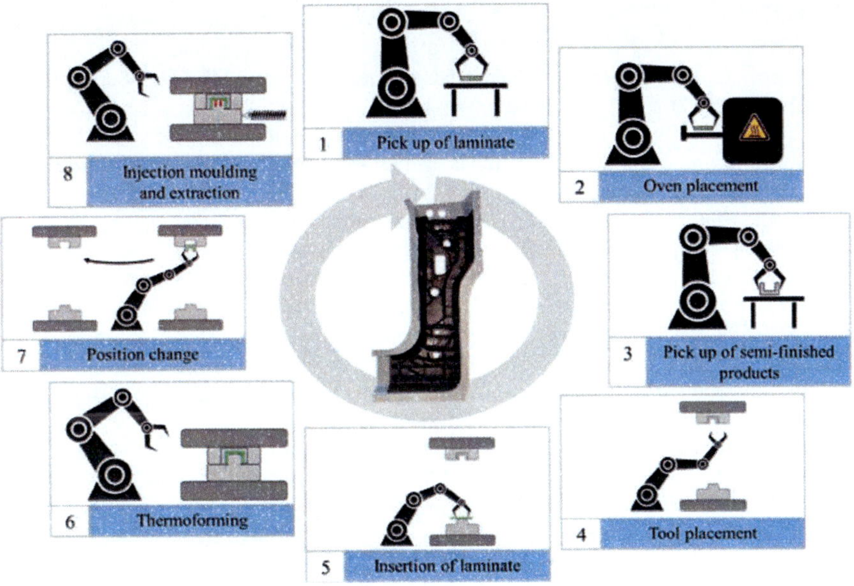

Fig. 5. Production cycle for the manufacturing of hybrid A-pillar reinforcement

3 Conclusion and Further Research

Due to the joint expertise of the project consortium, a hybrid structural component in the form of the A-pillar reinforcement of a Volkswagen Golf has been successfully investigated. Compared to the series production component in monolithic steel design, a 20% reduction in component weight was demonstrated while maintaining the same component performance. In addition, the number of individual components has been almost halved through function/component integration while maintaining the same functionality.

The process technology plays a key role in the implementation. The use of the organic pre-coated sheet metal can increase the bond strength compared to conventional bonding agents and does not show any significant damage after the forming process. The integrated manufacturing process allows the production of a hybrid A-pillar reinforcement in two directly successive process steps, thus eliminating the need for downstream assembly steps of the assembly. The investigated process can be transferred to series production by a simple modification of the periphery in the field of automation and heating technology as well as a reduction of the facility size used.

Acknowledgements. The authors would like to thank the institute directors Prof. Dr.-Ing. Thomas Vietor (IK), Prof. Dr.-Ing. Klaus Dröder (IWF) and Prof. Dr.-Ing. Klaus Dilger (ifs) as well as the partners of Volkswagen AG, Thyssenkrupp Steel Europe AG and BASF SE for supporting and financing the multilateral project "Use of organically coated semi-finished sheet metal products for multi material lightweight applications".

References

1. Friedrich, H.E.: Leichtbau in der Fahrzeugtechnik. ATZ/MTZ-Fachbuch. Springer, Wiesbaden (2017)
2. Liu, H., Lei, C., Xing, Z.: Cooling system of hot stamping of quenchable steel BR1500HS – Optimization and manufacturing methods. Int. J. Adv. Manuf. Technol. **69**, 211–223 (2013)
3. Drummer, D. (ed): Handbuch Kunststoff-Metall-Hybridtechnik. Erlangen (2015)
4. Kuhn, C., Klaiber, D., Altach, J.: Die Faser-Kunststoff-Metall Kombination, Kunststoffe, Ausgabe 3/2019, pp. 68–71. Hanser, München (2019)
5. Bader, B., Türck, E., Vietor, T.: Multi material design. A current overview of the used potential in automotive industries. In: Dröder K., Vietor T. (eds.) Technologies for economical and functional lightweight design. Zukunftstechnologien für den multifunktionalen Leichtbau, pp. 3–13. Springer Vieweg, Berlin (2019)
6. Kroll, L.: Technologiefusion für multifunktionale Leichtbaustrukturen. Springer, Berlin (2019)
7. Schongen, F., Patberg, L., Suenkel, R., Ferkel, H.: Coil-Based Hybrid Semi-Finished Product for Innovative, Firmly Bonded Structural Steel/Plastic Components, Conference Proceedings "Fazination hybrider Leichtbau 2018". Wolfsburg (2018)
8. Bader, B., Berlin, W., Demes, M., Dröder, K., Vietor, T.: Setzeffekte in Hybridbauteilen verhindern. Oberflächenstruktur sorgt für sichere Verankerung von Gewindeeinlegern in Organoblechen, Kunststoffe, 05/2020, pp. 58–61. Hanser, München (2020)
9. Berlin, W., Demes, M., Beuscher, J., Dröder, K.: Variothermie im Wandel der Zeit. Kunststoffe, 8, pp. 20–23. Hanser, München (2019)
10. Hopmann, C., Menges, G., Michaeli, W., Mohren, P.: Spritzgießwerkzeuge. Auslegung, Bau, Anwendung. Hanser, München (2017)

Net Shape Stacking and Consolidation of Thermoplastic Composite Tapes

Paul Zwicklhuber(✉) and Norbert Müller

Center for Lightweight Composite Technologies, ENGEL AUSTRIA GmbH,
Steyrer Str. 20, 4300 St. Valentin, Austria
{paul.zwicklhuber,norbert.mueller}@engel.at

Abstract. Thermoplastic composite tapes are produced with a specific width, e.g. 300 or 600 mm. A common production method is first to slice the tape rolls into narrow tapes bands with a width of e.g. 50 mm. These narrow tapes are used for stacking operations. Upon using such a technology, it is necessary to combine several narrow tapes to end up with a single layer. The outer edges of the tape stack normally extrude beyond of the area needed for the part, meaning, a cutting operation is necessary, either right after the stacking or after the consolidation. A beneficial alternative approach starts with large cutouts from the rolls, which are obtained by efficient stamping. These cutouts cover as much area as possible in one piece. The idea is to use the full tape width, if possible. The cutouts are stored in automatized magazines and become stacked efficiently employing a pick-&-place approach. The use of image processing enables precise adjustment of the gaps and overlaps.

This net-shape stacking technology leads to stacks that consist of a minimum number of tape cutouts. Even different types of tapes can be combined. The net-shape stacking is followed by an also net-shape consolidation in a heating-&-cooling unit. Both, the stacking unit as well as the consolidation unit are capable producing stacks and blanks within a typical injection moulding cycle time of one minute. The consolidated blanks fulfil the requirement of a best fit outer contour, which can be draped into the desired shape of the final part. Furthermore, the tailored composite blanks may expose the distinct variations in thickness and fibre orientation, which is optimized for the final part's specific loading conditions.

Keywords: Thermoplastic composite tape · Stacking · Consolidation · Image processing

1 Introduction

Thermoplastic composite tapes enable lightweight parts with superior performance. They are used, for example, for small robots, which perform part removal operations in injection moulding, Fig. 1.

K. Dröder and T. Vietor (Eds.): *Technologies for economic and functional lightweight design*, Zukunftstechnologien für den multifunktionalen Leichtbau, pp. 25–34, 2021. https://doi.org/10.1007/978-3-662-62924-6_3

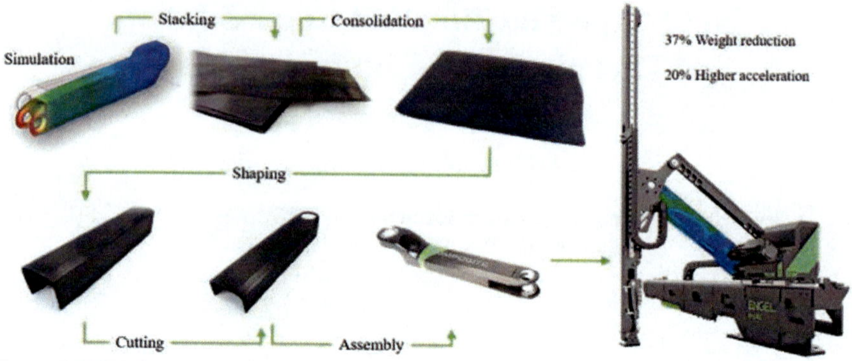

Fig. 1. Schematic representation of the production of a rotational axis for a small robot.

Due to the innovative manufacturing technique, a 37% weight reduction of the robot's rotational axis was achieved [1]. This leads to a rise of 20% of the robot's acceleration capabilities. Furthermore, the reduced weight yields energy saving in operation.

However, in order to reach from thin thermoplastic composite tapes to a final robot component, one needs to run through several steps. For the part, tapes with 0.14 mm of thickness, carbon fibres as reinforcement, and polyamide-6 as matrix are used. From the tapes, first stack that consists of 0°, 90° and ± 45° cutouts are produced. Extensive use of 0° layers alongside the parts length delivers pronounced stiffness in the necessary orientation. These stack needs to be consolidated into stiff and solid blanks with a thickness of 2.8 mm. In the next step, the thermoplastic composite blanks are heated above the melting temperature of the matrix material. In a shaping operation, a u-shape component is produced, which then is cut to the final outer dimensions. Two of such components are combined to a rotational axis of the robot by means of adhesive bonding. For the mechanical interface to adjacent components, machined aluminium parts are combined with the composite fraction.

The described approach - where also manual operations are necessary, especially with the handling and the assembly - can be used successfully for a limited number of parts, e.g. several hundred composite parts per year. The intention of the further development was to establish processing technology that is founded on the same conception as for the robot's rotational axis, but, in addition, enables fully automated production and the chance to utilize injection moulding for the functionalization. Then, the efficiency of the production of composite parts gets close to the levels known from injection moulding.

2 State of Research an Applications

Processing of prepregs and tapes with an epoxy resin as matrix is a well-established technology in the production of aerospace parts. The processes are not time-critical since curing of the epoxy matrix is activated after the stacking by means of autoclave processing. Switching from thermoset tapes to thermoplastic tapes is associated with benefits, but also with new challenges. Thermoplastics offer the chance to be processed significantly faster, since melting and solidification can be carried out in minutes. There are two general routes for the processing of thermoplastic tapes, either stacking with spot welding and subsequent consolidation or in-situ consolidation during the tape stacking, e.g. by means of laser of infrared irradiators as heat sources. With the in-situ consolidation, it is possible to manufacture tailored thermoplastic composite blanks without the need for post consolidation [2].

The tape stacking that starts with tapes of a specific constant width normally leads to a stack that extrudes beyond the outer contour which is necessary for the part. This means, a cutting operation usually is necessary after the stacking and consolidation. If wider tapes are used, then less tape strips are necessary for a specific stack, but the scrap from the following cutting is quite high. If narrower tapes are used, then the cutting scrap decreases, but the number of tape strips to be placed increases significantly [3]. A higher number of tape strips within a given tape stack increases the time necessary for the stacking. I.e., there are two competing optimization targets, reduction of number of tape cutouts in a stack and reduction of cutting scrap. With stacking concepts that are founding on tapes with constant width, on optimum is reached only when a multitude of the tapes width fits very well to the stack's width. However, only with few applications, this is the case.

Moving from thermoset tapes, with constant width, to thermoplastic tapes, again with constant width, was evolutionary development that did not question whether the constant-width-route is still the most appropriate way for the thermoplastic materials. There are arguments in favour of cutting or stamping of the tape-cutouts precisely to the net-shape directly from the roll: optimum use of the full tape width, minimization of scrap rate by optimum nesting, reduction of the number of tape cutouts in a specific stack, and no need for an additional cutting operations after stacking and consolidation. For actual applications, as for example an automotive centre tunnel, also stacks with a combination of glass- and carbon-fibre reinforced tapes are considered. A determining benefit of the laminate is that not only the fibre orientation can be chosen, but also different types of materials can be combined [4].

3 Investigations and Results

The primary objective of the development work was to establish manufacturing technology for the processing of thermoplastic composite tapes that supports both, high productivity and high precision of the stack. For that reason, a pick-&-place approach was chosen for the stacking and a heating-&-cooling approach was utilized for the consolidation. In order to achieve superior precision in the stacking operations, a high precision vision measuring system was employed. The presented system is distinguished by that fact that - on an industrial standard - it makes extensive use of vision control for the purpose of in-line optimization of the accuracy upon tape stacking.

3.1 Processing Sequence Employing Injection Moulding

For the processing of thermoplastic composite sheets, where woven fabrics are incorporated in a thermoplastic matrix material, the well-established approach is first to cut the sheet to the necessary outer shape. This is a net-shape outer contour, which fits very well into the cavity after draping. This means, after shaping, there is no need for a cutting operation at the three-dimensional composite shell.

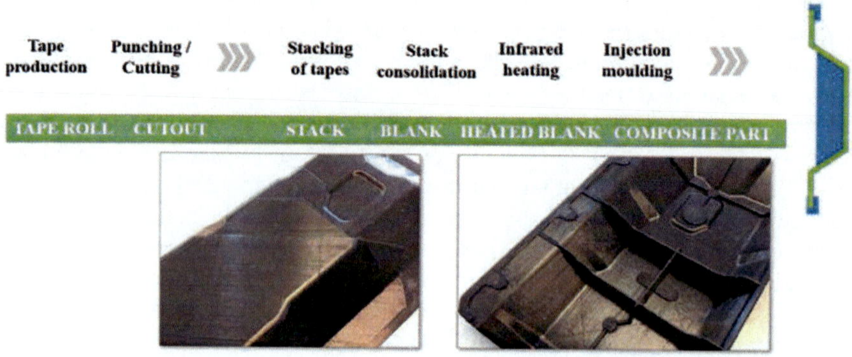

Fig. 2. Processing sequence for the production of parts from thermoplastic composite tapes with net-shape stacking and consolidation.

Consequently, the shaping and the subsequent injection moulding, which delivers the ribs and the geometric details, can be carried out in a one-step processing. The heated thermoplastic composite sheet is shaped, reconsolidated, and overmoulded in the same cavity of an injection moulding mould.

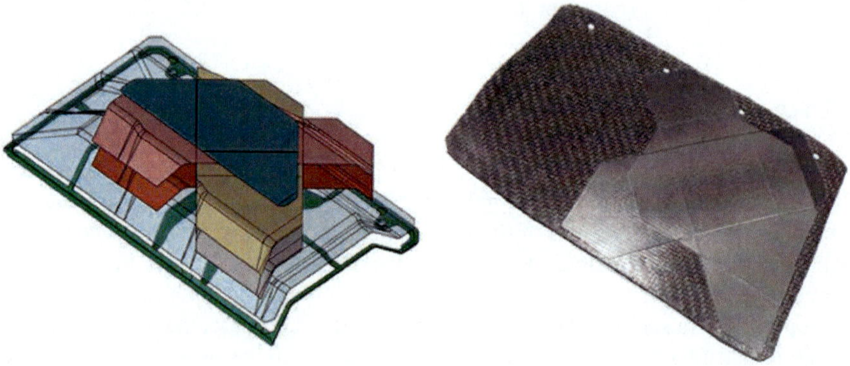

Fig. 3. Part for testing of materials and processing (left). Corresponding net-shape stack with base layer and tape reinforcement (right).

These benefits, with respect to efficiency, are to be transferred to a processing sequence for the production of parts from thermoplastic composite tapes, Fig. 2. However, with a switch from using thermoplastic composites sheet with woven fabrics to employing composite blanks made from continuous fiber composite tapes, as series of changes is associated [5]. An important aspect already is to start with tape cutouts as large as possible and with net contour, as it is actually needed for the product, Fig. 3.

3.2 Tape Stacking with Pick-&-Place

For the stacking, the tape cutouts are loaded to automated magazines, Fig. 4. Each magazine is capable of independently transferring a single tape from the stack in the magazine to a separate provision table. This is associated with the important benefit for the articulated robot, to pick the tape very quickly. The position and angular orientation of the tape is not that crucial at that point.

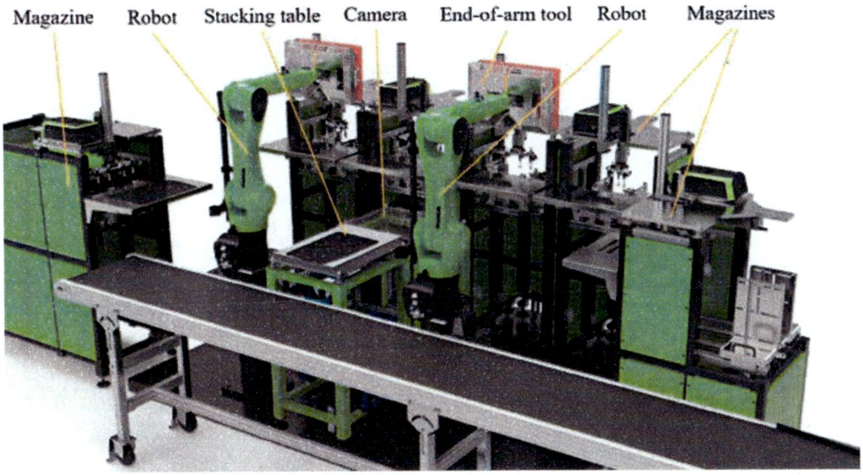

Fig. 4. Tape stacking unit for stack dimensions of 460 × 360 mm consisting of six automated magazines, two articulated robots with end-of-arm tools, camera system, stacking table, and conveyor belt.

The robot grabs the tape cutout with suction cups at its end-of-arm tool and transfers it to the glass platen of the camera system. The end-of-arm tool is equipped with an illumination for the improvement of the optical measurement. With the digital camera, that is placed below the glass platen, a high-resolution image is taken. Using an image processing software, information is gained on the actual location and angular orientation of important outer edges of the tape cutout. With that information, the position of a relevant corner point is calculated.

Whilst the robot is moving from the camera table to the stacking table, information from the vision measurement is supplied to the robot's control system. The placing and angular orientation of the tape is corrected in order to minimize the width of gaps and overlaps between tapes that are placed next to another, Fig. 5.

The stacking table is a vacuum platen. This ensures that the first layer is kept in place. The first layer can either be a thin thermoplastic composite tape, or a base layer, which might already have a higher thickness than a tape, e.g. 1.5 mm or 2.0 mm. The second and the subsequent layers usually are thermoplastic composite tapes with a thickness between 0.1 mm and 0.3 mm. Also, combinations from different materials can be processed, e.g. carbon fibre reinforced tapes on a base layer with a glass fabric inside.

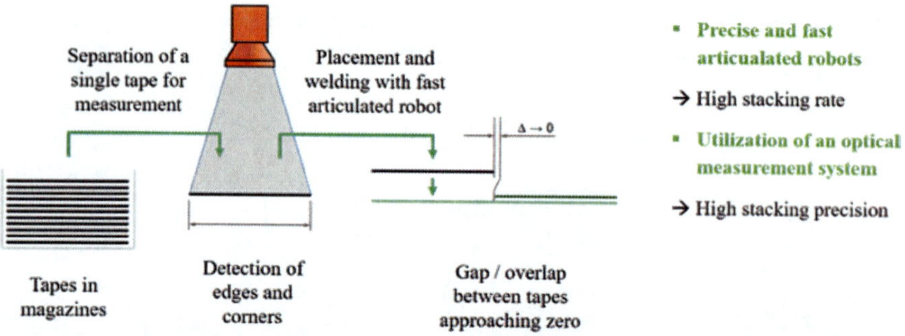

Fig. 5. Precise control of gap and overlap upon stacking of thermoplastic composite tape cutouts.

For the fixation, spot-welding with heated pins is employed. The tape cutout are spot-welded at least at two positions onto the layer below. This ensures that the stack stays as required with respect to the position and angular orientation of each individual tape cutout contained. The fixation is intended to maintain until the consolidation of the stack is finalized.

With the optical measurement system in service, it was proven that it is possible to achieve a stacking precision of less than 0.5 mm of gap or overlap between two tapes stacked adjacent to each other. A direct comparison between precision with and without the vision system in use not intended. This is because the precision of the position of the tape cutouts in the magazines and on the provision tables is not maintained. Since the optical measuring system makes corrections for each and any of the tape cutouts, it does not matter if the correction is just a few tenth of a millimetre, or several millimetres. I.e., the system is set up to be operated only in conjunction with the optical measuring device. It is the characterising feature that the system permanently heads for minimum gap and overlap width.

3.3 Consolidation with Heating-&-Cooling

The consolidation takes place between thin sheet metal moulds, Fig. 6. For each individual stack, the entire consolidation process takes several minutes, typically between 2 to 5 min. However, several moulds are in operation within the consideration unit. Therefore, the equipment is capable delivering a consolidated blank in less than one minute.

The mould with the stack inside enters the heating module and the stack is heated up above the material's melting temperature. This is done under low pressure. When the stack's centre plane is molten, the mould is transferred to the cooling module. There, the mould and the stack within are cooled under moderate pressure. The mould and the stack don't need to be cooled to room temperature. It is sufficient, when the temperature drops distinctly below the crystallization temperature of a semi-crystalline matrix material, or the solidification temperature of an amorphous matrix materials, resp. When this is achieved, the consolidated blank can already be removed from the mould. Since the mould is still at considerably high temperature even after removal of the consolidated blank, the next blank, which is loaded to that mould, automatically is preheated during the shuttling to the heating module, Fig. 7.

The determining factor for the cycle time of the consolidation unit usually is the duration of the heating. The cooling as well as the unloading, loading, and shuttling of the stack normally takes considerably less time. For a 2.8 mm thick stack, which consists of 20 layers of carbon fibre reinforced polyamide-6 tape, a duration for the heating of 25 s was measured. This is the time in the heating module for the centre plane of the stack to reach the melting temperature of 223 °C. This means, an overall duration of less than 45 s is reasonable, where also the necessary time for the unloading, loading, and shuttling is taken into account. The subsequent processing of the consolidated blanks in an injection moulding machine is performed in a similar fashion as the functionalization of organic sheet materials [6].

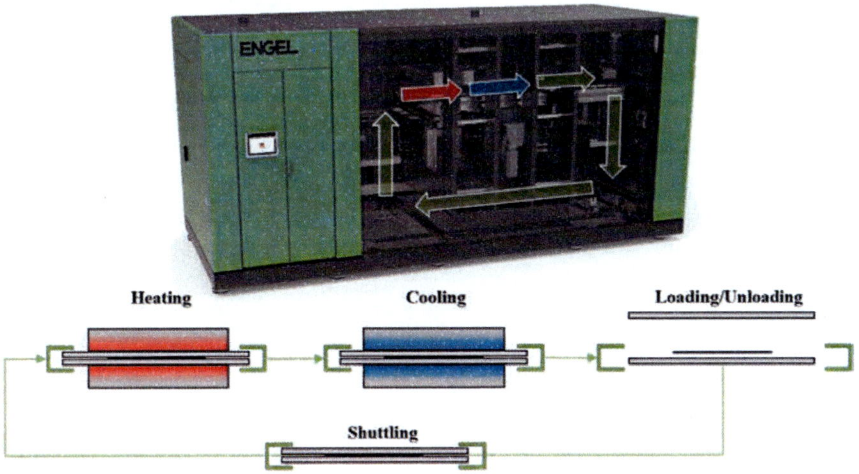

Fig. 6. Fully automated consolidation with heating and cooling of stacks from thermoplastic composites tapes between thin sheet metal moulds.

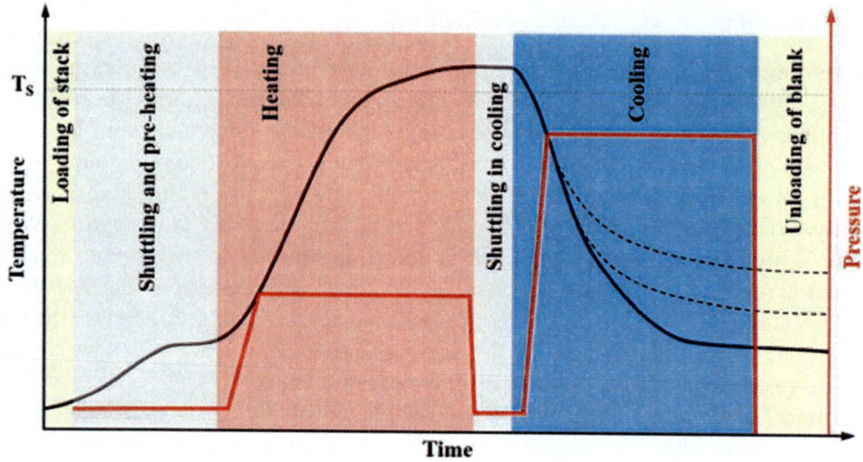

Fig. 7. Schematic representation of the course of temperature and pressure during fully automated consolidation.

The approach for the consolidation with heating-&-cooling between thin sheet metal moulds enables composite blanks that have intended thickness variations. Furthermore, the net-shape outer contour, which is already established during the stacking, is maintained during the consolidation. Since there is only very little pressure applied during the heating, whilst low to moderate pressure acts during the cooling (a maximum of 1.0 MPa usually is completely sufficient) only very limited effects of fibre displacement and disalignment are seen during the consolidation.

Despite of the low pressure acting, the technology delivers consolidated blanks with extraordinarily low remaining porosity. At adequate process settings, actually no remaining porosity is found between the tape layers amalgamated during the consideration. However, remaining porosity that is already existent in the tape is an issue. Whatever porosity already is present in the bulk of the tape, still can be found in the blank, i.e., after stacking and consolidation - at least partially. Therefore, the quality of the tapes, which are used for the stacking and the consolidation, is crucial for the remaining porosity of the final product.

Measurements with computer-tomographic means have shown that a remaining porosity of far below 1.0% can be achieved using the described approach. This is expected to be a fully satisfying performance characteristic for the majority of technical applications.

4 Conclusions and Outlook

The presented approach for the production of thermoplastic composite tape blanks using a pick-&-place stacking and a consolidation with heating-&-cooling delivers several benefits over alternative processing routes, Fig. 8. The utilization of large net-shape cutouts from the tape roll allow extraordinarily high output rates. The effect becomes as more predominant, as bigger the entire stack is. In case of a large stack, machinery that is using tapes with a specific width (e.g. 50 mm) requires numerous operations in order to finish a single layer. In contrast, with large net-shape cutouts and a pick-&-place technology, a single layer is finalized normally with just one to max. three stacking operations.

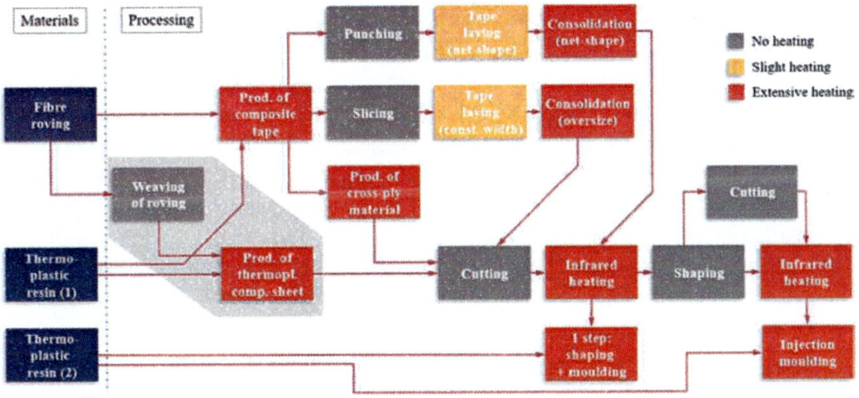

Fig. 8. Various processing routes for the production of composites parts based on thermoplastic composite tapes; in grey: processing route for composite parts based on sheets with woven fabrics inside.

Already during the design of the part and the stack, an optimum utilization of the tape roll with respect to resultant clippings is planned. E.g., the cutouts for the 45°-layers are nested as triangular shape, parallelogram shape, or trapezoidal contours in the tape roll, which leads to minimum scrap and most efficient use of the full tape width.

The machinery for the stacking as well as for the consolidation are designed to deliver final stacks and blanks within a typical cycle time of an injection moulding process. The intention is, to have the productivity of any module for the thermoplastic composite part in a similar range as is the case in injection moulding with a single cavity mould. This means, the tape stacking as well as the consolidation deliver as many stacks and blanks within a shift, as the injection moulding machine, equipped with an infrared oven, is able to shape and overmould in the same time. The existent lab equipment allows to perform a stacking operation in less than 4.0 s. Therefore, with a reasonable limitation of the number of cutouts contained in a stack, equal productivity in stacking, consolidation, and injection moulding can be achieved.

Since the stack is produced with net-shape outer contour, and this net-shape contour is maintained during the consolidation, no cutting operation is necessary at the finalized thermoplastic composite blank. After the consolidation, the blank might directly be heated in an infrared oven, and subsequently shaped as well as overmoulded in an injection moulding mould. A key requirement to force the way of (multi-material) lightweight constructions and design into automotive large-scale production is a continuous and automated process chain [7]. However, the fully interchained production, where the remaining heat in the blank from the consolidation is maintained and used, is an aspect of current investigations. The direct interconnection of the consolidation in the infrared heating will allow to take advantage of further energy saving potentials. This is of concern, not only from the cost viewpoint, but also for the minimization of the carbon footprint of the resultant composite part.

The current lab system for the tape stacking is capable of producing stacks with maximum dimensions of 460 × 360 mm. The consolidation unit is designed for stack dimension of 860 × 360 mm. Within the scope of the outlook of the ongoing R&D-work, the achieved performance characteristics are to be transferred to larger dimensions of the product. For the stacking, the next size is 1,100 × 600 mm of maximum outer dimension. Amongst the most important benefits of the system is, that is can make use of the full width of the tape roll. With the next size of the machinery, tapes rolls with 600 or 1,100 mm width, respectively, might be used to cover a layer in just a single stacking operation. This increases the output rate in terms of kilograms per hour significantly. Together with the precision, that is obtained from the utilization of the vision measuring system, the larger machinery is supposed to also end up with both, high productivity and, at the same time, outstanding accuracy. These have been and stay the paramont performance criteria.

Acknowledgements. The laboratory capabilities for the processing of thermoplastic composite tapes were established in close cooperation with FILL Gesellschaft m.b.H., Gurten (A) and the LIT-Factory at the Johannes Kepler University Linz, Linz (A). The authors gratefully acknowledge the contributions of the partners.

References

1. Müller, N., Egger, P., Zwickhuber, P., Steinbichler, G.: Layer-for-layer precision plus high performance. Kunstst. Int. 10/2016, 6–9 (2017)
2. Janssen, H., Peters, T., Brecher, C.: Efficient production of tailored structural thermoplastic composite parts by combining tape placement and 3D printing. In: 1st Cirp Conference on Composite Materials Part Manufacturing, pp. 91–95 (2017)
3. Kropka, M., Mühlbacher, M., Neumeyer, T., Altstädt, V.: From UD-tape to final part – a comprehensive approach towards thermoplastic composites. In: 1st Cirp Conference on Composite Materials Part Manufacturing, pp. 96–100 (2017)
4. Kuhn, C., Klaiber, D., Altach, J.: Lightweight design in the vehicle structure with the example of a center tunnel for the porsche boxster. Kunstst. Int. **4**, (2019)
5. Müller, N.: Manufacturing of load optimized structural parts from thermoplastic tapes with focus on flexibility and efficiency. In: 4th International Conference & Exhibition on Thermoplastic Composites (ITHEC), Conference Proceedings, Messe Bremen, Bremen (2018)
6. Müller, N.: Converging technologies - manufacturing of thermoplastic composites and injection molding of structural parts. In: 18th Automotive Composites Conference & Exhibition (SPE ACCE), Conference Proceedings, Society of Plastics Engineers, Novi, USA (2018)
7. Schnurr, R., Beuscher, J., Dietrich, F., Müller, A., Dröder, K.: Process design concept for automated pre-assembling of multi-material preforms. In: 21st International Conference on Composite Materials, Xian, China (2017)

Thermoset Technologies for Cost Efficient Production of Lightweight Composites

Lars Moser[1](✉), Sigrid Heide[2], Ian Swentek[3], Uwe Schmidt[4], and Manuel Seiz[4]

[1] Hexion GmbH, Varziner Str. 49, D-47138 Duisburg, Germany
lars.moser@hexion.com
[2] Hexion B.V., Seattleweg 17, 3195 ND Pernis, Netherlands
sigrid.terheide@hexion.com
[3] Hexion Inc., 180 East Broad Street, Columbus, OH 43215, USA
ian.swentek@hexion.com
[4] VOTTELER Lackfabrik GmbH & Co. KG, Schwieberdinger Strasse 97/102, 70825 Korntal-Münchingen, Germany
{u.schmidt,m.seiz}@votteler.com

Abstract. The automotive industry is evolving rapidly with increasingly rigorous emission targets and leaps toward electrification and autonomous driving. These forces continue to trigger lightweight solutions, advancing the adoption of novel composite materials for diverse applications. Composite materials have been used for body parts in sports and luxury vehicles for a long time, enabling design freedom, astonishing aesthetics and leading-edge driving performance. This extended abstract discusses state of the art composite manufacturing processes enabling cost-efficient high-quality parts production.

Keywords: Lightweight composites · Cost efficiency · HP-RTM · SMC · Epoxy · Phenolic · Class A · Fire resistance

1 Introduction

Thermoset composites are well suited for automotive lightweighting as their high specific strength and anisotropic nature enable high weight saving potentials, but traditionally, their use has mainly been limited to the premium segment. In order to make this technology available to other segments, Hexion has developed materials that address the needs for volume production, such as short cycle times, compatible mold release agents, robust processes, or superior performance. This abstract presents an overview of selected application examples where these new materials are the key enabler for high volume production.

K. Dröder and T. Vietor (Eds.): *Technologies for economic and functional lightweight design*, Zukunftstechnologien für den multifunktionalen Leichtbau, pp. 35–41, 2021. https://doi.org/10.1007/978-3-662-62924-6_4

2 HP-RTM Processing of Composite Leaf Springs

High Pressure-RTM technology has been developed to manufacture high-quality parts at short cycle times. The technology supports consistent production and provides flexibility to produce complex composite shapes that simply are not possible with conventional processes. The technology is very suitable for high volume production of composite leaf springs. Epoxy material systems have been tailored to address challenges in preforming and molding of thick laminates. Results show a high process output through short cycle times and elimination of post curing. Further process automation enables cost effective parts production.

Hexion's fast cure epoxy systems for suspension applications feature a unique set of properties which enables fast and robust processing. Opposed to typical cure times of 30 min or longer in prepreg or filament winding processes, this has been decreased to 5 to 7 min, depending on the part geometry, ensure high productivity. These fast-cure epoxy systems developed for HP-RTM processing exhibit thermo-latent behavior: the extended period of time for a mixed epoxy at the cure temperature to maintain a low viscosity. A large amount of resin is injected at low viscosity and after impregnation, rapid curing of a three-dimensional cross-linked network takes place. This allows large, multi-cavity tools to maintain the low effective curing time per part. Internal mold release agents guarantee good demolding behavior and make the per-shot application of external mold release agents unnecessary. Also, epoxy resin systems are insusceptible to humidity, which significantly contributes to robust processing. An example of an in-house molded, 30 mm thick laminate is presented in Fig. 1.

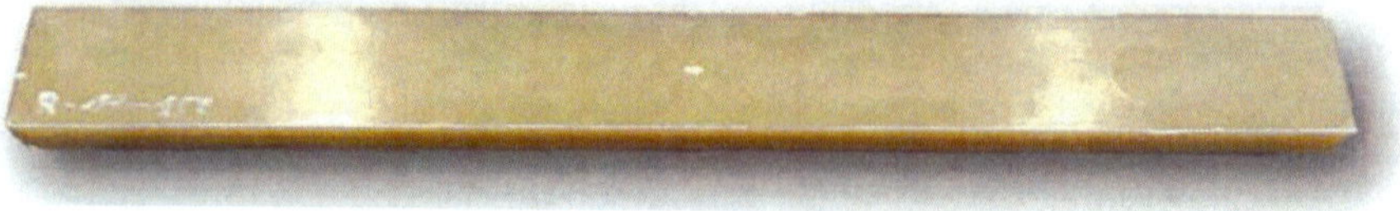

Fig. 1. Representative thick-sectioned HP-RTM leaf spring demonstrator.

Binder stabilized fabrics are easier to handle and position in the mold which is particularly important when large ply stacks need to be handled like in the manufacturing of leaf springs. Depending on the required level of fabric stabilization, Hexion offers both non-reactive and reactive epoxy binders fully compatible with fast-cure epoxy systems.

Additional benefits of epoxy systems are the absence of GADSL-listed materials, lower VOC, and long shelf life in the range of 2–3 years. Epoxy systems can be designed with a higher Tg for applications where the mechanical performance must be maintained under elevated temperature conditions. Table 1 shows the fast cure properties of the Hexion EPIKOTE Resin TRAC 06150 and EPIKURE Curing Agent TRAC 06150. The system allows for fast conversion out of the mold eliminating the need for post cure. In a typical production setting with a multi-cavity tool, this translates to per-part cure times well below one minute.

Table 1. Conversion of glass fiber reinforced Hexion TRAC 06150 epoxy resin in thick components (30 mm thickness, 58% fibers by volume).

Molding temperature [°C]	Cure time [s]	Degree of conversion [%]
105	600	98.5
120	300	99.2

3 In-mold Coating for High Volume Production of Class A Surfaces

In the production of Class A parts, the so-called fiber print-through effect most often forces part makers to apply expensive surface preparation and processing steps in order to achieve acceptable surface quality. This includes molding at lower temperatures to decrease the thermal shrinkage and limit the effects of different coefficients of thermal expansion of carbon fiber and epoxy resin. The resulting cycle times (typically 15 min or more) do not allow to apply this technology for volume models.

The new PuriCoat platform combines Hexion's fast-cure resin system suitable for High Pressure -RTM with Votteler's PU coating system applied by an in-mold coating (IMC) process. The part-to-part cycle time with this single-cavity one-shot process is only 3 min. A simple polishing step is sufficient to achieve an excellent Class A surface with minimal or without additional surface preparation or sanding enables a part cost reduction of up to 50% compared to parts produced using a traditional painting process. Overall, the system offers a lower release of VOC.

The development focused on ensuring that a durable interface is achieved between the carbon fiber-reinforced polymer (CFRP) part and the coating, while optimizing cycle time as well as surface quality. The surface topography and the yellowing tendency of the composite were also evaluated to assess the impact of climate, temperature and visible & UV light.

The effects of varying epoxy chemistries, internal release agents, process parameters (e.g. cure cycle) and surface preparation were evaluated. Optimal adhesion was reached utilizing the EPIKOTE Resin 06000, EPIKURE Curing agent 06130 and internal mold release agent HELOXY Additive TRAC 06805 system, combined with the VOTTER PURIFLOW® PU911IR system.

Despite the fact that both systems are self-releasing, it was possible to determine the factors that influence interface durability and to optimize the process to achieve consistently good results regarding surface quality and adhesion (see Fig. 2). In the development, especially light fastness (PV1303) was as critical test identified.

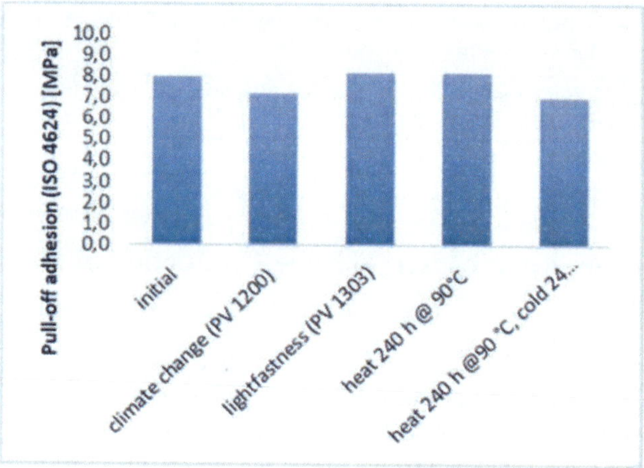

Fig. 2. PuriCoat interface durability.

4 Phenolic Sheet Molding Compound for Battery Enclosures

Another important industrial manufacturing constituent is sheet molding compound (SMC) due to its widespread use and high-throughput. Hexion has introduced a new phenolic system suitable for sheet molding compound processing. The basic components are EPONOL™ Resin TRAC 06921 paired with a suitable latent acid EPIKURE™ Catalyst TRAC 06921 and an internal mold release agent HELOXY™ TRAC 06951. Fillers are not required to achieve any of the mechanical or thermal performance but can be added to improve surface finish, color, and/or other functionality as needed.

Validation of the commercial formulation was conducted at the Fraunhofer Project Center. A series of increasingly larger batches were compounded to check the handle-ability of the material and the and maturation robustness. Without any changes to the equipment, and without any heating or other specialized functionality, the phenolic resin was able to compound 4800 tex glass and 50 k carbon fiber at relatively high basis weights of 5000 and 3000 g/m^2 respectively. In electric vehicle applications, the glass fiber is the more important design choice since it represents the lowest cost solution, though there are several niche applications for the carbon fiber variant. Unidirectional fiber variants were also run in order to understand the options of hybridized and reinforced designs. Much of the success of the compounding is due to the lack of fillers in the paste which enable the low viscosity resin to easily wet out high fiber content SMC (Fig. 3).

Fig. 3. Phenolic SMC being fabricated for scale-up testing

The compound was matured at room temperature for a period of 3–5 days until the complex viscosity of the compound reached at least 40 million centipoise via B-stage reaction. The ambient temperature during the first 12 h after compounding is the most important to minimize the initial exotherm and can be done by keeping the roll or festoon at 20 °C before completing the maturation cycle.

Panels were molded at 135 °C for 5 min, though the cure time was subsequently reduced to 2 min by tuning the formulation and molding parameters. Square plaques (457 mm^2, Fig. 4) of a variety of thicknesses and mold coverages were produced to check the mold flow within the tool, the cure speed, the surface finish, and document various molding parameters. A complex geometry battery box was also produced for demonstration and for customer testing. Mold coverages down to 40% were obtained with tuned process conditions.

Fig. 4. Sample phenolic/glass SMC plate molded at the Fraunhofer Project Center, dimensions 457 × 457 × 2.5 mm

While the phenolic resin formulation is designed to be user friendly, there are several important industrial considerations to use the material successfully:

- The SMC carrier film needs to be compatible with the resin system such that it will not disintegrate in contact with an acidic system and will release once the phenolic has properly matured.
- The mold surface needs also to be resistant to corrosion so as not to foul the tool.
- Quality control methods for the resin and compound will need to align to the unique properties of this formulated system.

The molded phenolic SMC plates were fully characterized and compared to current materials typically used in fire retardant applications. The tension testing follows ISO 527-4, the fiber content analysis came from ISO 7822 testing, and the impact analysis was conducted in accordance with ISO 6603-2. Panels were molded with charge patters designed to achieve uniform, planar isotropic properties. The two commercial incumbent materials also commonly seen in lower-performance battery protection applications are an ATH-filled (Aluminum TriHydrate) unsaturated polyester/vinylester resin blend and A356 cast aluminum (Table 2).

Table 2. Selected material properties of phenolic/glass SMC versus incumbent materials

Parameter	Unit	PF SMC	PF Prepreg	UPR/VER SMC w/ATH	A356 cast aluminum
Fiber weight fraction	%	56	60	41	N/A
Density	g/cm^3	1.78	1.85	1.8	2.7
Tensile strength	MPa	220	630	90	240
Tensile modulus	GPa	16	30	11	74
Impact energy absorption	kJ/m^2	68			45
Melting temperature	°C	N/A	N/A	N/A	600

The phenolic sheet-molding compound (SMC) technology achieves a 2-min cycle time and addresses the unique requirements in an electrified vehicle architecture. This drop-in replacement includes all the industrially relevant considerations including material processing, shelf life, and surface finish. A demonstrator battery cover highlights the superior fire resistance, impact resistance, and light weighting that is achieved with this technology. The material economically meets the most restrictive automotive fire performance standards such as ECE R100 and GB/T 31467.3.

5 Conclusion

New thermoset materials expand the applications of thermoset composites to volume models. The needs for short cycle times have been implemented and robust processes have been developed.

References

1. ter Heide, S., Moser, L., Swentek I.: Epoxy matrix technologies enable cost-efficient mass production of composite leaf springs. In: Society of Plastic Engineers Automotive Composites Conference and Exhibition. SPE ACCE 2017, Novi, Michigan, 2017
2. Friedrich, L., Defoor, F., Seiz, M., Kretzschmar, D.: A new platform for producing class A surface quality components. In: JEC Magazine N°121 May–June 2018
3. Swentek, I., Defoor, F., Ball, C.: Phenolic SMC for automotive fire resistance. In: Society of Plastic Engineers Automotive Composites Conference and Exhibition. SPE ACCE 2019, Novi, Michigan, 2019

Factories of the Future

Contribution to Digital Linked Development, Manufacturing and Quality Assurance Processes for Metal-Composite Lightweight Structures

Daniel R. Haider(✉), Fabian Folprecht, Johannes Gerritzen, Michael Krahl, Sebastian Spitzer, Andreas Hornig, Albert Langkamp, and Maik Gude

Institute of Lightweight Engineering and Polymer Technology (ILK), Technische Universität Dresden, Holbeinstr. 3, 01307 Dresden, Germany
{daniel.haider,fabian.folprecht,johannes.gerritzen, sebastian.spitzer,andreas.hornig,albert.langkamp, maik.gude}@tu-dresden.de, michael.krahl@tu-dreden.de

Abstract. More and more effort in development processes is required to meet the constantly increasing demands on technical components. Especially in hybrid structures, the high number of degrees of freedom in development due to the different material properties and interactions between them leads to complex and multi-disciplinary processes. Digitalisation is one option to meet the increasing demands for shorter development times and more efficient products. In addition, hybrid structures require a generally applicable procedure and guidelines for the design and preliminary dimensioning to support the engineering. A novel approach is presented to digitally link the development steps to create an interactive development process and a structure for a holistic data analysis. This is exemplary shown for an open, ribbed profile made of hybrid metal-composites in a module-based scheme. For a pre-dimensioning solution of the cross-section profile, analytical models, linked to adaptable numerical models, have been build up and transferred in the process model. Moreover, experimental validation concepts for the bending properties of a car body component is presented.

Keywords: Hybrid structures · Lightweight design · Development methods · Manufacturing technologies

K. Dröder and T. Vietor (Eds.): *Technologies for economic and functional lightweight design*, Zukunftstechnologien für den multifunktionalen Leichtbau, pp. 45–58, 2021. https://doi.org/10.1007/978-3-662-62924-6_5

1 Introduction

Due to their outstanding mechanical properties, fibre reinforced polymers (FRP) are becoming increasingly important for use in mobile systems. The combination of FRP with metallic components allows the realisation of new hybrid metal-composite structures (MCS), which offer advantages compared to classical solutions in terms of degree of function, design space and weight [1]. Innovative MCS are predestined for highly stressed structural elements such as body parts of cars [2]. On a prototype scale, such hybrid structures have been already designed, manufactured and tested successfully [1–4]. Some of the hybrid solutions were successfully transferred to application [5].

The transfer of MCS into industrial application is currently still challenging. The multitude of adjustable and interacting parameters in the areas of design, dimensioning, manufacturing and quality assurance leads to a complex and multi-disciplinary development, validation and manufacturing process [4]. To handle this, a clear and structured procedure is required, for which examples are given in some VDI standards. The VDI-guideline 2221 systematise a product development process for classical structures in main phases and have been established itself as a standard procedure [6]. For FRP in detail, VDI 2014 describes a structured procedure for design and dimensioning [7–9]. Furthermore, the importance of an interactive development process for FRP is shown in Helms [10]. Especially for FRP, Fig. 1 shows the interaction of the key steps "design", "dimensioning" and "manufacturing", which can hardly be realised with an iterative approach. In this way, production-related restrictions and validation steps are taken into account in an early stage of development. First design guidelines for ribbed profiles are given in Zhou for the combining of metal with injection moulding as well as in Müller and Faber for the combining of FRP with injection moulding [11–13]. Regarding a structured detailed development process, Lauter presents a module-based scheme for the engineering of hybrid structures to visualise the complex and interacting development procedure as a fundamental approach without discussing a detailed solution for the linking of the data [14]. A procedure for MCS, how the specific design steps can be linked to a uniform data level in the design process and models for pre-dimensioning, are not known. In this novel approach to obtain a digitally linked process model, the basic procedure is divided into three main phases.

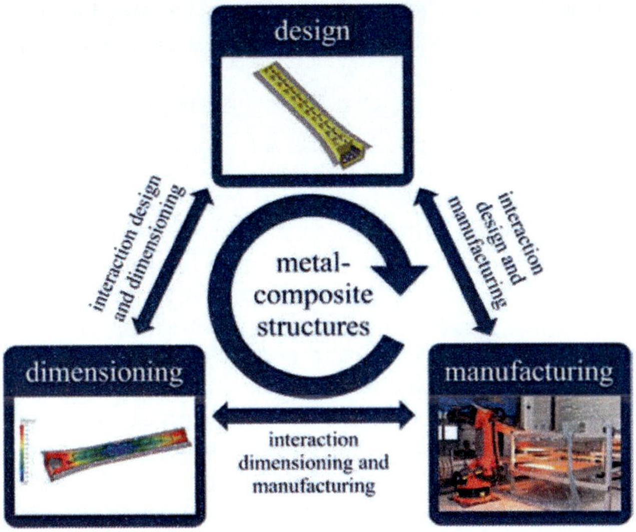

Fig. 1. Interaction of the engineering disciplines in the development procedure of a metal-composite structure according to Helms [10]

In the first phase a concept for a digital linked process model is developed (Fig. 2). The basic structure which consists of model, method and data level is based on FELDHUSEN and is combined with the methodology of VDI 2221. The process model is build up with the individual development steps for MCS. This procedure is shown for the pre-dimensioning as an example in chapter 3.

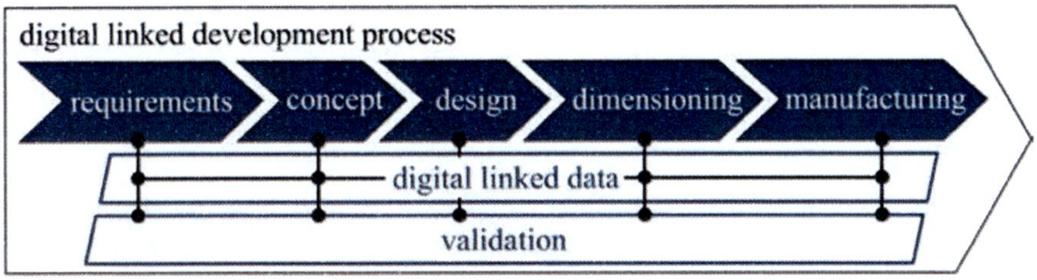

Fig. 2. Concept of digital linked development process

In a second phase, detailed concepts are developed on how to link the individual steps at the data level. Furthermore, the manufacturing process and validation should also be transferred into the basic structure of the process model and linked. In the third phase the developed solutions will be implemented to a digital linked development process. At the data level, the individual steps are linked to each other, creating interaction between them and enabling a holistic data analysis. This kind of digital linked development process is intended for support in development phase by structuring of relevant product data and provide engineering knowledge as well as support in integration of quality assurance by identifying and observing the most relevant parameters.

In addition to the concept for digital linking, models are also needed for dimensioning the components. Especially for open, ribbed profiles in MCS design, no analytical solutions for pre-dimensioning are available. A similar pre-dimensioning for open, ribbed profiles made of FRP but without metal profile has been already presented in Faber [13]. In this paper two approaches of preliminary design concepts are investigated for MCS. Thus a robust and efficient implementation of hybrid lightweight structures shall be enabled by an accompanying validation. In later phases of the design process, this approach allows the identification of critical parameters, which can strongly influence the function and the structural integrity of the component from the beginning of the development process. The linked development approach further allows the identification of quality management methods to monitor the identified key parameters. A first validation method for hybrid structures was already investigated in [15–17] as an online quality assurance system in component manufacturing, which enabled continuous data measurement of the production parameters in all process steps. These data were linked with the extracted data of a continuous virtual process chain in a database. They are used to increase the process understanding and as validation possibilities of simulation results. The current work for this digital linked approach in the early stages of development process is presented for the first phase of the linked development process.

2 Concept of an MMD-based Engineering Process Model

An efficient and robust development and manufacturing process can be supported by a deep understanding of the relevant parameters and their interaction with regard to the critical functions of the component. By measuring the time and effort spent for the different sub process steps the efficiency can be benchmarked. A well-structured and consistent process description is therefore the fundament. As part of the first phase for the digital linked process model a concept for the description of the development process is presented. Such a process description is as detailed as necessary to map all relevant development steps in which decisions with an impact to the product definition are made. Due to the repetition of development steps for similar structures such a development process description can be used as a template for subsequent product developments, whereby the development cycles can be supported in additional.

For the description of the development process an approach of Feldhusen and Grothe is used, that every process step consists of an executive method, an associated model and a set of data as shown in Fig. 3 [10]. Such a model represents an aspect of a physical issue with regard to the product information and can describe specific characteristics or the functional relationship of the parameters, a method is a systematic proceeding to reach a technical objective [18, 19]. Parameter values or sets are defined as data as base for information or knowledge.

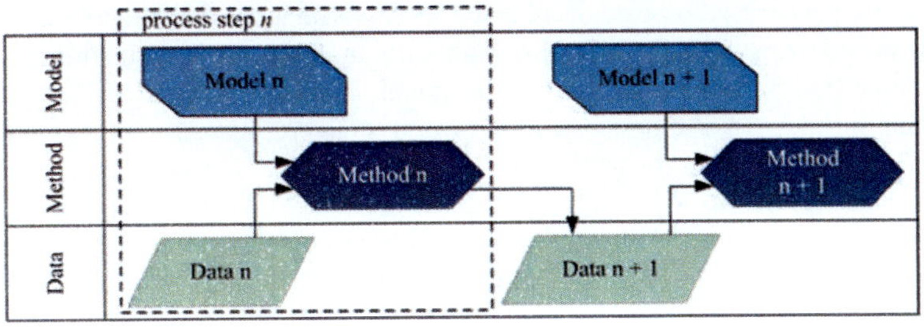

Fig. 3. Interaction of model, method and data in process steps according to Feldhusen [10]

This approach of method, model and data (MMD) is implemented in a process model for the consistent mapping of the development procedure. According to the development process, the model is structured with regard to the phases (PH) in VDI 2221, the phases consist of process steps (PS) in different detailing levels (L0+n) as shown in Fig. 4. Every step can be specified with a set of sub-steps, if a level of higher detailing is useful. In order to achieve an unambiguousness the steps are numbered in order of action with hierarchical numbered sub-steps. The highest detail level of the development plan is described by method (Me), model (Mo) and a set of data (Da). For allocation in nomenclature Me and Mo succeed the numeration of the associated process step. Data are not necessarily related to a single step and need an independent numeration. The understanding of the development process is supported by visualisation of the process model that is based on a standard for visualising workflows according to DIN 66001 and done in MS Visio 2016 [20]. There are shapes for the illustration of process steps as well as different kind of data described. Figure 4 shows in addition to the general shape for data shapes for methods and models.

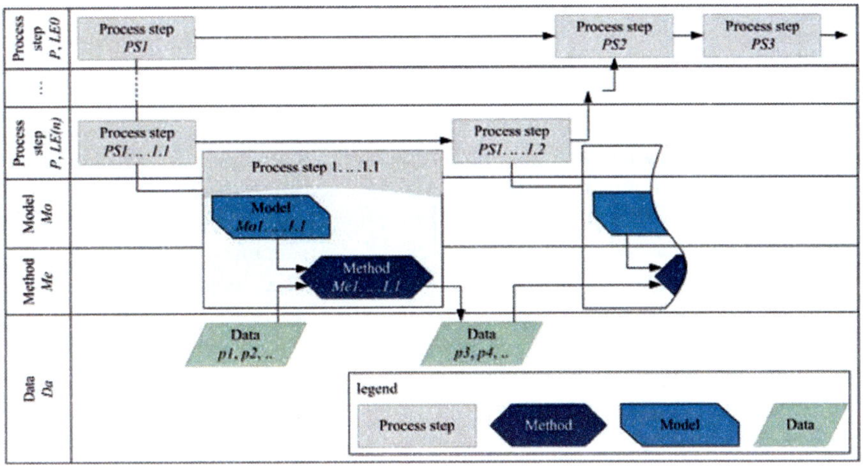

Fig. 4. MMD-based engineering process model as a map for the development process

Complex process steps can be realised either with complex models and simple methods or vice versa. The benefit of designing complex models with simple methods over running complex methods on simple models can be estimated by recording the time of an initial and an update run of a process step. Additionally, other step-related data are recorded, e.g. the ownership of the step, required experts or associated costs. A documentation of those parameters enables benchmarks for the development process. The models are stored in a catalogue with short description and intended use. Furthermore, an instruction for the method is stored in the same way, both catalogues are interlinked.

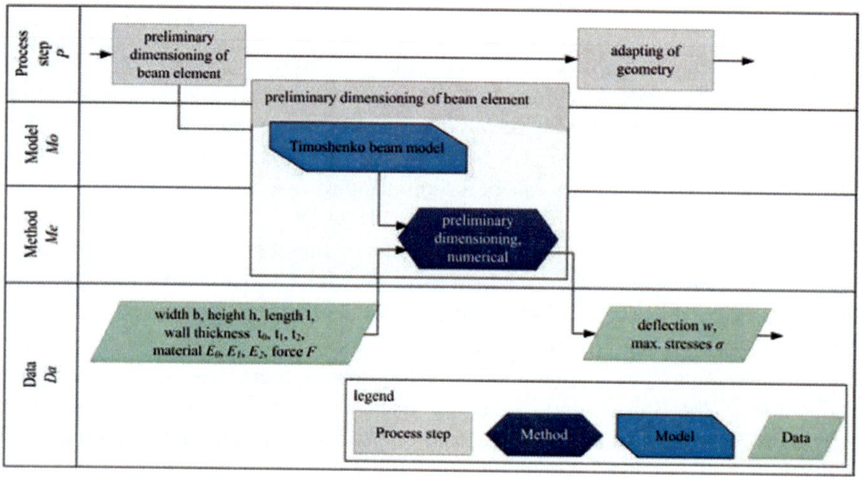

Fig. 5. Exemplary step “preliminary dimensioning of beam element” in the process model

The description of methods and models enables a flexible re-usage in other teams or projects and can support in quality assurance. In the development of hybrid structures the process model is exemplary shown on the preliminary dimensioning of a generic hybrid component as a representative sub-structure of a car body structure. In this step visualised in Fig. 5 the structure in 3D-Hybrid design [2, 4] is simplified as beam model in the material combination with high-strength steel, continuous fibre reinforced thermoplastic sheets and short fibre reinforced thermoplastic injection moulding compound. The model is based on a Timoshenko beam model and contains geometrical attributes like width b, height h and wall thickness t0, t1, t2 as well as properties of the different materials. Additional input data contain material parameters and boundary conditions like a load F. The associated method is a numerical dimensioning and includes different sub-steps for building up sub-models with parameter values from the input data. Additional sub-steps are to run these models and extract the parameter values of interest like deflection w and maximum stresses σ for the output data.

3 Exemplary Application of a Preliminary Design

In early stages of the design process, high fidelity methods in combination with robust and adaptable models enable an efficient study of the effect of parameter changes. In this way, it is possible to estimate the sensitivities to a geometry change for open, ribbed profiles and connect this data with the design process. Two approaches of preliminary design concepts are investigated and subsequently compared to a high-fidelity method in terms of prediction accuracy, sensitivity and adaptability. As preparation, the structure is abstracted to the model shown in Fig. 6 with two distinct cross-sections. The load case is adapted accordingly.

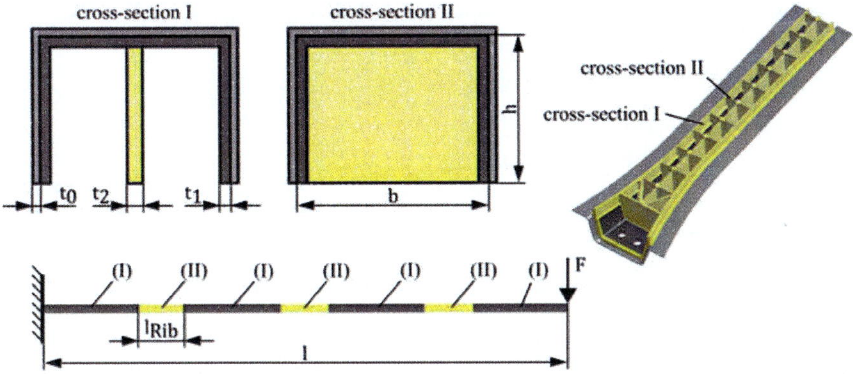

Fig. 6. Simplified model of the generic hybrid metal-composite structure as a beam element

For the first approach, a beam model in ABAQUS is chosen. Given the hybrid nature of the cross-section, preceding analyses of the occurring cross-sections, shown in Fig. 6, is required. In order to ensure adaptability, the entire build-up of the model is scripted, taking the geometric parameters b, h, t_0, t_1, t_2 as well as the material parameters $E_0, E_1, E_2, Mat_0, Mat_1$ as input parameters and returning an input file for analysis. For the analysis the option "*BEAM SECTION GENERATE" is used, which leads to the effective stiffness properties of the analysed profile to be calculated and written to a file, which can be used in future analyses [21]. This procedure is carried out for both cross-sections, whereby the script gathers from the input parameters if the profile is a transverse rib and adjusts the model build-up accordingly.

To enable the consideration of a variable number of transverse ribs along the length, the beam model is build-up by a second script. This script takes the total length of the beam l, number of transvers ribs n_{Ribs}, and thickness of these ribs l_{Rib} as input parameters and returns an input-file, ready to be solved. By running the script the previously written files are linked, assigning them to the respective region. The generated input-file is submitted to Abaqus/Standard for numerical analysis. The results for an example with the parameter values in Table 1 are shown in Fig. 7. This approach is suitable to predict the structure's deflection. For comparison with the other methods, the deflection at the loading point, which amounts to 1.11 mm, is chosen.

Table 1. Exemplary parameter set for the generic structure

Geometric parameter	b	h	t_0	t_1	t_2	l	l_{Rib}
Value in mm	37	28.5	1.5	2	3	300	3
Further parameter	E_0	E_1	E_2	F			
Value	210,000 MPa	22,000 MPa	11,000 MPa	410 N			

Since the profile was not explicitly part of the second simulation, the cross-section cannot be displayed directly within its results. It is rather necessary to transfer deflection and deformation results to the result of the cross section itself. This can be achieved by using the ABAQUS internal script "compositeBeam.py". The longitudinal stress at the clamping point for the example beam is shown in Fig. 7. For further comparison, the respective maxima and minima in each of the materials are chosen. The determined values for the example are given in Table 2.

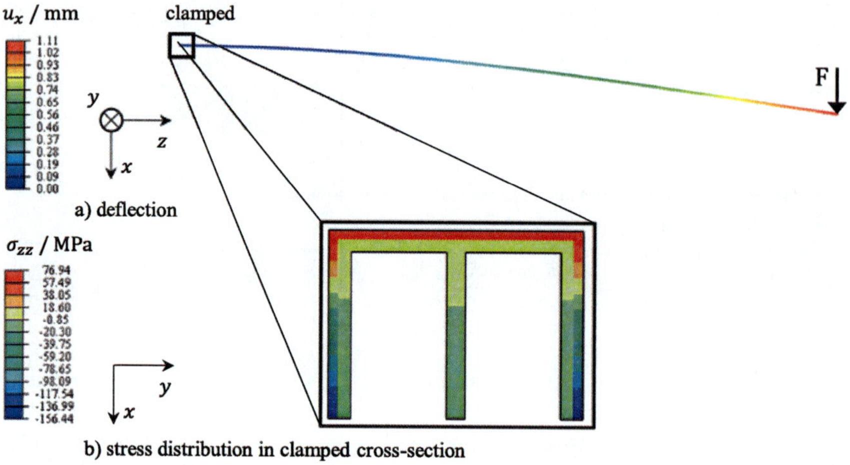

Fig. 7. Results from beam analysis in ABAQUS

A stress and deformation analysis using a finite element (FE) approach with a beam model for this hybrid structure takes several minutes. This is mainly due to the number of sub-steps, which partly have to be carried out manually every time.

Table 2. Stress extrema at the clamping point

Stress component	$\sigma^+_{zz,0}$/MPa	$\sigma^+_{zz,1}$/MPa	$\sigma^+_{zz,2}$/MPa	$\sigma^-_{zz,0}$/MPa	$\sigma^-_{zz,1}$/MPa	$\sigma^-_{zz,2}$/MPa
Abaqus Beam	76.94	6.84	2.60	−156.44	−16.39	−8.19
Analytical	77.07	6.85	2.61	−156.70	−16.42	−8.21

As alternative to the numerical approach, an analytical one is derived. A beam theory, based on the works of TIMOSHENKO, applicable to the multi-material-beam is proposed. The total deflection at the loading point w_F as a result of the applied force F and the resulting moment M_b can be found as

$$w_F = \sum_{i=1}^{m} \int_{l_i} \underbrace{\frac{M_{bi}}{(EI)_i} \frac{\partial M_{bi}}{\partial F}}_{Bending} + \underbrace{\frac{F_Q}{(GA)_i} \frac{\partial F_Q}{\partial F}}_{Shear\ Force} dz_i. \tag{1}$$

Equation 1 further contains the decomposition of the beam into m sections with a respective length l_i, equivalent bending stiffness $(EI)_i$ and equivalent shear stiffness $(GA)_i$ [22]. It must be noted that $(GA)_i$ does not contain additional information on the distribution of shear stress across the cross-section and the resulting influences on the deformation resistance. Such effects are neglected since the shear deformation ratio is small compared to the bending ratio. The respective stiffnesses are determined by

$$EI = \int_A E(x, y)y^2 dA \quad \text{and} \tag{2}$$

$$GA = \int_A G(x, y)dA \tag{3}$$

for both cross-sections. The stress distributions at a given point along the beam can be obtained by

$$\sigma_{zz}(x, y, z) = \frac{M_b(z)}{(EI(z))} \cdot E(x, y) \cdot y \quad \text{and} \tag{4}$$

$$\tau_{xz}(x, y, z) = \frac{F}{(GA(z))} \cdot G(x, y). \tag{5}$$

Substituting the values from Table 1 into (4) and (5) yields the respective extrema given in Table 2. Given the assumptions of TIMOSHENKO beam theory, shear deformation is constant across the cross-section. This leads to a more conservative design since maximum values of normal and shear stress coincide, which they actually do not [23, 24]. Furthermore, the analytical approach enables direct determination of sensitivities to determine the influence of the parameter to the properties of the structure. Two examples are given in Table 3 for the sensitivities of deflection and maximum tensile stress in Mat_1 on the cross-section parameters.

Table 3. Sensitivity of structural reactions on cross-section parameters (5 ribs)

Geometric parameter	b	h	t_0	t_1	t_2
Sensitivity w_F in mm/mm	−0.01	−0.10	−0.70	−0.05	−0.01
Sensitivity $\sigma^+_{zz,1}$ in MPa/mm	−0.07	−0.75	−6.01	−0.38	−0.08

As a reference for the preliminary design strategies, high fidelity FE-models in ABAQUS/STANDARD are used. The geometry is discretised by 3D-volume elements. To prevent numerical locking, second order elements are used. At least three elements are used along the smallest occurring edge in the model. Due to the jumps in stiffness along the beam, only deflection values are comparable since beam theory predicts unreasonable homogenous stress distribution along each section along its axis. To deal with the amount of possible combinations, the model build-up is scripted. The deflection predictions of the different models are shown in Fig. 8 with increasing number of ribs. As can be seen, all models predict very similar values for up to 25 ribs, showing less than 5% deviation. With increasing number of ribs, the aforementioned inadmissible homogenisation of stress in the respective cross-sections leads to unrealistic stiffening of the beam structure and thus increasing deviation.

Overall, the developed models show good applicability for early stages of the design process due to accurate predictions as well as low response times. Especially the analytical model offers great utility since it yields sensitivities of all target variables on all input variables. An efficient identification of relevant parameters with regard to the critical functions of a hybrid structure is possible and therefore enables targeted adjustments if needed. This exemplary process step for the preliminary design has the aim to support the engineering in an interacting digital way. In further steps several models with different levels of detail must be built up in the process model.

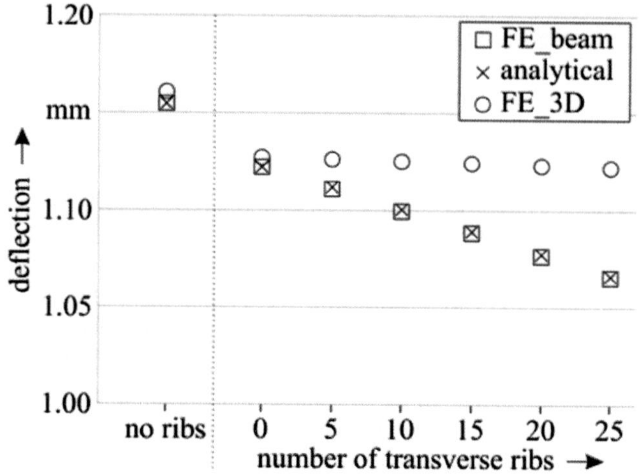

Fig. 8. Deflection predictions for the proposed models

4 Validation Steps in Early Stages of the Development Process

An essential step towards the suitability of this new group of materials for mass production is the early assurance of component quality and process reliability already in the development process. This requires a holistic and continuous validation strategy from the material to the entire system. By a continuous use of virtual and experimental validation methods, there is a constant comparison between the goals or requirements and the actual state of development. In this way, development times and costs can be reduced while at the same time ensuring that the quality requirements are met.

In an early phase of the validation, material data are determined using simple and standardised test specimens (tensile, bending and single-lap shear specimens, etc.) in order to create a material card. In the next phase, test specimens enable to identify geometry or process influences such as the specifications of rib elements in rib peel tests [25, 26]. Most of these test specimens are not standardised, which complicates comparability. Due to the fact that the technology validation of prototype components is a process involving high costs and time, many research projects are working with a simplified geometry. The experimental validation is presented Fig. 9 for a complex car body structure (A-pillar) in 3D-Hybrid technology [3, 4].

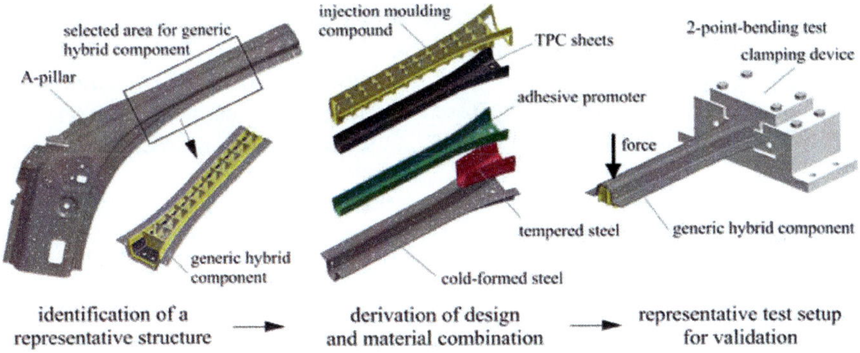

Fig. 9. Validation of the bending properties using a generic hybrid structure derived from an A-pillar

During one of the main load cases, car-rollover, the component is bended. The investigated metal-composite structure represents the main features of the reference car body component as a generic structure. The geometry was simplified and designed symmetrically, while the respective thicknesses of steel and thermoplastic composite (TPC) sheet as well as the material combination were acquired from the A-pillar. As a representative test setup, the bending properties were determined in a quasi-static 2-point-bending test. This enables the investigation of the complex interactions between geometry, material and manufacturing process influenced parameters and properties in a simplified and easier manageable specimen. The data from the validation must also be transferred into the above described structure of the process model.

5 Conclusion

To create an interactive development process and a structure for a holistic data analysis, a procedure with three phases is presented. As part of the first phase, a process model for the consistent description of the development process is introduced. This model describes the process by process steps and sub-steps in different detailing levels. Steps in highest level of detail are described by an approach of FELDHUSEN and consists of methods, models and data, which can be stored in a catalogue for the transfer to further iterations and projects. This process model is exemplary shown by the process step of preliminary dimensioning of a beam structure in 3D-Hybrid design. Two approaches are investigated and subsequently compared for the dimensioning of the ribs. All models predict very similar values for up to 25 transverse ribs, showing less than 5% deviation. Moreover, the analytical approach enables direct determination of sensitivities to determine the influence of the parameter to the properties of the structure. In a further step, these input and output data can be transferred to a next process step, e.g. a stability design, and thus be linked together for dimensioning. A strategy for technology validation of prototype components with a simplified generic geometry is demonstrated. This allows the hybrid metal-composite design to be tested on a near-series component structure and complements the standardised test specimens. By a method description these evaluation steps can be used effectively in the process model and integrated in parallel in the development of MCS in 3D-Hybrid design.

Acknowledgments. This research has received funding under the grant number 100339955 ("robust EVP 4.0" project) by the European Regional Development Fund (EFRE) and the German Federal State of Saxony.

References

1. Modler, N., Adam, F., Maaß, J., Kellner, P., Knothe, P., Geuther, M., Irmler, C.: Intrinsic lightweight steel-composite hybrids for structural components. Mater. Sci. Forum **825–826**, 401–408 (2015)
2. Kellner, P.: Zur systematischen Bewertung integrativer Leichtbau-Strukturkonzepte für biegebelastete Crashträger. Dissertation, Technisch Universität Dresden, Dresden (2013)
3. Haider, D.R., Krahl, M., Gude, M., Kellner, P., Knötschke, D.: Quality-assured process chains for the production of highly loaded lightweight structures in metal-FRP design. In: Plastics in Automotive Engineering, 14.–15.03.2018, Mannheim (2018)

4. Gude, M., et al.: Qualtitätsgesicherte Prozesskettenverknüpfung zur Herstellung höchstbelastbarer intrinsicher Metall-FKV-Verbunde in 3D-Hybrid-Bauweise. In: Plattform FOREL Abschlussbericht Q-Pro (2018)
5. Wawers, U., Stein, S.: Leichtbau und Digitalisierung im Entwicklungs- und Produktionsprozess am Beispiel Porsche 911. In: 23. Internationales Dresdner Leichtbausymposium, Dresden 27./28.06.2019
6. Verein Deutscher Ingenieure: VDI 2221 Part 1 – Design of Technical Products and Systems – Model of Product Design. Beuth, Berlin (2019)
7. Verein Deutscher Ingenieure: VDI 2014 Blatt 1 – Entwicklung von Bauteilen aus Faser-Kunststoff-Verbund; Grundlagen. Beuth, Berlin (1989)
8. Verein Deutscher Ingenieure: VDI 2014 Blatt 2 – Entwicklung von Bauteilen aus Faser-Kunststoff-Verbund; Konzeption und Gestaltung. Beuth, Berlin (1993)
9. Verein Deutscher Ingenieure: VDI 2014 Blatt 3 – Entwicklung von Bauteilen aus Faser-Kunststoff-Verbund; Berechnungen. Beuth, Berlin (2006)
10. Feldhusen, J., Grote, K.-H.: Pahl/Beitz Konstruktionslehre – Methoden und Anwendung erfolgreicher Produktentwicklung, 8th edn. Springer Vieweg, Berlin (2013)
11. Zhao, G.: Spritzgegossene, tragende Kunststoff-Metall-Hybridstrukturen. Dissertation, Universität Erlangen-Nürnberg, Erlangen (2002)
12. Müller, T.: Methodik zur Entwicklung von Hybridstrukturen auf Basis faserverstärkter Thermoplaste. Dissertation, Universität Erlangen-Nürnberg, Erlangen (2011)
13. Faber, J.: Beitrag zur konstruktiven Gestaltung offener, verrippter Profile aus Faser-Thermoplast-Verbunden mit lastpfadgerechten Verstärkungen. Dissertation, Technische Universität Darmstadt, Darmstadt (2016)
14. Lauter, C.: Entwicklung und Herstellung von Hybridbauteilen aus Metallen und Faserverbundkunststoffen für den Leichtbau im Automobil. Dissertation, University Paderborn, Paderborn (2014)
15. Haider, D. R., Krahl, M., Gude, M, Hengstmann, R., Titze, S., Haupt, M.: Continuous data measurement and analysis in automated manufacturing processes for hybrid lightweight structures. In: SAMPE Conference, 14.–16.10.2017, Stuttgart (2017)
16. Jakubik, J., Maron, B., Maaß, J., Gude, M., Masseria, F., Bublitz, D., Ropers, S.: Virtual and experimental analysis of a continuous 3D hybrid metal-composite process for automotive applications. In: Advanced Metal Forming Processes in Automotive Industry, 28–29.06.2016, Wroclaw, Poland (2016)
17. Wollmann, J., Haider, D.R., Krahl, M., Langkamp, A., Gude, M.: Linked a priori and a posteriori models of composite manufacturing process chain. In: Proceedings of the 14th International Scientific Conference: Computer Aided Engineering, pp. 823–828 (2018)
18. Grote, K.-H., Feldhusen, J., Beitz, W., Pahl, G.: Konstruktionslehre – Grundlagen erfolgreicher Produktentwicklung – Methoden und Anwendung. Springer, Berlin (2007)
19. Stachowiak, H.: Allgemeine Modelltheorie. Springer, Wien (1973)
20. Deutsches Institut für Normung: DIN 66001 – Informationsverarbeitung – Sinnbilder und ihre Darstellung. Beuth, Berlin (1983)
21. Dassault Systemes Simulia Corp.: SIMULIA User Assistance 2017 (2016)
22. Balke, H.: Einführung in die Technische Mechanik: Festigkeitslehre, 3rd edn. Springer, Berlin (2014)
23. Spura, C.: Technische Mechanik 2. Elastostatik: Nach fest kommt ab. Springer Fachmedien, Wiesbaden (2019)
24. Gross, D., Hauger, W., Schröder, J., Wall, W.A.: Technische Mechanik 2: Elastostatik, 13th edn. Springer Vieweg, Berlin (2017)

25. Liebsch, A., Kupfer, R., Krahl, M., Haider, D.R., Koshukow, W., Gude, M.: Adhesion studies of thermoplastic fibre-plastic composite hybrid components part 1: Thermoplastic-thermoplastic-composites. In: Hybrid Materials and Structures, Bremen (2018)
26. Haider, D. R., Krahl, M., Koshukow, W., Wolf, M., Liebsch, A., Kupfer, R., Gude, M.: Adhesion studies of thermoplastic fibre-plastic composite hybrid components part 2: Thermoplastic-metal-composites. In: Hybrid Materials and Structures, Bremen (2018)

Thinking Innovation Ahead – Joint Semantic Modelling for Integrated Product and Production at the Research Campus Arena2036

Dr. Clemens Ackermann[1(✉)], Manuel Fechter[2], and Peter Froeschle[1]

[1] ARENA2036 Research Campus, Pfaffenwaldring 19, 70569, BW, Stuttgart, Germany
{clemens.ackermann,peter.froeschle}@arena2036.de
[2] Fraunhofer IPA, Nobelstr. 12, 70569, BW, Stuttgart, Germany
manuel.fechter@ipa.fraunhofer.de

Abstract. The goal of the Research Campus ARENA2036 is, based on excellent, interdisciplinary basic and applied research, to produce potentially disruptive and leap-frog innovations, to transfer them to industry, and thus to contribute to the active shaping of work, mobility, production of the future, and digitization. The seamless transfer of research results into industrial application is intended to increase the competitiveness of the business location Baden-Württemberg and to enable the creation of novel business models – especially for SMEs. An essential component here is the interdisciplinary and trans-institutional approach of various fields of science and application, which is reflected in the close cooperation of all actors under the umbrella of ARENA2036. Based on this basic idea, current research work is carried out in the field of data interoperability in the domains of product development, and production system design. The goal is to achieve a significant overall reduction of product development and market introduction times. This topic is particularly significant due to the currently ongoing transformation processes in the automotive industry, which question the prevailing product and production patterns and require an increasing flexibility of manufacturing processes. New product and production technologies have to be incorporated into serial production at ever shorter intervals, which poses great challenges for both product design and corresponding production systems.

This paper conceptually approaches the basic ideas in innovation design incorporated at the Research Campus ARENA2036 as a research platform that allows for joint research in a precompetitive environment thus enabling all partners to think innovation ahead. One example for this, is the holistic semantic modelling of integrated product and production development in the research projects Fluid Production and Digital Fingerprint. Both projects conceive an approach to production that emphasizes the need for constant flexibility qua anthropocentric reconfigurability.

K. Dröder and T. Vietor (Eds.): *Technologies for economic and functional lightweight design*, Zukunftstechnologien für den multifunktionalen Leichtbau, pp. 59–68, 2021. https://doi.org/10.1007/978-3-662-62924-6_6

Keywords: Assembly administration shell · Data interoperability · Industry 4.0 · Fluid production

1 Introduction – A Titan of Research

The joint EUREKA-project PROMETHEUS, initiated by Daimler AG, is considered a blueprint for forward-looking and at the same time sustainable corporate research. Not least because it shows how cooperation between the scientific and industrial sectors can help to leverage undreamt-of potentials. Walter Ziegler, project manager of the "**Pro**gra**m**me for a **E**uropean **T**raffic of **H**ighest **E**fficiency and **U**nprecedented **S**afety", summarizes that "there can only be one solution to the increasing traffic problems": "new technologies – especially microelectronics, sensor technology, telecommunications and information processing – [must] be integrated into road traffic as comprehensively as possible". The project started on 01 October 1986 and almost all the technologies developed in PROMETHEUS are now in series production or – 25 years after the project was completed – will be soon.

Both the central questions and also the main approaches to solving them have changed almost exclusively in terms of their granularity over the last decades. The decisive change took place in another, structurally systemic area. According to Daimler AG's 2018 Annual Report, approximately 25,600 people are currently employed in the Research and Development division of Daimler AG, which in turn is divided into the following departments: Corporate Research & Mercedes Benz Cars Development, Daimler Trucks, Mercedes-Benz Vans, and Daimler Buses. The announced goal of the Corporate Research is to "offer customers fascinating products and tailored solutions for needs-based, safe and sustainable mobility." The structural systemic difference between this postulate of the 2018 Annual Report and the research work carried out within the PROMETHEUS project lies in the detail. Today, almost without exception, research and development is understood to be inextricably linked with one another, with the main obligation to ensure the most seamless transfer possible to the customer, whereas the horizon of past decades was much deeper. By way of comparison, in its 1986 Annual Report Daimler AG stated with regard to the PROMETHEUS project that "its realization will certainly extend well into the next century." By contrast, actual research projects are not even mentioned in current Annual Reports.

But in view of this development and in addition to the sustainable mobility qua product currently being striven for, how can it be ensured that there is actually sustainable research that will create solutions that reach well into this century?

This much is certain: holistic research must not only consider the product. But instead, it has to think production and product simultaneously – one actor alone will not be able to achieve all this. In the past, the corporate research of the OEM was the North Star, which Tier 1 and Tier 2 suppliers could orient themselves to. This was the direction in which the journey would take in the long term, and it was in accordance with this direction that the suppliers developed and produced. If this perspective "far into the next century" is lost (1986), the only thing that remains at Tier 1 and Tier 2 level is development close to the "fascinating product and tailor-made solution" for the customer (2018). It quickly becomes clear that this development is at its core contrary to the dogma of sustainability, which is why it is essential to find alternatives that promise a truly long-term perspective.

2 Forethought Instead of Hindsight

The hunt for the next disruptive technology – for the next What3Words[1] – and the next record quarter increasingly forces the hunt for all those technologies that already exist and find their way into the product almost immediately. Once such a position has been taken, one quickly finds oneself in the role of the person who is forced to take the next step always only after the fact. The role of Prometheus was exchanged for that of his brother Epimetheus: today, one literally turns into the one who thinks afterwards – i.e. in this case: after the emergence of a technology. In a nutshell, this means that one has become involved in outsourcing one's innovative potential to the indefinite. But how is it possible today to become part of an innovative network with a truly long-term perspective? How is it possible to think ahead?

Parts of a possible answer to this question are already concealed in PROMETHEUS: actual innovation is not shaped hierarchically within a silo, but in the pre-competitive area and in interdisciplinary as well as transinstitutional alliances. It is precisely for this purpose that new types of ecosystems need to be created that enable us to work together on future-oriented topics. The research campus ARENA2036 – the **A**ctive **R**esearch **E**nvironment for the **N**ext Generation of **A**utomobiles – undertakes such an attempt to think ahead together. The horizon – the year 2036 – not only resembles the pre-competitive character of research, but also marks the 150th anniversary of the automobile.

2.1 Questions Instead of Answers

It has become obvious over the past years that long-term, IP-driven silo research to achieve unique selling points that ought to secure a patent-based business model cannot keep pace with the cycles of the tourism-oriented investor market. Of course, many reasons for this are more far-reaching than can be outlined in just one sentence, above all, since they often lie outside the sphere of influence of the corporate strategists. Accordingly, however, it is all the more important to find new ways to increase one's own innovative strength again. It is also clear, however, that there is no silver bullet, but that the key to long-term success lies in a clever combination of the most diverse and the development of completely new routes.

The research campus ARENA2036, founded by Daimler AG, Robert Bosch GmbH, Fraunhofer Gesellschaft and the University of Stuttgart, among others, and funded by the German Federal Ministry of Education and Research (BMBF), outlines such a possible path to increased innovative strength by providing a platform where

1 www.what3words.com

more than 40 partners from industry and science can jointly exchange ideas and try out new ideas in a pre-competitive environment. The proximity created in the 10,000 m^2 research factory and the resulting interdisciplinary linking of the most diverse stakeholders ensures that the above outlined stalemate between OEM and suppliers can be resolved. The combination of the various stakeholders and the recombination of the resulting ideas is a value in itself, but not the only unique selling point of the Research Campus. It is rather, the distribution of risk among all participants that creates an opportunity to once again research topics that do not (necessarily) have to be included in the next model cycle, but which have the potential to 'reach far into this century'.

For this purpose, ARENA2036 navigates the thin line between basic and applied research in the fields of production, mobility, future work, and digitization. All these areas are thought of as interdependent and thus in close connection to each other and are dealt with through interlinked collaborative research projects, of which the Fluid Production project described at the beginning of this article marks the cornerstone regarding a versatile, reconfigurable vehicle production. Over the past six years, it has been shown that cooperative research between science and industry can not only offer solutions for incremental, iterative improvement of existing products and provide answers to familiar questions, but also that, based on trans-institutional proximity, new dimensions can be opened up and thus generate added value for all those involved.

A good example is the vehicle concept FlexCAR, which is currently being developed at ARENA2036. FlexCAR is a new type of mobile platform that deals with a mobility concept in transition and is able to create added value for participating partners already at the time of its design, system integration and later operation.

3 Joint Research Activities to Gain Synergies in Automotive Production

In the field of research, it is often not the isolated solutions and precise answers that ensure progress on the chosen route towards an overarching vision, but rather the questions that arise from cooperation of different knowledge domains that make the next step possible. An exemplary challenge is the joint development of product and production information models and semantics at the ARENA2036 Research Campus. Although the parallel exploration of production engineering and product design efforts have been postulated in research and teaching for decades (cf. systems engineering [1]) and the benefits of continuous data chains are highly appreciated, there is still a major break in the information flow between product development and production conception [2].

Particularly serious is the fact that existing model-based approaches for component design, simulation or analysis are not transferred purposefully into production development [2] as well as built-in sensors of today's smart products are not beneficially used in production application. A continuous information exchange and consolidated semantic base to describe items during all steps of product lifecycle for fast system integration and short production ramp-up times is needed. Manually performed information exchanges and not well-defined data interfaces might lead to hierarchically separated, monolithic implementations of product design and production control, which are not capable of performing volatile changes as desired by market demands [3–5]. For sustainable economic value creation, the information models and relations of the product and production domain must be available and comprehensibly described for all systems at all times. This is the only way to enable reconfiguration of the production systems in a short period of time in line with volatile market demands.

In order to handle the above mentioned challenges of collaborative product and production design with focus on interoperable data handling, the researchers at ARENA2036 work on two major research projects: the *Digital Fingerprint* and the *Fluid Production*. Both projects closely interact with the already mentioned *FlexCAR* project of an all-new vehicle concept. All projects directly refer to use-cases from the automotive industry either in the area of modular and reconfigurable final assembly or the draft, storage and exchange of product data during testing, design, operation and recycling.

3.1 Joint Semantic Modeling for I4.0 Component Description

During workshops and joint technology sprints between the projects, the consortia partners agreed to use common data modeling and standardized communication parameters for holistic data interoperability and transparency. Uniform information models and semantic parameters are used to describe the different application domains within the context of the automotive industry use-case. [6] Standardized interfaces – so called asset administration shells (AAS) – extend the conventional physical assets for an interconnected production scenario [7]. In doing so, the available data of all products and production resources gets accessible by every asset and user to be working within the ARENA2036 Research Campus. The so gained data transparency and interoperability enables new data-driven business models in smart production or new marketplaces for virtual data properties in the product domain.

Figure 1 illustrates the described approach. Every physical asset during product lifecycle, whether product or production resource, is extended by an AAS. The union of physical asset and virtual shell can be considered to be an I4.0-Component. Each I4.0-Component is individually aggregated of standardized sub-models from joint sub-model repositories.[2] To conclude the meaning of every individual semantic module, the descriptions can be referred to standardized dictionaries in the context of automotive product and production design.

[2] https://github.com/boschresearch/assets2036-submodels

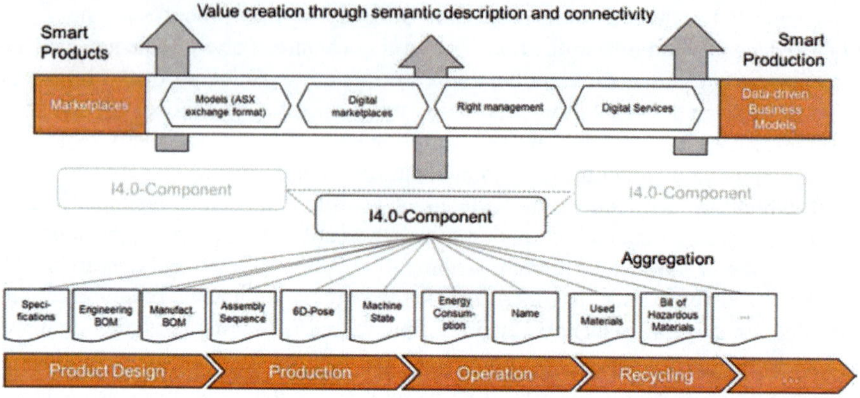

Fig. 1. Semantic definition of I4.0-components by aggregation of sub-models covering all stages of product lifecycle

3.2 Cyber-Physical Production Systems

Fluid Production addresses the challenges in production engineering design in reference to long-term investment, planning and configuration decisions of conventional, rigidly linked production lines. These more or less static production systems are being questioned, as they do no longer meet the requirements of volatile demands, increasing numbers of product variants, ever-shorter product life cycles and uncertain technological developments in course of transformation of the automotive industry [8]. Figure 2 highlights the differences between common production systems, e.g. the widely known dedicated manufacturing lines (DML), reconfigurable manufacturing systems (RMS), such as the Matrix Production System [9, 10] and the targeted Fluid Production System (FLMS) [11].

The guiding principle of *Fluid Production* can be described in analogy to the physical behaviour of a fluid drop. The production system should be capable to adapt itself in accordance to the occurring boundary and environmental conditions regarding form and position of the elements in the layout as well as competencies and capabilities of the individual process modules aggregated from various cyber-physical production modules (CPPS). FLMS are always an aggregation of individual process modules, consisting of multiple I4.0-components, e.g. production resources, varying in their capability to perform the production tasks in line with the process requirement from product domain. The context of resource allocation and utilisation is thus defined just before the start of production (SOP) by comparing the functional capabilities and characteristics of the available operating resources with the requirements of smart products to be manufactured.

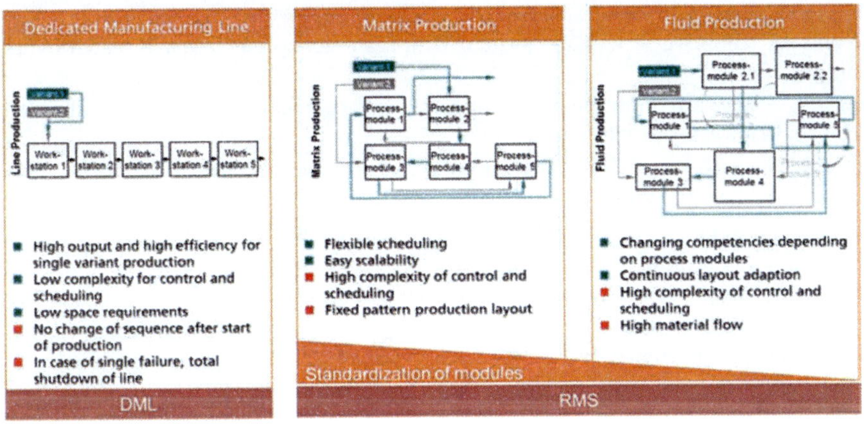

Fig. 2. Conventional production systems in automotive manufacturing in comparison to fluid production

The identified degrees of freedom of the Fluid Production system consists of a flexibilization of work content beyond state-of-the-art production systems, as well as new approaches of task scheduling and resource allocation for advanced routing flexibility and demand-based production scenarios beyond predefined flexibility corridors. Furthermore, the value-adding, primary production processes can merge directly with secondary, supporting processes as they can be found in logistic operations.

The overarching goal within *Fluid Production* is to postpone design and therefore production setup decisions as close to the SOP as possible and to dramatically reduce the system complexity of parameterization, integration and commissioning to a competency level every blue-collar worker can cope with [12]. This implies that the efforts for system integration and setup time have to be decreased drastically to perform economically suitable production scenarios for every lot-size and product variant. Every blue-collar worker in production should be capable to implement, commission and qualify assembly processes with the help of new approaches in work organization and the comprehensive use of data-driven assistive systems. The so-called anthropocentric production system within the Fluid Production should be capable to react to changing production environments within the time span of a few hours without the need for additional expert knowledge from specialized departments inside the organization or outside the manufacturing company.

3.3 Cyber-Physical Product Design

Based on the described semantic models and I4.0-Component definition, the *Digital Fingerprint* project focuses on new approaches to continuously describe product parameters from the early stage of initial draft and design up to continuous data collection during production with integrated quality inspection, as well as on-board data acquisition during operation to gather information on possible product improvements for further product generations. Ideally, the product describes its properties in such a comprehensive way, that in – combination with smart production resources – a versatile and self-adaptive production system within a heterogeneous resource pool can be achieved.

As a result, all production resources are connected and semantically modelled within the described AAS approach, the setup and commission time to start a manufacturing process only refers to a resource allocation, parametrization and linkage of process modules for a dedicated production scenario. By shifting the system design and setup as close as possible to SOP, the FLMS approach additionally minimizes uncertainties regarding forecasts in market demands and applied technologies. This way, the FLMS production system actively encounters possible discrepancies between intended product design and the real production case, which might appear during long-term parallel product and production development as it is performed today.

3.4 Technology Transfer to an All-New Modular, Cyber-Physical Vehicle Concept FlexCAR

The described cyber-physical product and production concept will show the greatest impact, when the modelling of product and production system are going to join forces in a shared product, such as the modular vehicle concept *FlexCAR*. The first stage of this project is intended to set up and roll out the rolling chassis (RC) platform, which will be a universal, modular base for future mobility applications. By merging product design, system integration and production, alongside the use and benefits of cyber-physical systems at every stage of product lifecycle engineering, the described RC is role model of the described semantic modelling approach inside ARENA2036 research campus. No matter which period of the RC lifecycle will be addressed in future use, the comprehensive data transparency and interoperability is a key enabler to guruantee comprehensive product understanding and design, production setup and integration as well as knowledge acquisition during operation of the RC.

Furthermore, the use of RC is not just defined by its final purpose of being a customized road vehicle in the B2B or B2C market. Early after the installation of first sensors, the basic energy supply and electric powertrain components, the RC might also serve as a resource inside the production setup, delivering goods to workstations or providing energy supply to remote production applications. The modular RC concept thus allows new solutions for transport and production inside an automotive factory. The specialization of smart components inside the factory, whether it is used as a product to be manufactured or a production resource to create added-value, thereby gets more and more fuzzy and use-case dependent.

4 Concluding Remarks – The Missing Link

One major drawback experienced during the first year of project work can be referred to the difficulties in developing new business models and marketplaces. Since these business models often do not result from the purely functional and causally driven research perspective, an external, economic way of thinking and consideration is necessary to unfold full potential. Solution- and application-oriented thinking patterns, as they prevail for example in engineering sectors, must therefore be combined with entrepreneurial action to generate new, goal-oriented value propositions. The awareness of entrepreneurial action and the value-adding transfer of results into new value creation systems is a critical and often missing element, which has been identified and will be tackled in the upcoming work packages.

While the close cooperation between industry and science in the pre-competitive environment already showed first promising results, it quickly became clear that innovations often take place in the niches in between. In order to tap this potential, Daimler, the University of Stuttgart and ARENA2036 founded the start-up accelerator STARTUP AUTOBAHN. The accelerator, which is also located under the roof of the Research Campus, not only guarantees that the start-up scene in one of Germany's strongest industrial areas is (re-)supported, but also that the innovation potential of the inventors is seamlessly integrated into high-tech research.

The dynamics created at the research campus ARENA2036 is thus the result of the proximity of basic science, application-oriented industrial research and the tinkering spirit of the start-up scene. It is precisely this combination of different institutions that makes it possible to recombine the most diverse competencies, which in turn makes it possible to actually develop innovative potential in such a way that disruption becomes conceivable.

5 Sumary and Outlook

This paper outlines the basic research ideas and structure behind ARENA2036 research campus to overcome the obstacles and impediments perceived in the past. The research focuses on the automotive industry which was in constant change over the past years due to changes in product portfolios, volatile demands and current technology pushes in area of electrification and autonomization.

The described approach of joint semantic modelling of product and production data during all stages of product lifecycle management, making use of the comprehensively available IT technology, can be considered to be a key-enabler for accelerated product development, faster production setup and therefore short time to market periods. The work details are part of the ongoing research endeavors. Future publications will focus on the benefits, impediments and lessons learned into the comprehensive use of cyber-physical product and production systems within the automotive industry.

Acknowledgements. The authors would like to thank the Federal Ministry of Education and Research (BMBF) for funding the research project Fluid Production (funding code: 02P18Q620-02P18Q629) as well as the project participants Fraunhofer Gesellschaft, Pilz GmbH & Co. KG, Mercedes-Benz AG, EntServ Deutschland GmbH, BALLUFF GmbH, Robert Bosch GmbH, Universität Stuttgart, BÄR Automation GmbH, Schunk GmbH & Co. KG, and KUKA Systems GmbH.

References

1. Winzer, P.: Generic Systems Engineering. Springer, Berlin (2016)
2. Ehrlenspiel, K., et al.: Kostengünstig entwickeln und konstruieren. Springer, Berlin (1998)
3. März, L., Von Langsdorff, P.: Flexibilität und Marktorientierung in der Montage. In: Montageplanung—effizient und marktgerecht, pp. 3–10. Springer, Berlin (2001)
4. Koren, Y.: The Global Manufacturing Revolution: Product-Process-Business Integration and Reconfigurable Systems. Wiley (2010)
5. Berkholz, D.: Wandlungsfähige Produktionssysteme – der Zukunft einen Schritt voraus. Wandlungsfähige Produktionssysteme: Heute die Industrie von morgen gestalten, pp. 13–18 (2008)
6. IEC White Paper: Semantic Interoperability: Challenges in The Digital Transformation Age. Geneva (2019)
7. Wagner, C., et al.: The role of the Industry 4.0 asset administration shell and the digital twin during the life cycle of a plant. In: 2017 22nd IEEE International Conference on Emerging Technologies and Factory Automation (ETFA). IEEE, pp. 1–8 (2017)
8. Dietz, T., Fechter, M.: Einleitung. In: Entwicklung, Aufbau und Demonstration einer wandlungsfähigen (Fahrzeug-) Forschungsproduktion, pp. 5–9. Springer Vieweg, Berlin (2020)
9. Foith-Förster, P., Bauernhansl, T.: Changeable assembly systems through flexibly linked process modules. Procedia CIRP **41**, 230–235 (2016)
10. Greschke, P.: Matrix-Produktion als Konzept einer taktunabhängigen Fließfertigung. Dissertation. Technische Universit ̈at Braunschweig (2016)
11. Fries, C., et al.: Fluide Fahrzeugproduktion. VDI-Z Jg. 161, VDI Fachmedien (2019)
12. Fechter, M., Dietz, T.: Zusammenfassung und Ausblick. In: Entwicklung, Aufbau und Demonstration einer wandlungsfähigen (Fahrzeug-) Forschungsproduktion, pp. 145–158. Springer Vieweg, Berlin (2020)

Integrated Factory Modelling – Enabling Dynamic Changes for the Factory of the Future at the Example of E.GO Mobile AG

Peter Burggräf(✉), Matthias Dannapfel, and Sebastian Patrick Vierschilling

Chair of Production Management, Department of Factory Planning, Laboratory for Machine Tools and Production Engineering (WZL) of RWTH Aachen, Campus Boulevard 30, 52074 Aachen, Germany
office@peterburggraef.com,
{M.Dannapfel,S.Vierschilling}@wzl.rwth-aachen.de

Abstract. Fast-moving changes in products, materials and process technologies require factory planning processes and procedures to be flexible and dynamic. Today, most factory planning projects are missing their budget (72%) and time targets (60%). To reduce these deviations, digitalization is key to success, but in current approaches, coordination between different planning disciplines is missing as well as different technology maturity levels prohibit automated interfaces. The Integrated Factory Modelling (IFM2) is an interdisciplinary planning approach for Green- and Brownfield factories coordinating all planning disciplines from infrastructure to process planning across the factory lifecycle. Therefore, the Integrated Factory Model (IFM) as a single dataset is established for all planning participants, accessible everywhere and on every device. Collaboration is enhanced by the working mode with an agile factory scrum process. Based on the IFM user-specific smart expert tools have been developed supporting planners and managers.

As a result, planning processes could be improved significantly, reducing costs by 20–30%, saving one-year planning time for a Greenfield and reduce planning failures significantly. IFM was initially applied and introduced to the e.GO Mobile AG, which is used as an example showing real-life use cases, challenges as well as next development steps.

Keywords: Factory of the future · Factory planning · Coordination · Single data source · Agile planning

1 Introduction

Manufacturing companies need to be adaptive and dynamic to solve future production challenges. Shortened product life cycles, e.g. as a result of new digital functions or new production technologies caused by new materials and technologies, lead to

K. Dröder and T. Vietor (Eds.): *Technologies for economic and functional lightweight design*, Zukunftstechnologien für den multifunktionalen Leichtbau, pp. 69–81, 2021. https://doi.org/10.1007/978-3-662-62924-6_7

frequent changes in factory requirements and require a continuous factory planning [1]. Additionally, the use of a wide range of IT tools in factory planning has become essential to manage complexity but also, factory planning is expensive and often time-consuming. Time and financial budget targets are usually not met as a result of various challenges in the planning process [2]. Frequent changes in requirements during the planning process lead to time loss and repeated planning efforts. In addition, the strong parallelization of product development and factory planning often forces planners to make assumptions, even about basic product characteristics such as material and production time. It is also associated with various challenges such as shortened planning phases, increased frequency and complexity of factory planning tasks. The main challenges facing factory planning today are [3]:

(1) Need for dealing with various planning tasks simultaneously involving multiple stakeholders
(2) Limited availability of information and resources
(3) High expectations in limited time and at an effective cost
(4) Need to keep up with continuous reorganization of production flows and layouts

In addition, Schenk [4] counts the requirements for future factory planning and the factory operation to be the follows:

(1) high planning speed and reliability;
(2) reliable knowledge already at the planning stage for the operation of a factory before product start-up;
(3) change of understanding from the unique, project-related to the permanent planning of individual phases of the factory life cycle (holistic project management) and the permanent management task thus required;
(4) integration of new participative planning and control methods and tools in connection with the unity of planning and control of dynamic processes and factory systems;
(5) holistic consideration of the planning objects across different object structure levels;
(6) extension of the consideration levels (from the market, company, process perspective);
(7) change of the planning objects by cross-linking of enterprise, factory and competence networks.

Looking especially towards Brownfield planning, the integration of new and innovative materials is challenging planners significantly. For example in automotive production, new production processes change the value streams significantly and lead to new and additional equipment which needs to be fit into the existing factory.

To summarize these requirements, new factory planning approaches need to be collaborative and cross-discipline beside of knowledge-based, dynamic and fast. This leads to the assumption that the today's planning challenges are party result of too many interfaces as well as to less cross-discipline understanding. To solve this and fulfill the named requirements named above, a new factory planning approach is required to ensure a collaborative planning process.

2 State of the Art

2.1 Factory Planning

In a cross-sector definition the Association of German Engineers (Ger. Verband Deutscher Ingenieure (VDI)) defines factory planning as a

> "Systematic, objective-oriented process for planning a factory, structured into a sequence of phases, each of which is dependent on the preceding phase, and makes use of particular methods and tools, and extending from the setting of objectives to the start of production [5]."

Spur and Schmigalla give another definition. They differentiate themselves from most other definitions due to their broad understanding by considering and focusing on the convertible and flexible factory. Thus, considering the frequent changes in their environment and consequently their repeated adjustments. According to their definition [6], factory planning

> "is the predestining design of factories. The factory is to be planned according to economic goals as well as according to the requirements of the working people and the environment."

According to them, factory planning contains of the analysis, goal setting, function determination, dimensioning, structuring, integration, and design of factories as systems as well as their subsystems, elements, substructures, and processes. [6] To better understand factory planning on greater detail, the key subject and objects in the planning process, the factory and the factory planner, shall be explained as well.

2.2 The Subject: Factory

Bergholz classifies the objects of factory planning as a cube with three dimensions. All objects of a production system are allocated within categories (organization, processes and resources), hierarchic levels and functions (processing, transport, storage and supporting functions). Due to external and internal impulses, the factory as an object is continuously under pressure, which results into the following examples of driver of change [7]:

- global linked production of goods;
- individualization of the products;
- integrated products and services;
- local changes in environment and society.

Those drivers make it essential to continuously adapt and re-plan the factory and therefore necessitate the process of factory planning.

The staff responsible for operating the factory planning project is key for the success of the project. The project team can be divided into the internal project team, which are responsible for the operational planning and coordination during the project, and external partners and influencers to the team [8]. As an example, participants in a project can be listed as following:

An internal project team consisting of (extract)

A. Project coordinator
B. Production and process planner
C. Production IT planner
D. Layout planner
E. Quality planner
F. Logistics planner

(1) Additional external influences on the project are provided by (extract)

G. Stakeholder/Management
H. Engineering department
I. Factory Operator
J. Architect, External Building teams and General contractor
K. Equipment supplier

Considering the preceding description of the complexity of the stakeholders, it becomes clear that coordinating them with the use of a classical project/team structure can lead to significant communication and information coordination problems. In worst case part of the project team loses time and ability for actual planning due to the high coordination effort. A study by Bracht stated that the average planner just uses 20% of his working time for conduction planning activities. For this reason, a new type of project structure must be sought, as shown in Fig. 1, to efficiently including a growing number of experts [9].

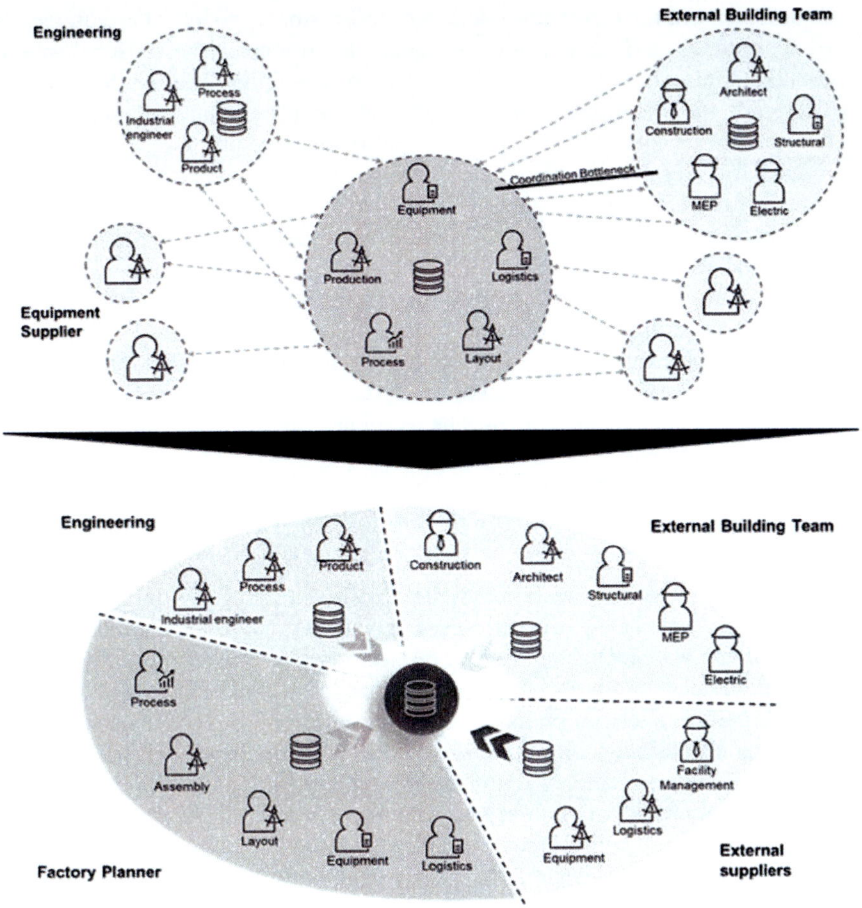

Fig. 1. Factory planning project teams: from traditional centrally organized planning teams structures to data based collaborative planning teams

2.3 Digital Factory Planning

Digital factory planning is the concept of carrying out planning processes with the assistance of various IT tools. It can be applied to both, greenfield and brownfield, factory planning projects. The procedure for the digital planning of a factory starts with the creation of 2-dimensional basic layouts and ends with 3-dimensional CAD models of the entire factory. Digital tools will lead to major changes in the way organizations function and in the content of work for many employees in industrial companies [10]. Various IT tools support factory planning and can be summarized as digital factory planning tools. Digital factory planning is characterized by three elements [9]:

(1) Data management is digitally integrated into a single model: The required data from products, product programs, processes, and resources, etc. must be gathered, created, adjusted and provided at different places during the planning process. Therefore, the task of data management is to provide and maintain an up-to-date and continuous data structure during the entire planning process
(2) Exemplary representation: With the increasing performance of computer systems, complex graphical representations as user interfaces in digital factory planning have become state-of the art. With this development, it is possible to carry out and visualize planning steps in a realistic computing environment. This enables factory planner to check and improve the dimensioning of product plants and process designs during the planning phase in a digital environment.
(3) Simulation as a forecasting instrument: By using three-dimensional simulation applications, the behaviour of complex systems such as a factory can be reproduced. The planning result can thus be tested and optimized before its realization. In particular, spatial processes such as collision tests for robot programming or the design of assembly processes can be performed. Furthermore, process flows, product dimension fittings or value streams can be mapped and simulated or tested.

The choice of IT specialist tools for digital factory planning has increased significantly. The tools must be coordinated in such a way that their functionalities build on each other or complement each other without causing interface problems. In this way, the entire digital planning process can be implemented within a single software architecture and inefficiencies can be avoided. The development of such a software architecture, which also displays the entire factory with all participants, is still the essential basis of digital factory planning. Furthermore, the software architecture should consider the three elements of digital factory planning discussed above [9].

In recent years, the construction industry faced similar challenges to those faced by factory planning and developed a. To address these challenges, the concept of building information modelling (BIM) was developed. The model is defined by international standard ISO 29481-1:2010 (E) as a "digital representation of physical and functional characteristics of any built object that forms a reliable basis for decisions" [11]. BIM is realized with object-oriented software and consists of parametric objects representing building components [12]. A completed building information model contains precise geometric and informational data, which is required to support the design, manufacturing and construction activities, needed to realize a building [13]. BIM can be viewed as a virtual process that incorporates multiple attributes of a facility within a single virtual model which can be used by the stakeholders from various disciplines to collaborate [14].

For this reason, the BIM approach can be used as basis for a digital and networked planning process, which is supplemented by the requirements of digital factory planning.

3 Integrated Factory Modelling

The methodology of Integrated Factory Modelling (IFM) is designed to meet these challenges. Therefore, ideas of the BIM methodology have been enhanced by production needs with the target of enhancing the planning collaboration of the project team.

The IFM works with the idea of five main project phases: During the first phase, the so-called planning phase, process concepts, an initial rough 2D layout and the basic material flows are developed. During the design phase, a 2D & 3D CAD model of the factory is created, which serves as a data basis for the further planning process. The next step is to use technologies like virtual reality for design reviews of the CAD-Modell and carrying out a clash detection and issue Management as part of the validation phase. The fourth phase, the build phase, is where the planning is transformed to stone and concrete and the model is used for a rigorous site management and progress monitoring. Therefore, regular plan/is comparisons are conducted. During the operating phase a predictive maintenance approach is used. In addition, the generated data from the operate phase is made available as a data basis for the future re-planning as well as continuous time (4D-planning) and scenario planning (5D-planning). This basically enables a continuous improvement process and reduces the information collection time and effort along the factory lifecycle.

To ensure that all stakeholder are properly aligned during the process, the IFM is using a factory scrum process. Each planning sprint has a duration of about two to four weeks, which the planning participants use for enhancing their planning maturity. During each sprint meeting at the end of a planning period the coordinated factory model is reviewed, changes are discussed and documented in the model as well as new tasks for the sprint are defined. Additionally, daily spring meetings can be arranged in the planning groups to align in small teams.

Therefore, the three main characteristics of the IFM methodology, which are displayed in Fig. 2, are now explained in detail:

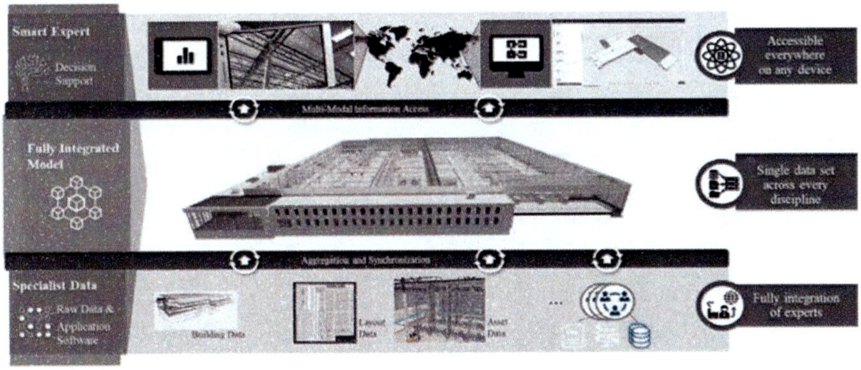

Fig. 2. Interrelations of the IFM model

(1) One single dataset across every discipline

All planning activities are operated and documented in a centralized planning model. The factory model always reflects the current planning status of all planners, making changes and errors immediately visible by automated collision control. Errors are thus detected and eliminated early in the planning process. Since this is done purely digitally and virtually, error costs are dramatically reduced. To achieve all this, IFM must be integrated in all planning phases of the factory planning process, from the design of 2D layouts to the virtual reality representation of the model in a 3D environment, as well as all disciplines. All planning activities carried out by internal and external parties must be aligned and synchronized with the platform. This also includes the product development team which connects its current product model with the IFM model. Due to this layout relevant changes can be recognized and implemented at once. For direct integration of supplier, IFM provides flexible extensions of the platform. The suppliers will get their own enclosed space on the platform and will be connected to the processes of IFM without sharing of all planning information (see Fig. 3). In the case of IFM, the 3D model of the factory represents the central interface to the planning process, which can be used throughout all life cycle phases of the factory

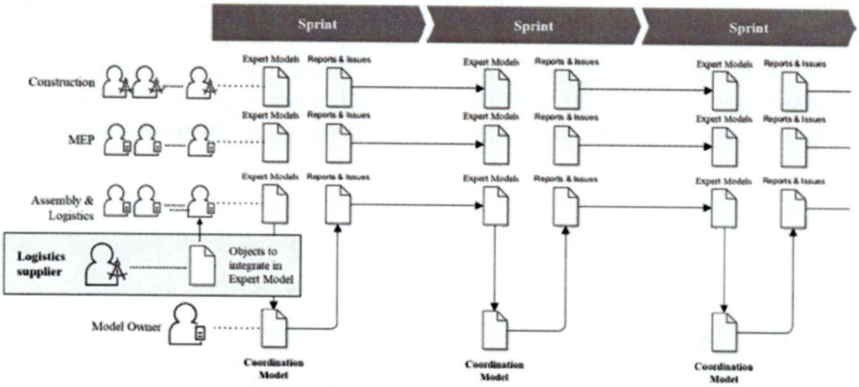

Fig. 3. Integration of specialist planners and expert processes of IFM

(2) Fully integration of experts

The CAD factory model consists of many individual part models, which in turn are divided into further part models down to the level of individual objects, as can be seen in Fig. 3. Typically, the production is split into assembly, logistics, quality and more areas of the factory are modelled as a separate sub-model. The validation of the overall model is done by coordinating the sub-models and subsequent analyses, which can be partially automated. Nevertheless, every planner can use his or her specialist tool for the planning task but must connect his results to the overall factory model. Therefore, an inter-operationally between different planning tools need to be considered in the software architecture selection.

(3) Smart Experts accessible everywhere on any device

As already explained, the main idea of IFM is to create a common platform for all project participants to interact and exchange data. To achieve the overall communication between all necessary stakeholders, an implemented cloud service is used. The cloud service integrates both the factory model and the planning process to ensure a more connected and effective workflow. To secure the explained synchronized working, the platform must synchronize data from multiple sources and provide consistent, up to date information to every stakeholder. Therefore, IFM includes a centralized document management system that bundles the available information at all necessary levels of detail within in the 3D factory model. Furthermore, individual applications, so called Smart Experts, can be implemented based on the planning model. These smart experts can start can be used for example in the building stage as model overlay to the building site to prove correct building. Similar to the use of BIM models, this step enables a target-performance comparison of the virtual model and the factory currently under construction. This allows to prevent errors in the early phase and to reduce the error costs significantly. Another smart expert could be a material flow simulation based on the model details or a virtual reality process validation. The number and application of smart experts is depending on the planning case and user specific.

As an example of the usage of IFM the equipment supplier model delivery process which was implemented at the e.GO Mobile AG is presented. The e.GO Mobile AG is a young automotive start-up founded in 2015 by Prof. Günther Schuh in Aachen. The founding ide of e.GO was to produce electric small-size vehicle for less than 10.000€. Furthermore, the product design logic was changed from customer-orientated design to production-orientated design. Additionally, the factory and production planning was parallelized to the product design, leading to a factory planning of just 18 months and just two months lead time from production design freeze and first product build. The e.GO Life, a small-sized two person electric vehicle with a range of around 150 km, is the first product of the e.GO Mobile AG using an aluminum-hybrid structure. The e.GO Mover, the second product, is planned to be a partly autonomous shuttle bus for up to 25 people. The e.GO Mover is planned have a hybrid body structure of carbon and aluminum. Both, the e.GO Life and the e.GO Mover, are produced in two manufacturing and one body-shop factory in low volume in Aachen, Germany.

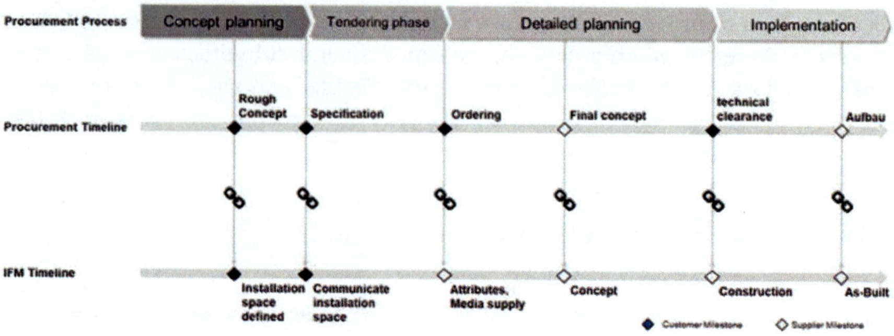

Fig. 4. Connected IFM and procurement timelines

Key to success for the e.GO Mobile AG was the fully integration of product, production and procurement processes. This connectivity was particularly crucial in the successful implementation of the change from an L6e vehicle classification to M1. Due to the digitally connected processes, changes in the product, like dimensional or material changes directly were communicated production planner. In case of dimensional changes, various measures are possible to ensure producibility, e.g. editing the assembly station design or the line side space requirement for the part's container. For material changes, existing machining capabilities and capacities needed to be checked. Nevertheless, due to the direct information and possibility to review every change, the production could quickly update their planning or hand feedback to the development team to review their planning due to restrictions in the production. In total this leaded to a reduced product and production development time, leading to a quicker time to market, by around 30 percent in comparison to automotive industry standards. A very important cooperation for the production planning were the integrated factory and procurement processes, which can exemplarily be seen in Fig. 4. A five step stage-gate process was developed by e.GO to align on the one hand factory planning and procurement milestones and on the other hand integrate the supplier data directly into the factory model. Key of the stage-gate process is the so-called Level of Detail (LoD) which define model requirements similar to deliverables in BIM. Therefore, LoD 100 to LoD 500 are indicating the degree of maturity of the digital planning status. The extension is especially used for the awarding of contracts for larger plants or for areas to be contracted externally. Each step of the stage-gate process will be explained following. Additionally, Fig. 5 gives an example regarding the model maturity at each stage.

As first step, the equipment planner starts a supplier award process, defines and specifies the available/planned space which is intended for the equipment to be purchased, the so called LoD 100 model, and provides a specification sheet for the supplier. All of these is published to selected suppliers as an Request for Quotation (RfQ). As initial feedback, the interested suppliers provide first requirements for any media supplies, bearings and other general requirements. This results into a parameterized model which is called LoD 200. After these are aligned, the supplier specifies his model and creates a first conceptual presentation of the solution of the equipment, which is called LoD 300. Based on these modes, the equipment and processes with it will be closely evaluated and validated. At the end of this step, the equipment planner finally decides on a supplier with whom he will complete the project. Following the general planning and the selection of suppliers, the supplier develops a detailed plan including a business case. In addition, the supplier creates an interface examination. This allows an examination of possible material flows or even logistics processes. The model is refined and detailed until a complete engineering representation, the so called LoD 400. Based on the engineering representation model the final order confirmation and approval for production. After the equipment is installed at the facility, a 3D scan of the finished area is created and compared with the 3D planning model of LoD 400. If the scan matches the model and any agreed changes have been implemented, payment to the supplier can be released and the process ends with an As-built model of the equipment, the so called LoD 500.

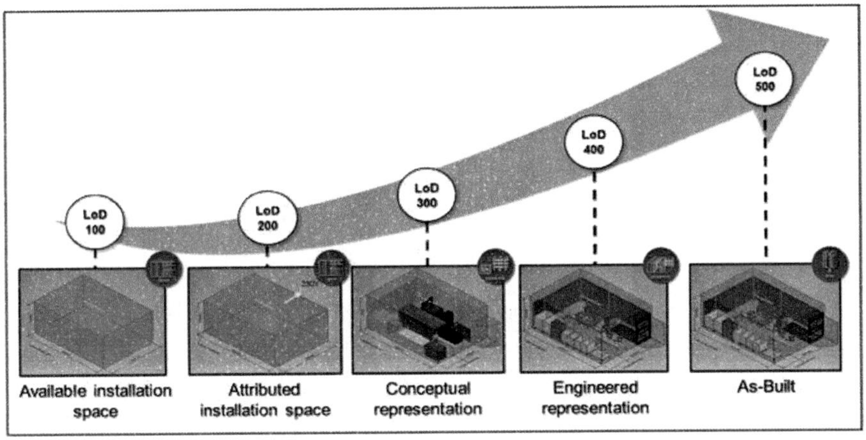

Fig. 5. IFM the five stages equipment supplier model delivery process

4 Outlook

By using integrated factory modelling (IFM) at the e.GO Mobile AG significant advancements in the area of factory planning could be achieved. A shortened planning time by 15 months as well as planning cost savings of roughly 33% in comparison to comparable projects can be named besides a higher quality of planning and accurate digital models. Nevertheless, additional functions and advancements in the methodology are possible and in research. Three major projects are shortly named:

The long-term goal should be the automation of the factory planning processes. Therefore, using methods of artificial intelligence (AI) should be considered as suitable approach for the stated goal with the digital factory model being a basis. This model already provides all required information in a machine-readable form which can be used by the algorithms of the AI. Similarly, AI can feed its results back into the digital factory model, allowing easy and efficient further use of the model.

Furthermore, it possible to integrate the so-called Generative Layout Design Approach which is based on the Aachen factory planning procedure and evolutionary algorithms. Based on the desired input parameters, a considerable number of layout alternatives are generated, and a possible solution space is defined. This space is systematically searched and automatically limited by a targeted evolution algorithm. This procedure enables to generate the best possible solution in a reasonable computing time. The rapid calculation of the evolutionary algorithm therefore ensures that changing parameters in the planning data which can be quickly evaluated and included in the calculation. This approach makes it possible to significantly reduce the planning time.

The IFM is also an important first step towards optimizing data management in factory planning. A study by Bracht stated out that a factory planner invests 23% of their time for searching information. Even if a piece of information is found, there is no guarantee that is up-to-date, complete or correct. The IFM already is provides a central platform with linked information for each object, for example the maximum permissible load of a forklift truck. Further information can be automatically added and updated using algorithms known from social media platforms. With automated information retrieval technologies combined with the central information database the value-added time of planners can be significantly increased and double efforts can be reduced.

References

1. Dombrowski, U., Schmidt, S., Wrehde, J.: Herausforderungen in Der Automobilproduktion. ZWF **101**(5), 254–259 (2006)
2. Schmitt, R., Schuh, G. (eds.): Advances in production research. In: Proceedings of the 8th Congress of the German Academic Association for Production Technology (WGP), Aachen, November 19–20, 2018. Springer International Publishing, Cham (2019)
3. Achim, K., Burggräf, P., Krunke, M., Becks, M.: Type-oriented approach for the value-optimized application of heuristics in factory planning. In: FAIM 2014: Proceedings of the 24th International Conference on Flexible Automation and Intelligent Manufacturing: Capturing Competitive Advantage Via Advanced Manufacturing and Enterprise Transformation, pp. 385–394. DEStech Publications, Lancaster (2014)
4. Schenk, M., Wirth, S.: Fabrikplanung und Fabrikbetrieb. Methoden für die wandlungsfähige und vernetzte Fabrik, 2nd edn. Springer, Berlin (2004)
5. Verein Deutscher Ingenieure (02.2011): VDI-Richtlinie 5200 (2011)
6. Schmigalla, H.: Fabrikplanung. Begriffe und Zusammenhänge, 1st edn. Hanser, München (1995)
7. Bergholz, M.: Objektorientierte Fabrikplanung, 1st edn. Shaker, Aachen (2006)
8. Burggräf, P.: Wertorientierte Fabrikplanung, 1st edn. Apprimus Wissenschaftsverlag, Aachen (2013)
9. Bracht, U., Geckler, D., Wenzel, S.: Digitale Fabrik. Methoden und Praxisbeispiele, 2nd edn. Springer Verlag, Berlin (2018)

10. Küpper, D., Kuhlmann, K., Köcher, S., Thomas, D., Bruggräf, P.: The factory of the Future. https://www.bcg.com/de-de/publications/2016/leaning-manufacturing-operations-factory-of-future.aspx. Accessed 5 June 2020
11. International Organization of Standardization (05.2010): ISO 29481-1:2010 (2010)
12. Cerovsek, T.: A review and outlook for a 'Building Information Model' (BIM): a multi-standpoint framework for technological development. Adv Eng Inform **24**(2), 224–244 (2011)
13. Eastman, C.M.: BIM handbook. A Guide to Building Information Modeling for Owners, Managers, Designers, Engineers, and Contractors, 2nd edn. Wiley, Hoboken (2011)
14. Azhar, S.: Building Information Modeling (BIM): trends, benefits, risks, and challenges for the AEC industry. Leadersh. Manag. Eng. **11**(3), 241–252 (2011)

Life-Cycle Engineering

Methodology for Assessing the Environmental Impact of Emerging Materials

Malte Schäfer[1](✉), Martina Gottschling[2], Felipe Cerdas[1], and Christoph Herrmann[1]

[1] Institute of Machine Tools and Production Technology (IWF) / Chair of Sustainable Manufacturing and Life Cycle Engineering, Technische Universität Braunschweig, Langer Kamp 19b, 38106 Braunschweig, Deutschland
malte.schaefer@tu-braunschweig.de

[2] Group Innovation – Center of Innovation Battery / Polymers / Sustainable Materials, Volkswagen AG, 1777-4, PO Box 011, 38436 Wolfsburg, Deutschland
martina.gottschling@volkswagen.de

Abstract. In order to reduce environmental impacts of product systems through material research and development, to identify mitigation potential, and to avoid problem shifting, information about the environmental impact of emerging materials is needed at an early stage. This information can support decisions on material selection as well as manufacturing process optimization. The goal is to reduce the impact of an emerging material so that it is lower than the impact of an established material. We propose a methodology to address this need. It consists of four steps: 1) reference LCA, 2) forecast LCA of emerging material, 3) scaling and 4) comparison. We apply the methodology to automotive seat cover materials such as bovine leather, faux leather and a fictitious flax-based material, where the production of the latter shares some production steps with leather. The results indicate how much environmental space the to-be-developed manufacturing process of the fictitious material can take up before it surpasses that of leather and faux leather. This way, the methodology can support the material research and development process in identifying and creating alternative materials with a lower environmental impact.

Keywords: Material selection · Material research · Life cycle assessment · Screening · Scaling

1 Introduction

Materials and the process of extracting, refining, using and disposing of them are highly relevant to the environment. Furthermore, humankind's demand for materials has been increasing rapidly over the past decades [1]. In this regard, the development of new materials requires systematic analysis to avoid unintended consequences.

K. Dröder and T. Vietor (Eds.): *Technologies for economic and functional lightweight design*, Zukunftstechnologien für den multifunktionalen Leichtbau, pp. 85–96, 2021. https://doi.org/10.1007/978-3-662-62924-6_8

Emerging materials currently in the research and development (R&D) phase could potentially reduce the environmental impact compared to established materials, and may eventually replace them. However, in order to inform the decision making process of material researchers and developers, they require reliable information on the environmental performance of both the established and the novel material. This includes both direct environmental impacts caused throughout the life cycle of a material, and the indirect effects which their material properties may have. For example, a material with a lower density may potentially reduce the weight of an aircraft, thus reducing its fuel consumption and air pollution. Most importantly, this information should guide material R&D efforts by quantifying the environmental impact across the complete life cycle and by identifying environmental hot-spots within this life cycle. This might contribute to select which materials to develop further, and how to prioritize their R&D efforts, in order to achieve the desired product environmental performance.

Performing a lifecycle environmental assessment for an emerging material is challenging. These emerging materials are typically produced at a lab scale or pilot plant scale, and therefore the relevant data and information on which the assessment relies, is not (publicly) available, or not representative of a potential commercial scale product system.

The problem statement can therefore be formulated as follows: a methodology is required to guide material research and development efforts, which pursue the goal of creating novel materials with a lower environmental impact than an established material. In this paper we present a methodology which addresses this issue. The primary intended application are materials for vehicle interiors. Specifically, we present an application of the methodology to leather, artificial leather, and an emerging material alternative based on flax fiber.

In Sect. 2, we present the current state of research on approaches similar to the one proposed here. In Sect. 3, we describe the proposed methodology. In Sect. 4, we apply the methodology and discuss the results. In Sect. 5, we close with a summary and conclusion.

2 State of Research

From the problem statement presented in the introductory section, we searched the literature to identify similar methodologies and their applications in the current literature. The following requirements were identified as the most important ones in this context, as they address the central issues in the assessment of environmental impacts for emerging materials and decision making in the material research and development process based on this assessment. The methodologies should consider:

1. The complete material life cycle including the product in which the material is used
2. Multiple types of environmental impacts beyond global warming potential (GWP)
3. Uncertainty and variability of data and models
4. Scaling effects from technology development, technological learning and technology diffusion

Not taking into account the complete life cycle exposes the risk of missing environmental hot-spots, and engaging in problem-shifting (e.g. from the production to the end of life stage). Similarly, assessing only the effects on climate change while neglecting other types of impact – such as land use, water use, or local air pollution – bears the risk of problem shifting from one type of impact to another. An example is the transition from vehicles with combustion engines to electric vehicles, which appears to correlate with a reduction in GWP, while increasing environmental impacts in other categories, such as human toxicity or metal depletion [2, 3]. For an assessment of an established material, uncertainty and variability are already challenging yet important to factor in – even more so for emerging materials. For an example of variability, see Ashby on the variability of embodied energy in aluminium [4, p. 125]. There it is shown how the embodied energy of aluminium can vary by a factor of four, depending on the data source. Finally, the decision to include scaling effects is based on the experience that the upscaling process tends to correlate with significant improvements in efficiency and yield, which translate into both cost savings and environmental impact reductions [5–7]. This factor needs to be accounted for when comparing emerging and established materials. The underlying effects of scaling can be decomposed into the three aspects of technology development, technological learning and technology diffusion, which are explained and reviewed in detail in [8].

In the research area of material selection, the most established methodology is probably the one developed by Ashby and colleagues [4, 9]. The methodology has been implemented in a commercial software tool for systematic material selection, which includes a vast material database [10]. Both the methodology and the software tool allow the user to include environmental aspects into their selection. While the complete life cycle is taken into account to some extent, the impact assessment is limited to GWP and embodied energy, a proxy variable with a high degree of correlation with GWP. Uncertainty and variability are not considered and neither are scaling effects. Therefore, and more importantly because it is a methodology for material selection, the methodology cannot be considered suitable for the environmental assessment of emerging materials nor for guiding material R&D efforts. However, the step-by-step approach, the decision making support using charts and diagrams and implementation into a software tool are helpful elements to consider in the development of a novel methodology.

Several other methodologies for material selection based on environmental criteria have been proposed [11–16]. Sun et al. propose a simplified method to evaluate the environmental impact during material selection, based on material grouping according to the type of material and its environmental impact [11]. The approach by Ribeiro et al. combines technical, economic and environmental aspects, and allows the user to weigh these against one another using so-called ternary diagrams [12]. The method by Qiu et al. integrates structural design and material selection into a hybrid optimization model with the goal of reducing environmental impact [13]. Broeren et al. propose a material selection framework for incorporating environmental criteria and a simplified cost analysis into early stage product (re)design, applied to the case study of printer panels from polymers and bio-polymers [14]. The methodology by Hallstedt and Isaksson supports material selection in early product design based on material criticality assessment for alloy materials [15]. The framework by Kappenthuler et al. for material developers supports the identification and prioritization of research projects related to the development of materials for sustainable construction [16].

While most of these approaches address the first two criteria stated above to some extent, only a few consider uncertainty and variability. Only one approach [16] considers scaling effects. However, it is intended for assessing construction materials, a fact which the domain-specific assessment criteria reflect. Applying the methodology to other types of materials would require major modifications. Also, the proposed assessment is not entirely quantitative, but semi-quantitative and qualitative in several categories. The quantitative categories do not allow for the calculation of a value that represents the absolute environmental impact, but rely on comparative ranking (e.g. against all other materials within the same material category). This requires extensive data on both the reference materials and the emerging material, which may not be available. While none of the approaches discussed in this paragraph completely meet the initially stated requirements, they contain some useful elements. These include the idea of grouping similar materials [11] and visualizing uncertainty [14], both of which may be beneficial for the development of our methodology.

Life Cycle Assessment (LCA) may be considered the furthest developed and most widely applied methodology for quantifying the environmental impact of products [17]. LCA, per definition, meets the first two criteria (complete life cycle and multiple impact categories) [17, 18]. A range of studies have been published on the issues of variability on uncertainty in LCA, including [19–22]. The fourth criterion – scaling – has been primarily addressed by studies on ex-ante LCA (alternative descriptions: *screening LCA, prospective LCA, anticipatory LCA, future-oriented LCA, LCA of emerging technologies*). This relatively recent sub-domain of LCA research seeks to develop approaches to apply LCA to technologies at an early stage of their development. The review articles by Cucurachi et al. [23], Arvidsson et al. [24], Buyle et al. [8], Moni et al. [25] and Thonemann et al. [26] provide an overview of the research on ex-ante LCA. Since the authors identify central challenges in conducting an ex-ante LCA to be largely congruent with the four criteria stated at the beginning of this section, we consider the ex-ante LCA to be of high relevance for the development of the methodology presented in Sect. 3. The fact that several variations of ex-ante LCAs have been applied to (nano-) materials supports this proposition [27–29].

We conclude that from both the research domains of material selection as well as life cycle assessment, valuable contributions have been published on the early stage material assessment based on environmental criteria. In the following section, we put forth a methodology that seeks to combine the most suitable elements from previous approaches in such a way that it may potentially solve the challenges of the task at hand better, easier, or both.

3 Proposed Methodology

The methodology we propose for the early stage environmental assessment of emerging materials is intended to meet the four criteria stated at the beginning of Sect. 2.

It combines several elements from other approaches. In this section, we initially describe the methodology in a general manner (see Fig. 1), before we apply it in Sect. 4.

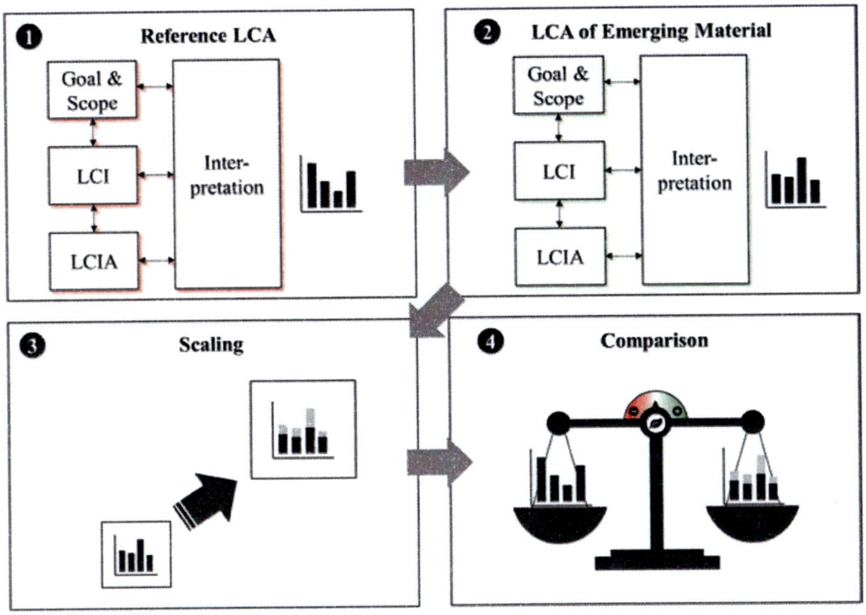

Fig. 1. The four steps of the proposed methodology

The *first step (reference LCA)* consists of conducting an LCA of a reference material that resembles the emerging material at hand as closely as possible. Characteristic features that define material similarity may include material properties (e.g. physical), material cost, origin of the raw materials, the configuration of the material production chain, or the location of the materials within an existing material classification system. A useful indicator may also be the first (commercial or pre-commercial) applications of a material. When exploring material alternatives for leather as seat cover materials (see Sect. 4), one may investigate whether novel materials have already been applied to other products often made from leather, such as footwear and apparel. Unsuitable indicators in this context would be annual production volume, technology readiness level (TRL) or other indicators that would render an emerging material quite dissimilar to an established material, due to their differing levels of development progress. This would, by default, disqualify most or all emerging materials. For the same reason, data on material properties and cost can only serve as a rough indication. The LCA can be based on primary data, if one has access to the material production facilities, or rely on published LCA studies that disclose details on the life cycle inventory (which is necessary to build one's own LCA model). Plenty of options are available for LCA software to choose from, both commercial (e.g. SimaPro, GaBi, Umberto) and free (e.g. OpenLCA, Brightway2). For the background system, it is likely that access to commercial databases (e.g. GaBi, ecoinvent) will be necessary. For guidance on which life impact categories to include in the life cycle impact assessment, see for example the Product Environmental Footprint Category Rules by the European Commission (EC), which now exist for a number of product categories. For products not included, published LCA studies may serve as orientation on which impact categories may be relevant. Both types of documents may also support the process of identifying a reference material for the emerging material of interest.

In the *second step (novel LCA)*, we conduct an LCA of the emerging material. The goal & scope, and therefore all assumptions, the functional unit, system boundaries etc. shall be the same as for the LCA of the reference material. However, the data availability for the emerging material is likely to be worse. Possible data sources are primary data from the lab or from a pilot production line, background data from LCA databases, expert information, published LCA studies on similar materials or on materials that make up part of the emerging material, public databases or producer information. For values that are uncertain, it may be viable to describe them using probability distributions (e.g. uniform, triangular, normal, log-normal), as demonstrated by Blanco et al. [30]. Using software packages such as Brightway2, these uncertainty distributions can be used to calculate probability distributions for the LCA results, which reflect the underlying uncertainty. This is possible for both the reference material and the emerging material, if the input parameters are specified accordingly.

In the *third step (scaling)*, we modify the results from the LCA for the emerging material to reflect the changes that are likely to happen once and if the material is scaled up to industrial production scale. In this case, the production process is likely to change drastically, and so are the costs and environmental impacts associated with it. Buyle et al. provide an overview of methods available to describe the upscaling of emerging technologies [8]. The ideal systems baseline (ISB) method is suitable for production systems driven by chemical reactions, and for which the environmental impact is governed by the yields and efficiencies of these processes. The problem solution space (PSP) is a creativity method relying on expert knowledge, and can be applied to ill-defined problem definitions and for exploring applications. The proxy technology transfer method for processes (PTT_p) works by means of analogy, partially replacing the inventory for existing production processes with novel elements. The proxy technology transfer method for the impact (PTT_i) does not rely on inventory information, but instead directly interpolates between the environmental impact at lab scale (for the emerging technology) and, for example, the ISB. Scaling and extrapolation (S&E) relies on empirical observations of scaling relationships for similar systems in the past, extrapolating current trends (e.g. based on cumulative production volumes) into the future for the emerging technology. Participatory methods (PM) describe the inclusion of expert knowledge in some form, e.g. using techniques such as the Delphi method. Depending on the emerging material at hand, one or several of these methods may be suitable to gather the information necessary to conduct an LCA. The result of the scaling step will be, depending on the method, either a scaled life cycle impact assessment (LCIA) result, or a scaled LCI which can then be used to calculate the LCIA results.

The final, *fourth step (comparison)* consists of comparing the results for both the established and the emerging material. The comparison shall take into account the greater amount of uncertainty attached to the emerging material. If Monte Carlo simulations are conducted for both materials, then the results of these can in part reflect the underlying uncertainty. However, the uncertainty attached to the scaling step will probably be difficult to quantify. Furthermore, there are likely to be so-called "unknown unknowns", which can neither be quantified nor anticipated. These may describe a new type of environmental impact that is currently not sufficiently characterized by existing LCIA methods, a new technology affecting the production process, or any other unforeseen changes that may have an effect on the result of the assessment. Therefore, experts judging the results should always assume the actual uncertainty to be greater than what can be quantified and described in the assessment.

4 Exemplary Application

In this section, we discuss how the methodology may be applied to the case study of materials for vehicle seat covers, with leather and faux leather as the established materials and a fictitious, flax-based material as the emerging material. Leather is a well-established material for seat cover, and its production process has been refined and optimized over the course of several centuries. However, its environmental impact is non-negligible [31]. Several established alternatives with potentially lower environmental impact exist, such as artificial leather ("faux leather") and textiles. Also, some novel alternatives are currently being explored, such as a synthetic leather with fibres from pineapple leaves [32].

For the reference LCA step, the goal and scope should be defined first [18]. The intended application is defined as a comparative assessment of the overall environmental impact associated with leather, faux leather and a fictitious, plant-based alternative. The functional unit is 1 m^2 of car seat cover lasting one vehicle lifetime (10 years/200'000 km). Attributional modelling is employed, and the impact categories GWP and FEP (freshwater eutrophication potential) are used in the life cycle impact assessment.

The life cycle inventory for leather is based on the data provided in [33] for the Italian case study. The background data is modeled using the ecoinvent 3.6 cutoff database, with some adjustments. Some chemicals not included in the database were replaced with suitable alternatives, and in one case, where the chemical could not be identified ("auxiliaries"), left out entirely. Also, the upstream process chain of cattle raising is included, as recommended in [31]. From the same reference, the allocation factors are taken into account, to allocate environmental impacts to leather and other by-products such as meat. The result of the LCIA, a GWP of 31.35–45.26 kg CO_2eq. per m^2, is in partial accordance with results from an internal document for leather seat covers (an environmental product declaration (EPD) by a leather manufacturer) and results from the GaBi database on seat cover materials [34]. The divergence in the results stems from assuming different origins of the bovine hide, where hide from Brazilian cattle results in a higher impact than South African based cattle hide. The reference results (GaBi and EPD) fall within the range of 45–47 kg CO_2eq. per m^2.

For the second step (LCA of an emerging material), we base the life cycle inventory on that of the leather case study. Assuming the fictitious material will undergo the same finishing process as leather, we leave all inputs and outputs for the finishing process unchanged. The wastewater treatment, chrome recovery, slaughterhouse and cattle raising are removed from the assessment. Instead, for illustrative purposes, we assume an input of one kilogram of flax fiber per kilogram of material output. Using this approach, we can define the minimum environmental input of a fictitious material based on flax fiber, which undergoes a similar finishing process step as leather. More process steps than just the provision of flax fiber and the finishing process will be necessary for a material that could replace leather, hence, the environmental impact of this simplified life cycle inventory constitutes a lower bound. The LCA results for leather can be seen as an upper bound which shall not be exceeded. Lower and upper bound combined define a "corridor" for the material manufacturing process development. As for the stated assumptions, the LCIA results indicate a GWP of 1.87 kg CO_2eq. per kilogram of flax fiber based material.

The third step is included in the second step, as for the LCA of the fictitious emerging material, we applied the PTT_p method to create a life cycle inventory from the inventory of the reference material, leather. However, this inventory is not based on a real material, and the manufacturing of such a material would certainly require additional manufacturing steps, in addition to the provision of the flax fiber and the finishing process. The LCIA results for the fictitious flax fiber based material can therefore be seen as a minimum.

In the comparison step, the results for the LCIA of the reference material (leather) are compared to that of the emerging material. Additionally, we include the LCIA results of faux leather, which is taken from an existing, commercial LCA database [34]. To include variability in the assessment, we compare different upstream processes (cattle origin for leather and base material for faux leather). To highlight potential goal conflicts, we extend the assessment beyond GWP, and include FEP (freshwater eutrophication potential) as well. The comparison is illustrated in Fig. 2.

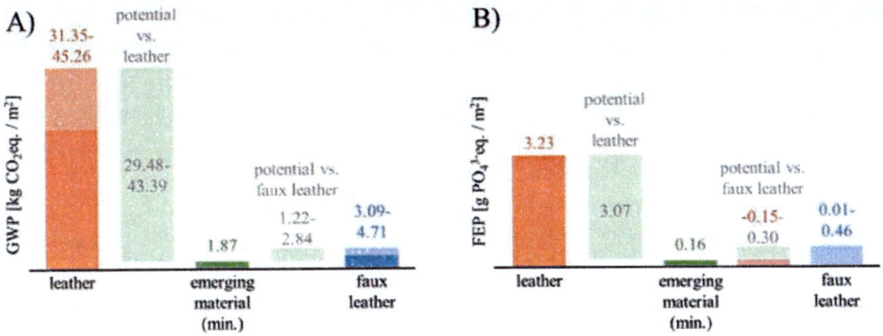

Fig. 2. Comparing GWP (A) & FEP (B) LCIA results for leather, faux leather and a fictitious, emerging material based on flax fiber. The potentials (light green/pink bars) indicate how much GWP/FEP the missing process steps in the material manufacturing process of the emerging material may cause before the environmental impact of the emerging material exceeds that of the reference material (pink bar: negative potential). The range of results for leather and faux leather can be explained by different upstream processes (origin of cattle for leather, base component for faux leather). For FEP, changing the upstream process from Brazilian to South African cattle does not cause a significant change in the results. For certain types of faux leathers (those using PET as a base component), the FEP is lower than that for the emerging material, hence the negative potential.

The comparison shows that for GWP, the resulting impact from the fictitious, emerging material based on flax fiber is the lowest. For FEP, some types of faux leathers have a lower environmental impact than the fictitious flax fiber material. Leather has the highest impact in both categories, regardless of variation in the upstream processes (cattle origin). The variability of results for faux leather stems from the base materials used (PVC, PUR or PET). The potential, or "corridor" for developing a material manufacturing process for a novel material based on renewable resources, indicates how much "environmental space" this process may occupy before it exceeds the impact of the reference material. The comparison shows that the extent of this corridor is, in this case, governed by the choice of the reference material. Since faux leather has proven its ability to fulfil the same function as leather in automotive seat cover applications, ambitious material research and development efforts should be directed at achieving an environmental impact that is lower than both leather and faux leather.

5 Discussion of the Methodology and Its Application

The methodology described here serves as a starting point for further research. Some need for improvement remains, which we briefly summarize here.

The most important step to develop the methodology further is to apply it to a real, emerging material. However, the challenge remains to get hold of relevant data that is not classified and can be used for research to be published. Also, for emerging materials, some information critical for an LCA may not yet be known, such as the expected lifetime of the material. Leather is known to last the complete lifetime of a car seat, for an emerging material this may not be known to a lack of long-term tests. Especially for bio-based materials, the expected amount of uncertainty will be high. Variations in growing locations, soil quality, use of fertilizers and pesticides, treatment of "waste" (incineration, composting) and other parameters have a major impact on the results, yet cannot always be determined with certainty. In this case, we recommend to conduct multiple LCAs with different assumptions, and compare these scenarios. Blanco et al. demonstrate how both parameter variations and more fundamental changes in the model can be addressed using Monte Carlo simulations [30].

Finally, since the assessment is future-oriented, changes in the background system should be included as well. Van der Giesen et al. discuss in their review the necessity for the adjustment of the background system (e.g. change in electricity mix), and list examples of studies which implemented these changes in LCA studies [34].

6 Conclusion & Outlook

The proposed methodology is a starting point for assessing the environmental impact of emerging materials. It is based on a review of the research literature on material selection and life cycle assessment, especially ex-ante life cycle assessment. The exemplary application has shown how the methodology could be applied to materials for car seat covers. Applying it to a real, emerging material may provide further insight into the strengths, weaknesses and opportunities for improvement of the methodology.

References

1. Allwood, J.M., Ashby, M.F., Gutowski, T.G., Worrell, E.: Material efficiency: a white paper. Resour. Conserv. Recycl. **55**(3), 362–381 (2011)
2. Hawkins, T.R., Singh, B., Majeau-Bettez, G., Strømman, A.H.: Comparative environmental life cycle assessment of conventional and electric vehicles. J. Ind. Ecol. **17**(1), 53–64 (2013)
3. Helmers, E., Dietz, J., Hartard, S.: Electric car life cycle assessment based on real-world mileage and the electric conversion scenario. Int. J. Life Cycle Assess. **22**(1), 15–30 (2017)
4. Ashby, M.: Materials and the Environment: Eco-informed Material Choice, 2nd edn. Elsevier/Butterworth-Heinemann (2012). https://www.elsevier.com/books/materials-and-the-environment/ashby/978-0-12-385971-6
5. Caduff, M., Huijbregts, M.A.J., Althaus, H.J., Hendriks, A.J.: Power-law relationships for estimating mass, fuel consumption and costs of energy conversion equipments. Environ. Sci. Technol. **45**(2), 751–754 (2011)
6. Villares, M., Işıldar, A., van der Giesen, C., Guinée, J.: Does ex ante application enhance the usefulness of LCA? A case study on an emerging technology for metal recovery from e-waste. Int. J. Life Cycle Assess. **22**(10), 1618–1633 (2017)
7. Piccinno, F., Hischier, R., Seeger, S., Som, C.: Predicting the environmental impact of a future nanocellulose production at industrial scale: application of the life cycle assessment scale-up framework. J. Clean. Prod. **174**, 283–295 (2018)

8. Buyle, M., Audenaert, A., Billen, P., Boonen, K., Van Passel, S.: The future of ex-ante LCA? Lessons learned and practical recommendations. Sustainability **11**(19), 1–24 (2019)
9. Ashby, M., Johnson, K.: The art of materials selection. Mater. Today **6**(12), 24–35 (2003)
10. Ashby, M. F., Miller, A., Rutter, F., Seymour, C., Wegst, U. G. K.: The CES Eco Selector – Background Reading. Design, (2009). https://www.researchgate.net/publication/268357513_The_CES_Eco_Selector_-_Background_Reading
11. Sun, M., Rydh, C.J., Kaebernick, H.: Material grouping for simplified product life cycle assessment. J. Sustain. Prod. Des. **3**(1/2), 45–58 (2004)
12. Ribeiro, I., Peças, P., Silva, A., Henriques, E.: Life cycle engineering methodology applied to material selection, a fender case study. J. Clean. Prod. **16**(17), 1887–1899 (2008)
13. Qiu, L.M., Sun, L.F., Liu, X.J., Zhang, S.Y.: Material selection combined with optimal structural design for mechanical parts. J. Zhejiang Univ. Sci. A **14**(6), 383–392 (2013)
14. Broeren, M.L.M., Molenveld, K., van den Oever, M.J.A., Patel, M.K., Worrell, E., Shen, L.: Early-stage sustainability assessment to assist with material selection: a case study for biobased printer panels. J. Clean. Prod. **135**, 30–41 (2016)
15. Hallstedt, S.I., Isaksson, O.: Material criticality assessment in early phases of sustainable product development. J. Clean. Prod. **161**, 40–52 (2017)
16. Kappenthuler, S., Seeger, S.: From resources to research—a framework for identification and prioritization of materials research for sustainable construction. Mater. Today Sustain, 100009 (2019)
17. Hauschild, M.Z., Rosenbaum, R.K., Olsen, S.I.: Life Cycle Assessment – Theory and Practice, 1st edn. Springer, Copenhagen (2018)
18. DIN EN ISO. DIN EN ISO 14040 – Umweltmanagement – Ökobilanz – Grundsätze und Rahmenbedingungen (2006)
19. Cox, B., Mutel, C.L., Bauer, C., Mendoza Beltran, A., van Vuuren, D.P.: Uncertain environmental footprint of current and future battery electric vehicles. Environ. Sci. Technol. **52**(8), 4989–4995 (2018)
20. Thonemann, N., Pizzol, M.: Consequential life cycle assessment of carbon capture and utilization technologies within the chemical industry. Energy Environ. Sci. **12**(7), 2253–2263 (2019)
21. Padey, P., Girard, R., Le Boulch, D., Blanc, I.: From LCAs to simplified models: a generic methodology applied to wind power electricity. Environ. Sci. Technol. **47**(3), 1231–1238 (2013)
22. Di Lullo, G., Gemechu, E., Oni, A.O., Kumar, A.: Extending sensitivity analysis using regression to effectively disseminate life cycle assessment results. Int. J. Life Cycle Assess. 25, 222–239 (2020)
23. Cucurachi, S., Van Der Giesen, C., Guinée, J.: Ex-ante LCA of emerging technologies. Procedia CIRP **69**(May), 463–468 (2018)
24. Arvidsson, R., et al.: Environmental assessment of emerging technologies: recommendations for prospective LCA. J. Ind. Ecol. **22**(6), 1286–1294 (2018)
25. Moni, S.M., Mahmud, R., High, K., Carbajales-Dale, M.: Life cycle assessment of emerging technologies: a review. J. Ind. Ecol. **24**(1), 52–63 (2020)
26. Thonemann, N., Schulte, A., Maga, D.: How to conduct prospective life cycle assessment for emerging technologies? a systematic review and methodological guidance. no. Lcc. **12**(3), 1192 (2020)
27. Simon, B., Bachtin, K., Kiliç, A., Amor, B., Weil, M.: Proposal of a framework for scale-up life cycle inventory: a case of nanofibers for lithium iron phosphate cathode applications. Integr. Environ. Assess. Manag. **12**(3), 465–477 (2016)

28. Gavankar, S., Anderson, S., Keller, A.A.: Critical components of uncertainty communication in life cycle assessments of emerging technologies: nanotechnology as a case study. J. Ind. Ecol. **19**(3), 468–479 (2015)
29. Wender, B.A., Seager, T.P.: Towards prospective life cycle assessment: Single wall carbon nanotubes for lithium-ion batteries. In: Proceedings 2011 IEEE International Symposium on Sustainable Systems and Technology ISSST 2011, pp. 1–4 (2011)
30. Blanco, C.F., et al.: Assessing the sustainability of emerging technologies: a probabilistic LCA method applied to advanced photovoltaics. J. Clean. Prod. 259, 120968 (2020)
31. De Rosa-Giglio, P., A. Fontanella, G. Gonzalez-Quijano, I. Ioannidis, B. Nucci, and F. Brugnoli. Product Environmental Footprint Category Rules – Leather (2019)
32. Nayak, R., Nguyen, L. V. T., Panwar, T., Jajpura, L.: Sustainable technologies and processes adapted by fashion brands. Elsevier Ltd (2020). https://www.sciencedirect.com/book/9780081028674/sustainable-technologies-for-fashion-and-textiles
33. Notarnicola, B., et al.: Life cycle assessment of Italian and Spanish bovine leather production systems. Afinidad **68**(553), 167–180 (2011)
34. van der Giesen, C., Cucurachi, S., Guinée, J., Kramer, G. J., Tukker, A.: A critical view on the current application of LCA for new technologies and recommendations for improved practice. J. Clean. Prod. **259**, 120904 (2020).

Systematic Design of Body Concepts Regarding Mini-Mal Environmental Impacts in an Early Concept Phase

Lars Reimer[1(✉)], Pavan Krishna Jois[1], Hartmut Henkelmann[1], Jens Meschke[1], Thomas Vietor[2], and Christoph Herrmann[3]

[1] Group Innovation, Volkswagen AG, Berliner Ring 2, 38440 Wolfsburg, Germany
lars.reimer@volkswagen.de
[2] Institute for Engineering Design, Technische Universität Braunschweig, Universitätsplatz 2, 38106 Braunschweig, Germany
[3] Chair of Sustainable Manufacturing and Life Cycle Engineering, Technical University of Braunschweig, Universitätsplatz 2, 38106 Braunschweig, Germany

Abstract. For internal combustion engine vehicles, the use stage dominates the life cycle emissions. In comparison, the life cycle emissions of battery electric vehicles highly depend on the electricity mix. With consideration of an European electricity mix the life cycle emissions split approximately equally between the production and use stage. Approximately 46% of these emissions is caused by the battery production. But the absolute and relative share of emissions from the vehicle production increase as well. Thus both stages have to be considered for the environmental assessment of body parts. Therefore the environmental impact of different material concepts as well as production and joining technologies are in focus of the development. A decision regarding environmentally optimized body concepts has to be made in the concept phase. A first approach provides mass indices from Ashby 1999. So, concepts made out of different materials can be developed in a given design space. These concepts are evaluated using a simplified life cycle assessment, which considers different body designs, mobility concepts and markets (electricity mixes). It can be shown that there is a large variance of greenhouse gas emissions for a given lightweight design potential. Hence, an optimization procedure to find concepts with the lowest environmental impacts is needed. In this paper a first approach for an optimization procedure concerning ecological aspects of body parts is described and demonstrated with an example application.

Keywords: Body parts · Optimization · Life cycle assessment · Life cycle engineering

K. Dröder and T. Vietor (Eds.): *Technologies for economic and functional lightweight design*, Zukunftstechnologien für den multifunktionalen Leichtbau, pp. 97–109, 2021. https://doi.org/10.1007/978-3-662-62924-6_9

1 Introduction

According to the UN Paris Agreement and for the limitation of the global warming, greenhouse gas (GHG) emissions have to be reduced. With approximately 18%, the transport sector causes a large part of these emissions [1]. The electrification of vehicles offers a potential for a mitigation. Depending on the electricity mix for charging, on one hand the emissions of battery electric vehicles (BEV) in usage decrease in part vastly in comparison to internal combustion engine vehicles (ICEV). On the other hand the emissions of BEVs shift into the production stage and consequently into the industry sector [2].

Applying the current European electricity mix the life cycle emissions of BEVs splits approximately equally between the production and the use stage. Thus both stages have to be considered for the assessment of BEVs [3]. Besides the battery production, which causes approx. 46% of the emissions in the production stage, the production of the vehicle body with approx. 43% offers a large potential for the reduction of the life cycle emissions of BEVs. [3] Therefore the impact of different body concepts in detail with various materials, production and joining technologies, are in focus of this research.

One method to find body concepts with minimal environmental impacts offers the life cycle engineering (LCE). LCE is a product based application for the analysis of ecological and economical properties also taking technical characteristics into account. The whole product life cycle is considered by a life cycle thinking approach [4]. Therefore LCE offers an approach for developing sustainable products with the premises of saving resources and protecting the environment [5]. In this context the life cycle assessment (LCA) offers an approach to quantify the environmental impacts of the product under development and their processes over the entire life cycle [6]. By means of LCA based LCE – an approach according to Herrmann et al. 2018 – the environmental hotspots of the product can be determined early in the product development process. Thus, measures for minimizing the environmental impacts caused by the product can be defined at an early stage [7]. Consequently LCA based LCE offers an approach for the development of body parts with minimal environmental impacts.

The development of body parts with minimal environmental impacts should be made in an early concept phase. Only this phase shows a sufficient design freedom. As a result, expensive change loops due to insufficient environmental compatibility of the body part can be avoided [8]. However, in an early concept phase various part designs are developed and evaluated. Within the design phase no detailed CAD- or FE-models are available. A simplified design methodology – for example according to mass indices, which are presented in Ashby [9] – shall be applied. Due to the absence of information regarding the production and use of the body part, different life cycle scenarios have to be analyzed. Consequently there are uncertainties with respect to the part design and the life cycle scenario. Nevertheless, in order to find the body concept with minimal environmental impacts, a simplified design methodology is needed. Based on estimating environmental assessment the design method should enable a fast selection of body concepts with minimal environmental impacts with regard to various material, manufacturing and joining technologies.

To support the simplified design methodology for the conceptual design of body parts with minimal environmental impacts the following three major requirements are derived:

I. **Requirements regarding the application procedure:** To guarantee the application in the concept phase, the design methodology has to be based on available information. The use of detailed CAD- or FE-models should be avoided. These models do not exist in an early concept phase. Therefore the design should be represented by simplified, analytically, computable mechanical analogous models. Due to the insufficient database, e.g. regarding the production and life cycle scenario of the body part, a simplification of the environmental assessment is needed. According to this, an environmental assessment methodology based on analytical equations with a few characteristics should be applied. Thus, trend statements regarding the environmental impact of different body concepts should be enabled.

II. **Requirements regarding the design methodology:** The design methodology should enable the design of body parts with arbitrary cross-section geometries. The design of multiple parts of one assembly in different materials should be enabled as well. In this context the mapping of the interfaces between the parts has to be considered. The design criteria should be the preservation of mechanical part properties, like equivalent bending stiffnesses. To guarantee a maximum potential of lightweight design, a shape optimization of the part geometries in a given design is recommended. In addition an intermediate coupling between the design and environmental assessment should be enabled for the aimed minimal environmental impacts.

III. **Requirements regarding the environmental assessment methodology:** The basic idea of an environmental assessment methodology according to DIN ISO EN 14040 is the life cycle point of view. The consideration of product caused environmental impacts along the entire life cycle, from the raw material extraction to the end of life processes, should be aligned with the environmental assessment methodology. In addition the assessment methodology should enable a simple consideration of various boundary conditions, like different mobility concepts and markets as well as future life cycle scenarios. Thus, first trends can be identified.

2 State of Research

Numerous publications address the conceptual design or the environmental assessment of body parts. But there are only a few studies, which consider the coupling between the conceptual design and the environmental assessment of body parts. In order to find the research gap, the common literature studies, which consider the coupling of design and environmental assessment were analyzed with respect to the described requirements. In this context, only studies with focus on the development of automotive components were considered. Table 1 show an overview on the identified literature studies. The black filling of the circles represents the fulfillment of the major tasks.

Table 1. State of research regarding the design of body parts with minimal environmental impacts

Literature study	**Ashby 1999**	**Kampe 2001**	**Ermolaeva et al. 2004**	**Giudice et al. 2004**	**Ribeiro et al. 2008**	**Wanner 2010**
Requirements regarding the application procedure						
Complexity of the design methodology	●	●	○	○	○	●
Complexity of the environmental assessment	●	●	○	○	○	●
Requirements regarding the design methodology						
Design of parts or assemblies with arbitrary geometries	◔	◔	◕	◕	◕	◔
Application of a shape optimization	○	○	◕	◔	◔	◑
Coupling with an environmental assessment	◕	◕	○	○	○	○
Requirements regarding the environmental assessment methodology						
Consideration of the entire life cycle	◔	◔	●	●	●	○
Consideration of various boundary conditions	◕	◕	◑	◑	◑	○

Literature study	**Mayyas et al. 2012a**	**Mayyas et al. 2012b**	**Poulikidou et al. 2015**	**Kleemann et al. 2016, 2017**	**Kaluza et al. 2016, 2017**	**Fröhlich et al. 2017**
Requirements regarding the application procedure						
Complexity of the design methodology	●	◔	●	●	●	●
Complexity of the environmental assessment	◑	◔	○	●	●	◕
Requirements regarding the design methodology						
Design of parts or assemblies with arbitrary geometries	◔	◕	◑	○	◑	◕
Application of a shape optimization	○	○	●	○	○	○
Coupling with an environmental assessment	◔	○	◔	◑	◑	◔
Requirements regarding the environmental assessment methodology						
Consideration of the entire life cycle	●	●	●	◑	●	◕
Consideration of various boundary conditions	◕	◑	◑	○	●	◕

Ermolaeva et al. 2004; Giudice et al. 2004; Ribeiro et al. 2008; Mayyas et al. 2012b; Kleemann et al. 2016/2017 focus the development of body parts with minimal environmental impacts basing e.g. on FE models for the design and complex LCA methodologies for the assessment of the environmental impacts. Due to this fact, an application of these methodologies is not possible in an early concept stage. A direct coupling between the disciplines design and environmental assessment is not investigated. However, Ashby 1999; Kampe 2001; Wanner 2010; Mayyas et al. 2012a focus at a design methodology based on mass indices. Due to this, these studies are not applicable for the design of parts or assemblies with arbitrary geometries. An application of an shape optimization of the part geometries is not possible. A direct coupling with an environmental assessment, which consider various life cycle stages and boundary conditions in the design process is basically possible. Only Poulikidou et al. 2015; Fröhlich et al. 2017; Kaluza et al. 2016/2017 enable a simplified design methodology regarding the design of parts or assemblies with arbitrary geometries and an environmental assessment, which consider various life cycle stages and boundary conditions.

Determination of the research gap

Due to different production technologies or part geometries the material yield $\eta_{RMat.}$ can vary vastly. The same applies for the assumed electricity mix $e_{Elec.}$ due to different markets. In addition, varying mobility concepts, like Mobility as a Service (MaaS) have great influence on the lifetime and driving cycle of the vehicle. All of these parameter have a high impact on the life cycle emissions. As a result of that, Fig. 1 shows the variability of the results of the environmental assessment regarding different boundary conditions and considering arbitrary lightweight design potential through body part concepts in different materials or production technologies. The hatched area in X-direction describes the solution space of possible results depending on the boundary conditions. It shows, that the results of the environmental assessment can vary vastly. Thus, a direct coupling between design and environmental assessment is needed for the determination of first trends for body part designs with minimal environmental impacts. Basically this requirement is fulfilled by e.g. Fröhlich et al. 2017 or Kaluza et al. 2016/2017.

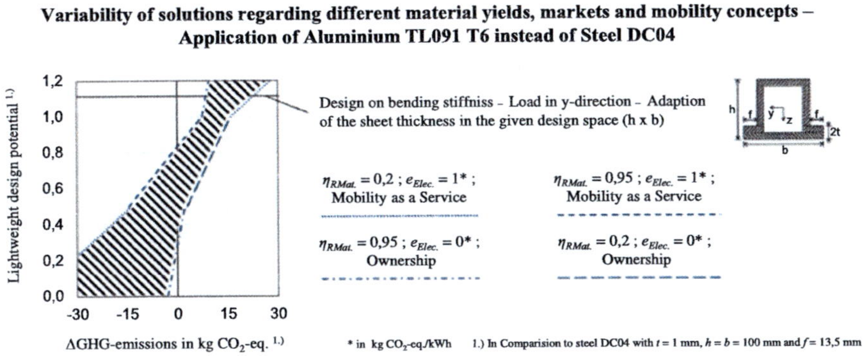

Fig. 1. Variability of solutions regarding different material yields, markets and mobility concepts; own calculation

The aforementioned studies takes only the adaption of one design parameter into account, like the sheet thickness of the investigated body part in different materials as the basis for the fulfillment of the mechanical properties. In this way, the lightweight design potential cannot be fully reached. For the achievement of the full lightweight design potential, an adaption of the whole geometry is needed. Hence, a shape and topology optimization of the geometry in a given design space, considering different materials and production technologies, is recommended. Accordingly, the solution space of possible results extends in Y-direction as well, Fig. 1. Thus, an optimization strategy to find the best combination of material and production technology for body parts with minimal environmental impacts is needed. Considering mechanical and environmental properties, the definition of an objective function for the calculation of the environmental impacts of a body part for an optimization of the body part can be stated as a research demand.

3 Conceptual Design of Body Concepts Regarding Minimal Environmental Impacts

3.1 Design Method

Figure 2 shows the concept for the design of body parts with minimal environmental impacts. Through this concept the conceptual design and environmental assessment are coupled. The methodology can be divided into the three pillar "Boundary conditions" (1), "Optimization procedure" (2) and "Solution" (3). The second pillar is subdivided into four main steps. These four steps include the concept derivation, the concept validation, the concept assessment and the final concept selection.

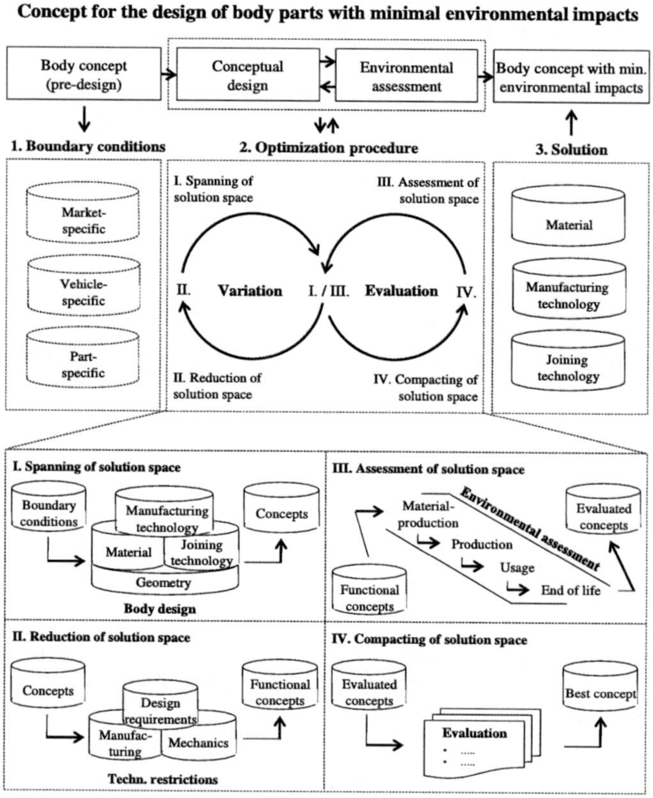

Fig. 2. Concept for the design of body parts with minimal environmental impacts

Boundary conditions

Initially the boundary conditions regarding the design of the part and the following environmental assessment have to be defined. Thereby it can distinguished between market-, vehicle- and part-specific influence factors on the life cycle ΔGHG-emissions of body parts.

I. **Market-specific:** Due to different electricity mixes, the ΔGHG-emissions of vehicles depend on the market, where the vehicles are produced or operated. The assumed future scenario also has an influence due to the electricity mix and following on the ΔGHG-emissions as well.

II. **Vehicle-specific:** Vehicles can be operated in an Ownership or Mobility as a Service (MaaS) use case. Vehicles with respect to MaaS are assumed to undergo extended mileages (600.000 km vs. 200.000 km in Ownership). At the same time, the driving cycle are different. Vehicle in a MaaS use case operate primarily in urban space. Conversely vehicle in an Ownership use case operate according to the average WLTC driving cycle. This has an influence on the energy reduction value of the vehicle. This value defines the energy saving through lightweight design.

III. **Part-specific:** Due to different geometries and loads as well as design criteria of the parts to be investigated, the lightweight design potential varies depending on e.g. alternative material concepts or production technologies. Through various manufacturing concepts the material yield can vary as well.

Optimization procedure

The optimization procedure to find the body concept with minimal environmental impacts in a given design space regarding the predefined boundary conditions can described in four main steps:

I. **Spanning of solution space:** Initially various body concepts are defined in a given design space. Here the material, manufacturing and joining technology as well as the part geometry can be varied.

II. **Reduction of solution space:** In the next step, the defined body concepts get analyzed e.g. regarding design requirements or mechanical and manufacturing restrictions. Only concepts, which fulfill these requirements and restrictions, are pursued.

III. **Assessment of solution space:** The functionally validated concepts get evaluated with a simplified environmental assessment methodology.

IV. **Compacting of solution space:** Finally the assessment results are analyzed and the functional concept with minimal environmental impacts get selected.

Solution

The result of the optimization procedure is the body concept comprising a combination of material, manufacturing and joining concepts, with minimal environmental impacts.

3.2 Application of the Optimization Procedure

To make a decision about body concepts with minimal environmental impacts, a strategy is applied to different scenarios, considering all the influencing factors, to find the best solution. The classical intuitive-iterative designing process is replaced by the mathematical design optimization process. The complexity of such an optimization increases drastically with increased variables and results in increased computational times. To steer clear of this issue, it is often better to choose an appropriate optimization method. In addition the part geometry can also be simplified as the study is conducted in concept phase. This includes the approximation of the geometry to analytically calculable profile cross sections as well as neglecting of elements such as radii and draft angles.

The objective function of a mathematical optimization is a simple quantification of a required result. Objective function $f(x)$ as shown in Eq. 1, calculates ΔGHG-emissions in different scenarios regarding the assumed boundary conditions. It is important to express emissions as function of the design vector: $x = \{h_i, b_i, t_i\}$. The parameter *hi*, b_i and t_i characterizes the dimensions of the body part. Figure 1 shows an example for that. The ΔGHG-emission calculations include a wide variety of combinations of production and joining technology to accommodate different profile designs with varying height, widths and thickness of the body part. The technical constraints regarding the design space are the lower and upper bounds of the optimization problem. The mechanical properties can be classified as the functional constraints of the part under study.

$$min.\ \Delta GHG - emissions\ as\ f(x)\ with \begin{cases} Mech.\ properties_{(Ref)}(x) - Mech.\ properties_{(Final)}(x) \leq 0 \\ h_{min} \leq h_i \leq h_{max};\ b_{min} \leq b_i \leq b_{max};\ t_{min} \leq t_i \leq t_{max} \\ for\ i = 1, \ldots,\ number\ of\ profiles \end{cases} \quad (1)$$

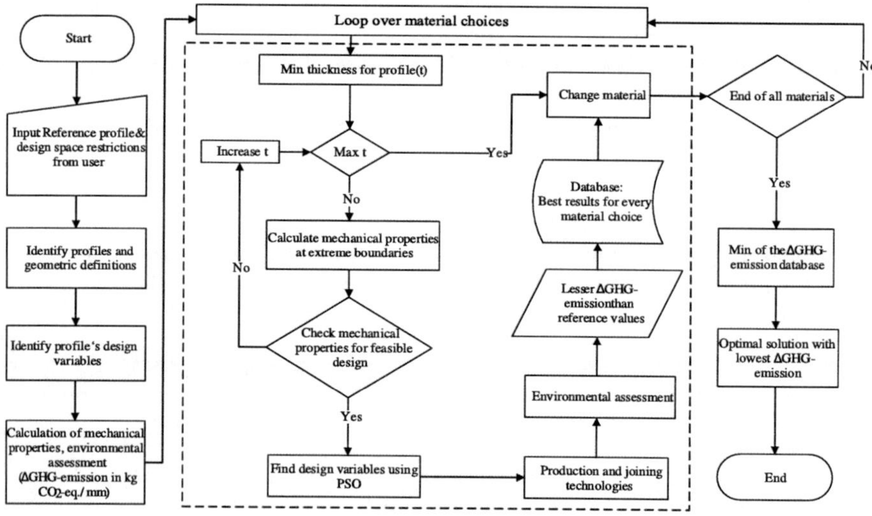

Fig. 3. Representation of the optimization procedure

The next step in solving the problem is to identify the optimization algorithm to be used. Given the fact that the function to calculate mechanical properties is highly non-linear, the analytical sensitivity cannot be derived "manually". Therefore, a gradient-based optimization method is inappropriate for directly solving this problem. To reduce computational costs and to make this complex problem tractable, a heuristics based evolutionary optimization method, e.g. a particle swarm optimization (PSO) is employed. The implementation of the optimization algorithm is based on the concept described in Fig. 2, which employs searching for an optimal solution in multivariable optimization case by spanning, reducing, assessing and compacting the solution space of the problem, also known as a domain reduction strategy. This works on the principle of eliminating less promising solution space, thereby enhancing the performance of optimization algorithm.

The developed algorithm is capable of handling multiple profiles with varying cross sections, different materials and manufacturing methods. Figure 3 illustrates one of such algorithms, e.g. with bending stiffness as design criteria for a one profile cross sectional shape. As previously mentioned, due to concept phase the profile is approximated to rectangular cross sections and height, width and thickness are considered as primary variables. The optimization methodology is illustrated in Fig. 3. The input design concept from user is marked as reference profile for the algorithm. The reference profiles are generally a commonly used, intuitive cross section of an automobile part under study. It is important for the user to accurately identify and define the cross sectional parameters as input. The reference material is chosen to be steel, so as to find a replacement for the most commonly used material in the automotive industry. Input also includes type of vehicle, its purpose and market scenario.

With the initial design, all the necessary parameters are calculated. The parameters are evaluated by multiple auxiliary functions, which work alongside the optimizer. These parameters include the geometrical definitions of the profile, like segment IDs, height, width, length of the walls of beam, area moment of inertia and others. A reference value for both mechanical property and environmental impact is computed. The optimizer is constrained to find a solution whose mechanical property is equal or higher than the reference mechanical property value and whose environmental impact is lower than that of the reference. Once the reference values are established, the next step in Fig. 3 is addressed. The core of the optimization strategy, as previously described, is to reduce the solution space efficiently. Given different material and manufacturing concepts chosen for the design evaluation, the solver evaluates all the profiles with multiples cross sections across all combinations. This forms the first important loop of the optimization strategy. As in this case, e.g. with 1 profile and 4 materials under investigation, there will be a total of 4 material choices and a single best possible solution in each case is found, if it exists. If multiple profiles exists in the input, the algorithm checks for all the material combinations instead. The simulation terminates, when all the material choices were checked.

In addition, at each iteration of the primary loop, optimal values of $x = \{h, b, t\}$ are to be found. The secondary loop runs over the thickness (t) of profile, as thickness can be pre-discretized due to profile's manufacturability restrictions. As an initial guess, the limits of the stiffness values are first checked for extreme values of the design variables in the design space. This possibility of a feasible solution is verified if the part profile with maximum boundary values for the height and width have a stiffness value higher or equal to the reference profile. With help of limiting stiffness values, the appropriate solution space for particle swarm optimization algorithm is fixed. Given a material choice, fixed thicknesses of the profile and a feasible solution space, the only two unknowns variables – height h and breadth b are found, while respecting the functional constraints. Until this point, the profiles, which respect the functional constraints are found. As a next step, with all the necessary information about the geometrical aspect of the profile, environmental assessment of the profile is done to check ΔGHG-emissions. The algorithm is capable of deciding new possible manufacturing method and adjust the environmental assessment values based on varying dimensions of the profiles. The database stores all emissions and other relevant information about each solution case like stiffness values, mass, etc. The algorithm is terminated when all the material choices are exhausted and the final design with the lowest environmental impact is chosen and displayed to the user as an output.

4 Verification of the Concept by Means of a Case Study

Figure 4 shows a door shell concept for an electrically driven concept vehicle. As a case example, the side crash beam (in blue), which is integrated into the door, is considered. Assumptions regarding the design concept, geometry and design space are shown in Fig. 4. Besides the requirement of the static and dynamic stiffness of the door an essential task of the side crash beam is the requirement of structural integrity to protect occupants from intrusions. A failure of the side crash beam should be avoided in case of a side crash. Due to these facts, the side crash beam is designed on bending strength.

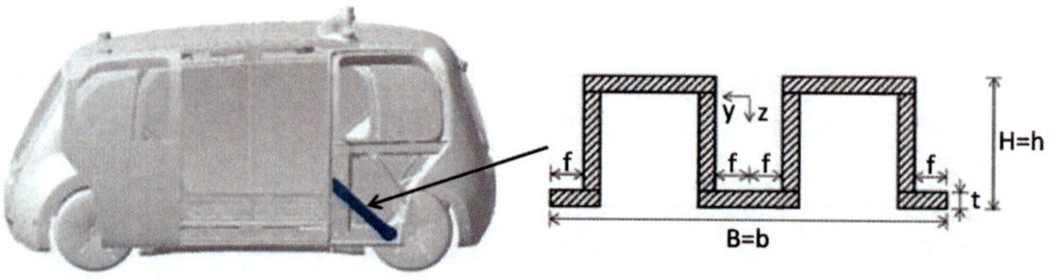

Design concept	Geometry	Design space	Design criterion
• Shell design	• $h = 25$ mm	• $H = 25$ mm	• Bending strength
• High ductility steel	• $b = 100$ mm	• $B = 100$ mm	
• Mass $m = 1{,}3$ kg	• $f = 10$ mm		
• Material yield $\eta_{R.Mat.} = 0{,}6$	• $t = 1{,}35$ mm		

Fig. 4. Definition of a case study

To find the concept with minimal environmental impacts, the methodology described before is applied. In this context different steel and aluminium concepts in shell and profile design get analyzed. Materials with a higher strength like the reference material are not considered, due to the fact that a higher strength in the door body would lead to a decreased deformation into the interior in case of a side crash. This results in an increase of the occupant accelerations and should be avoided.

Table 2. Results of the environmental assessment

Results of the environmental assessment										
Nr.	Design concept	Material	Geometry *h* [mm]	 *b* [mm]	 *f* [mm]	 *t* [mm]	Visualisation	Lightweight design potential	ΔGHG-emissions Ownership/ Europe[1.)] [Δ kg CO_2-eq.]	 MaaS/ China[2.)] [Δ kg CO_2-eq.]
R	Shell design	High ductility steel	25	100	10	1,35		1,0	0,0	0,0
1	Shell design	High ductility steel	25	100	11,9	1,2	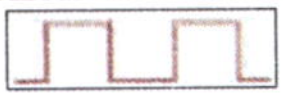	0,9	- 0,9	- 3,9
2	Shell design	Automotiv aluminium	25	100	10,6	3,9	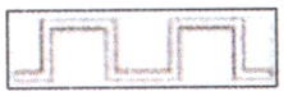	0,9	+ 13,8	+ 11,8
3	Profile design	Automotiv aluminium	25	100	0	1,9		0,7	+ 2,4	- 6,8

1.) Mileage s_{LC} = 200000 km, Emission in energy supply $e_{Elec.}$ = 0,42 kg CO_2-eq./kWh, Energy reduction value e_{ERV} = 3e-5 kWh/kgkm;
2.) Mileage s_{LC} = 600000 km, Emission in energy supply $e_{Elec.}$ = 0,84 kg CO_2-eq./kWh, Energy reduction value e_{ERV} = 5e-5 kWh/kgkm

Two different life cycle scenarios are defined. In the first scenario the vehicle operates in an Ownership use case on the European market. The second scenario considers a MaaS approach on the Chinese market. Both scenarios describe extreme cases regarding the variability of LCA results. Table 2 shows a selection of the results of the lightweight design potential and the ΔGHG-emission savings, which can be achieved in comparison to the reference concept (R). For example, a small lightweight design potential can be achieved by optimizing the cross-sectional geometry of the reference concept (R) while retaining the material. As a result, greenhouse gas emissions can be reduced slightly in the assumed life cycle scenarios.

Furthermore, a small lightweight design potential can be achieved with an aluminium concept in shell design (2). Due to the small lightweight design potential and the emission extensive material production of aluminium, this concept shows no potential for ΔGHG-emission saving in comparison to the reference concept. Therefore, this concept is not considered further. The aluminium concept in profile design (3) can achieve by a better utilization of the design space a much higher lightweight design potential. However, due to the emission intensive material production process of aluminium, this concept cannot achieve ΔGHG-emission savings compared to the reference concept in today's life scenarios. However, considering future life cycle scenario (MaaS/Chinese market) this aluminium concept in profile design (3) is advantageous regarding the life cycle ΔGHG-emissions. Due to higher mileages and energy reduction values as well as emissions in energy supply, the emission extensive aluminium production can be compensated through less emission in the use stage. Consequently the lightweight design potential becomes more important.

Finally, the results in Table 2 suggest that no ΔGHG-emission savings can be achieved with aluminium for structural parts with a medium strength in today's life cycle scenarios. If the assumed future scenarios are taken into account, potential ΔGHG-emission savings can be achieved with aluminium. Depending on the reached lightweight design potential, aluminium concepts can even be advantageous independently of the market. While the ΔGHG-emission saving potential for vehicles in Ownership is very low, a significant saving potential can be achieved in a MaaS vehicle operation. As a result, the use of aluminium in profile design for strength-relevant structural components with medium strength, like the investigated side crash beam, is to be recommended in the future.

5 Conclusion

In this work, a methodology is shown, which enables an optimization of body parts regarding minimal environmental impacts. The methodology is applied in the concept phase. In this context different material and manufacturing concepts are considered during the definition of possible concept designs. Different load cases and design criteria are considered as well. Furthermore an adaption of the geometry in a given design space is enabled to find concepts with the greatest possible lightweight design potential.

In the next step, the defined body part concepts are evaluated with a simplified environmental assessment. The assessment methodology is based on the calculation of the life cycle ΔGHG-emission and considers the variability of life cycle scenarios, like different mobility concepts or markets. Finally the body part concept with minimal environmental impact can be found with the described optimization procedure.

In order to assess the ΔGHG-emission saving through body concepts with minimal environmental impacts, the next step is the application of the described design methodology on a whole body of a vehicle concept. Due to that, potential ΔGHG-emission savings can be quantified and compared to other measures for the reduction of the greenhouse gas emissions in the life cycle of a vehicle.

References

1. Bundesministerium für Umwelt, Naturschutz und nukleare Sicherheit: Klimaschutz in Zahlen (2018). Accessed 17 Jan 2020
2. Cerdas, F., Egede, P., Herrmann, C.: LCA of electromobility. In: Hauschild, M.Z., Rosenbaum, R.K., Olsen, S.I. (eds.) Life Cycle Assessment: Theory and Practice, pp. 669–693. Springer International Publishing, Cham (2018)
3. Kreyenberg, D.: Fahrzeugantriebe für die Elektromobilität: Total Cost of Ownership, Energieeffizienz, CO2-Emissionen und Kundennutzen, Dissertation (2016)
4. Dieter, G.E.: Materials Selection and Design, 10th edn. ASM International, Materials Park, Ohio (1997)

5. Kaluza, A., Kleemann, S., Broch, F., Herrmann, C., Vietor, T.: Analyzing decision-making in automotive design towards life cycle engineering for hybrid lightweight components. Procedia CIRP **50**, 825–830 (2016). https://doi.org/10.1016/j.procir.2016.05.029
6. DIN EN ISO 14040:2009-11, Umweltmanagement_-Ökobilanz_-Grundsätze und Rahmenbedingungen (ISO_14040:2006); Deutsche und Englische Fassung EN_ ISO_14040:2006 (2009)
7. Herrmann, C., Dewulf, W., Hauschild, M., Kaluza, A., Kara, S., Skerlos, S.: Life cycle engineering of lightweight structures **67**(2), 651–672 (2018). https://doi.org/10.1016/j.cirp.2018.05.008
8. Ehrlenspiel, K., Meerkamm, H.: Integrierte Produktentwicklung: Denkabläufe, Methodeneinsatz, Zusammenarbeit, 6th edn. München, Wien, Hanser. www.hanser-fachbuch.de/buch/Integrierte+Produktentwicklung/9783446440890
9. Ashby, M.F.: Materials and the Environment: Eco-informed Material Choice, 2nd edn. Elsevier Butterworth-Heinemann, Amsterdam (2013)
10. Ashby, M.F.: Materials Selection in Mechanical Design, 2nd edn. Elsevier Butterworth-Heinemann, Amsterdam (2004)
11. Kampe, S.: Incorporating green engineering in materials selection and design. In ASEE Annual Conference Proceedings (2001)
12. Ermolaeva, N.S., Castro, M.B., Kandachar, P.V.: Materials selection for an automotive structure by integrating structural optimization with environmental impact assessment. Mater Des **25**(8), 689–698 (2004). https://doi.org/10.1016/j.matdes.2004.02.021
13. Giudice, F., La Rosa, G., Risitano, A.: Materials selection in the life-cycle design process: a method to integrate mechanical and environmental performances in optimal choice. Mater. Des. **26**(1), 9–20 (2005). https://doi.org/10.1016/j.matdes.2004.04.006
14. Ribeiro, I., Peças, P., Silva, A., Henriques, E.: Life cycle engineering methodology applied to material selection, a fender case study. J. Cleaner Prod. **16**(17), 1887–1899 (2008). https://doi.org/10.1016/j.jclepro.2008.01.002
15. Wanner, A.: Minimum-weight materials selection for limited available space. Mater. Des. **31**(6), 2834–2839 (2010). https://doi.org/10.1016/j.matdes.2009.12.052
16. Mayyas, A.T., Qattawi, A., Mayyas, A.R., Omar, M.A.: Life cycle assessment-based selection for a sustainable lightweight body-in-white design. Energy **39**(1), 412–425 (2012). https://doi.org/10.1016/j.energy.2011.12.033
17. Mayyas, A.T., Qattawi, A., Mayyas, A.R., Omar, M.: Quantifiable measures of sustainability: a case study of materials selection for eco-lightweight auto-bodies. J. Cleaner Prod. **40**, 177–189 (2013). https://doi.org/10.1016/j.jclepro.2012.08.039
18. Poulikidou, S., Schneider, C., Björklund, A., Kazemahvazi, S., Wennhage, P., Zenkert, D.: A material selection approach to evaluate material substitution for minimizing the life cycle environmental impact of vehicles. Mater. Des. **83**, 704–712 (2015). https://doi.org/10.1016/j.matdes.2015.06.079
19. Kleemann, S., Türck, E., Vietor, T., D. C. M.: 1.-1. 2. International design conference-design 2016, Towards knowledge based engineering for multi-material design (2016). https://doi.org/10.24355/dbbs.084-201704031138
20. Kleemann, S., Inkermann, D., Bader, B., Türck, E., Vietor, T.: Semi-formal approach to structure and access knowledge for multi-material-design (2017). https://doi.org/10.24355/dbbs.084-201708301114
21. Kaluza, A., Kleemann, S., Fröhlich, T., Herrmann, C., Vietor, T.: Concurrent design & life cycle engineering in automotive lightweight component development. Procedia CIRP **66**, 16–21 (2017). https://doi.org/10.1016/j.procir.2017.03.293
22. Fröhlich, T., Kleemann, S., Türck, E., Vietor, T.: Multi-criteria analysis of multi-material lightweight components on a conceptual level of detail (2017). https://doi.org/10.24355/dbbs.084-201709070942

Processing Capabilities for Thermoplastic Composites – Minimum Material Consumption and Recyclability

Norbert Müller(✉) and Philipp Seinsche

Center for Lightweight Composite Technologies, ENGEL AUSTRIA GmbH, Steyrer Str. 20, 4300 St. Valentin, Austria
{norbert.mueller,philipp.seinsche}@engel.at

ABSTRACT. An important measure for improved lightweight performance is utilisation of materials only to the extent, as it is necessary for the application. To achieve a minimum part weight in high volume lightweight applications, it is necessary to combine thermoplastic composite sheet materials exposing different thicknesses. Then, a minimum material consumption is gained. Such composite components have the capabilities of being fully recyclable. Used parts as well as scrap and cut-off that arises during production can be recycled and used for injection moulding of ribs and other geometry, even to become part of the same composite component later on. However, there are considerable challenges in the processing since different composite sheets need to be processed simultaneously. A processing unit was designed, build and tested that is capable to process thermoplastic composite sheets with three different thicknesses. The resultant part, a structural automotive door component, exposes areas were the composite sheet thickness is 0.6 mm, 1.0 mm, and 2.5 mm, respectively. Besides of the injection moulding, the processing requires two different kind of infrared ovens with a specialized software for the control of the heating and three articulated robots for the automated handling operations of the composite blanks.

Keywords: Processing of thermoplastic composites · Infrared heating · Injection moulding · High volume production

1 Introduction

On the one hand, by means of injection moulding, very complex geometries can be realized at reasonable cost. This, for example, is the case with automotive parts as air intake systems or oil pans. On the other hand, the production of composite structures with continuous fibre reinforcement enables parts with high lightweight performance that expose extraordinarily high strength and stiffness.

K. Dröder and T. Vietor (Eds.): *Technologies for economic and functional lightweight design*, Zukunftstechnologien für den multifunktionalen Leichtbau, pp. 110–121, 2021. https://doi.org/10.1007/978-3-662-62924-6_10

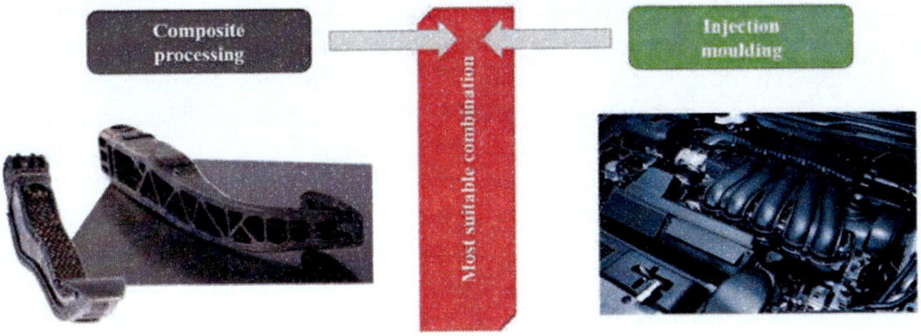

Fig. 1. Identification of the most suitable combination of design freedom plus functional integration (from injection moulding) and mechanical performance (from the utilization of thermoplastic composite materials)

The challenge, in order to establish a composite processing technology within the typical circumference of the injection moulding is, to combine the design freedom and cost effectiveness of injection moulding with the strength and stiffness of a continuous fibre reinforced composite (Fig. 1). Therefore, to find the most reasonable combination is the target of any product and processing innovation for high volume composite manufacturing. The key factors for success are use of thermoplastic composite materials and utilization of processing technologies that are fully automatable. The major intention of the processing of thermoplastic composite sheet materials is to reach the performance levels of continuous fibre reinforcement. This needs to be achieved in a typical cycle time of the injection moulding machine. Furthermore, extensive functionalization of the composite part is enabled by means of injection moulding [1].

2 State of Research an Applications

Common thermoplastic composite materials for the large volume applications consist of continuous glass fibres and a matrix from polypropylene or polyamide. The fibre reinforcement can either be one or several layers of woven fabric, or non-woven layers prepared from unidirectionally reinforced tapes. The subsequent processing of woven or non-woven materials is quite similar. The thermoplastic composite materials are delivered as sheets where the standard thicknesses range from 0.5 mm to about 4.0 mm. From these sheets, a net-shape cut-out is prepared normally by waterjet cutting. The cut-out is heated in an infrared oven fairly above the matrix polymer's melting temperature, transferred in an injection moulding mould, shaped by closing the clamp, and overmoulded with ribs and other geometry by means of injection moulding.

The standard processing setup is heating of a single cut-out from a thermoplastic composite sheet that has one specific thickness. Therefore, the entire production unit can be adjusted to that material processing conditions. In such a way, e.g. door-modules and frontend-beams are produced. The material's thickness affects the process conditions very intensely. As thinner the material is, as quicker the heating, but as less time is left for the transport of the heated cut-out to the mould.

The one-step process sequence described above is the most efficient way. Variants of the processing consider a two- or multiple-step approach. The most effortful sequence, which is reasonable for highly demanding parts as for aerospace applications, is first to heat a rectangular plate or a pre-cutout, shape this in a match mould, perform a three-dimensional cutting, re-heat the composite part and finally overmould that in a separate injection moulding mould. From the viewpoint of the resultant mechanical performance, the one-step route and the multi-step-version are supposed to be comparable. However, the multi-step approach can deliver improved surface quality and dimensional precision, in particular of the composite fraction [2].

Combining two or more cut-outs from the thermoplastic composite sheet can be beneficial for several reasons. It helps to reduce the scrap rate when an odd-shaped cut-out is combined from two pieces. The scrap rate with a conventional way of preparing thermoplastic composites for the processing often is seen above 25% [3]. Besides of scrap reduction, by patching the thermoplastic composite, a local reinforcement can be achieved and in the case of parts, where the load conditions differ significantly in the various regions, a distinct adjustment of strength and stiffness according to the actual loading is possible when using sheets with different thickness.

In this paper, a processing investigation on the manufacturing of a structural components that contains thermoplastic composite cut-outs with three different thicknesses is presented. The thickest sheet (2.5 mm) has more then 4 times the thickness of the thinnest sheet (0.6 mm) which is a novel combination for the processing. There is a third, intermediate thickness of 1.0 mm, which, as seen from the thickness viewpoint, is closer to the 0.6 mm sheet, but can be handled similar as the 2.5 mm sheet. This means, from a processing viewpoint, the exception is the 0.6 mm materials since is can be heated fast, even in a single sided infrared oven, but cools down also very fast. The latter is the actual challenge.

3 Investigations and Results

3.1 Processing of Thermoplastic Composite Blanks

A typical setup for the processing of thermoplastic composite materials consists of at least an injection moulding machine, an oven for the infrared heating, and an automation system, which is either a linear robot or an articulated 6-axis-robot. Dependent on the application and the machine setup, an infrared oven in horizontal or in vertical orientation is utilised (Fig. 2).

Thermoplastic composite materials with a thickness up to about 1.0 mm, usually can be heated quick enough by means of a single-sided infrared oven. Thicker materials require a heating of the blank from both-sides to remain with the heating and

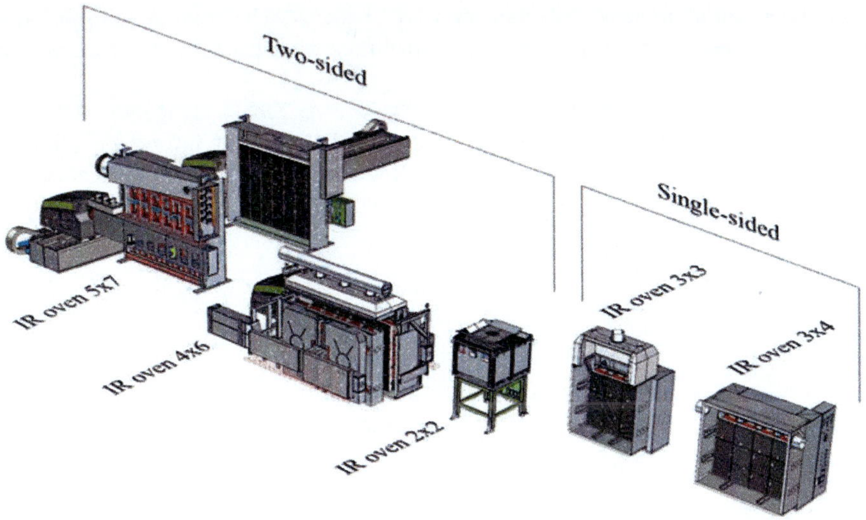

Fig. 2. Vertical infrared ovens for single-sided and two-sided heating of thermoplastic composite blanks

handling time within the injection moulding cycle. This means, the heating process should never become the determining factor for the overall cycle time.

In a one-shot composite moulding process, the thermoplastic composite sheet first is heated up in an infrared oven. With the automation system, the heated blank is transferred into the mould. A vertical clamping unit allows simplified placement of the heated blank on the lower half of the mould. When a conventional injection moulding machine is used, where the clamping direction is horizontal, some sort of a fixation of the heated blank in the mould is necessary.

This fixation can either be achieved with stationary pins, retractable clamping bolts, or mechanical grippers. After transfer of the heated blank, the mould is closed quickly, and the still hot thermoplastic composite sheet is draped into the final shape. The hot handling is critical with respect to time. Whilst a fairly thick heated thermoplastic composite blank, e.g. 2.5 mm allows a duration for the transfer of up to about 10 s, very thin thermoplastic composite blanks, e.g. 0.5 mm need to be placed in the mould and shaped within just a few seconds.

Another factor is the type of the fibres as well as of the thermoplastic resin used for the composite blanks. Since carbon fibres have a significantly higher thermal conductivity than glass fibres, they cool down a lot quicker in air. Furthermore, as higher the thermoplastic matrix needs to be heated for the shaping, as faster the cooling during the hot handling takes place. This is one of the reasons while the hot handling of e.g. polyamide-based thermoplastic composite materials is more time-critical than the handling of heated blank made with polypropylene polymer matrix.

The fibres within the sheet can either be woven fabric or a non-woven laminate. At current state, processing of thermoplastic composite sheet materials with imbedded woven fabrics inside has achieved the highest feasibility [4]. The non-woven cross-ply materials are an alternative that may offer cost benefits. The materials are produced from thermoplastic composite tapes. Such materials were used for the investigations (Table 1).

All materials were supplied by Kingfa Science & Tech, Guangzhou (CN). The thermoplastic composite sheets are balanced continuous fibre reinforced blanks with polypropylene as matrix. The injection moulding resin is a high flow long glass fibre polypropylene compound. The selection of the materials as well as of the processing methods were primarily cost driven. A cross-ply materials based on glass fibre reinforced tapes offers high structural performance at reasonable cost. The polypropylene matrix has a low density of 0.92 g/cm^3, which contributes to the overalls lightweight effect.

Table 1. Materials used for the investigations

Area/Purpose	Grade	Layers	Thickness
Centre region/main faces	Kingply KP-GFPP7004 BK001	2	0.6 mm
Window parapet/patch	Kingply KP-GFPP7004 BK001	4	1.0 mm
Window frame/structural	Kingply KP-GFPP7010 BK001	10	2.5 mm
Ribs/overmoulding	Kingfa GFPP-L30 HSBK001(RHM)	–	–

From the processing method's aspects, the intention was to generally stay with the one-shot approach that is most productive and cost effective. However, it was found that the shaping and overmoulding of the 2.5 mm thick cut-out is highly demanding since the requirement was overmoulding from both sides. Therefore, the method was changed from a typical one-shot process to a one-and-a-half-shot process, where the heated cut-out is shaped first in a separate draping-and-shaping cavity, but which is depicted in the same injection moulding mould.

During the heating, the tape based thermoplastic composite materials behave similar as materials containing woven fabric. However, they expose a different drapability and show an enhanced tendency to fill up rip cavities during the reconsolidation. Nevertheless, these issues can be solved with proper process settings and a precise match of cavity gap to the actual thickness of the composite blank.

3.2 Indicators for Performance Characteristics

In general, a lightweight component is characterised as relevant performance level, most time strength and stiffness, with respect to the component's weight. As higher such a reading is, as better the lightweight performance. However, in an environment, where cost and productivity are paramont, the approach normally differs. The intention is not to end up with the utmost lightweight performance, but to achieve a considerable improvement in the lightweight characteristics whilst maintaining the cost situations as well as the productivity that is present in the well-established production lines.

Reducing the composite thickness of large regions of a structural component is a contribution to both, less material utilization and less energy consumption for the heating. Therefore, during the design work and the structural simulation work, processing aspects already need to become involved. I.e., the structural component is set up in a fashion, were the processing can be expected to be still possible within a typical injection moulding cycle time. The relevant indicator for the performance of a processing approach is, if a stable, robust, and fully-automated processing can be achieved. The benchmark was the technology readiness level of existing one-shot processing of thermoplastic composites materials where only one cut-out is utilized, which is state of the art. The measure was, whether, with the solution proposed, the same technology readiness is possible to be achieved, i.e. when three different cut-outs are processed at once.

3.3 Variable Thickness of the Composite Within the Same Part

The use case for the investigations was the entire inner of an automotive door. In such a component, areas exist where primarily a thin, but crash resistant shell is necessary. Within the scope of the design work on the part's concept and geometry, which was performed by Brose Fahrzeugteile GmbH & Co. KG, Hallstadt (D) it was found that this requirement can be solved with a thermoplastic composite sheet of just 0.6 mm thickness (Fig. 3).

However, other areas, as for example the window parapet, require more strength and stiffness. Is was figured out that the local requirement might be solved with a patch having 1.0 mm of thickness. Since the patch is amalgamated with a 0.6 mm sheet, the resultant thickness of the window parapet region is 1.6 mm. A highly demanding region of a door inner component is the window frame. It needs to exhibit outstandingly high strength and stiffness. Furthermore, the window frame is a region that is visible for the passengers. Therefore, also optical quality characteristics were of concern. For the region of the window frame, a 2.5 mm thermoplastic composite blank was used.

As the project was a proof-of-concept, which had to show how the processing is set up appropriately, it was not necessary to produce the full door inner. A section of the intended

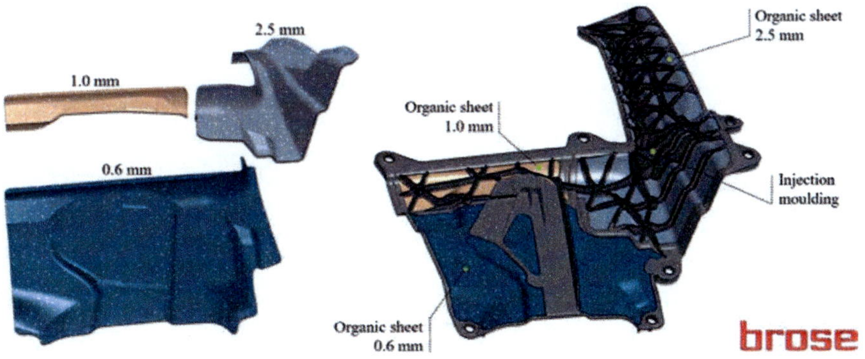

Fig. 3. Automotive door structure from thermoplastic composite materials and injection moulded fractions

part was chosen that contains the most critical regions and representative functional elements. These are the main door faces, the window parapet, and the window frame.

3.4 Combination of Horizontal and Vertical Infrared Ovens

The most challenging process steps for the production of the part were the heating and the handling of the thermoplastic composite sheets with three different thicknesses. For the 1.0 mm and the 2.5 mm material heating in a regular horizontal IR oven with a two-sided arrangement of radiators was employed. The heating of the 0.6 mm material took place in a separate, vertical IR oven that performed the irradiation single-sided (Fig. 4).

The 2.5 mm thermoplastic composite sheet requires the longest heating time. Additional soaking time is necessary to reach the required temperature range also in the material's centre region. The 0.6 mm material can be heated very rapidly, even when just single-sided irradiation is used. Corresponding to the thickness, the allowable handling time of the 2.5 mm organic sheet is much longer than that of the 0.6 mm material, which has to be placed in the mould very rapidly. For this reason, it is essential to locate the vertical oven as close as possible to the mould.

When the same heating setup is used for the 2.5 mm and the 1.0 mm material, then the 1.0 mm material heats up a lot faster. This is not intended, because the material would stay at high temperature longer than necessary. Therefore, an advanced IR oven control system was established that allows to automatically adjust the heating rate in each control region to the same value (Fig. 5). This means, the heating has the same slope at each point where the temperature is measured for control purpose. These are three spots on the top and three spots on the bottom of the blanks.

The system does not only level out the characteristic deviation with respect to the heating behaviour between the 1.0 mm and the 2.5 mm material. It also takes care for

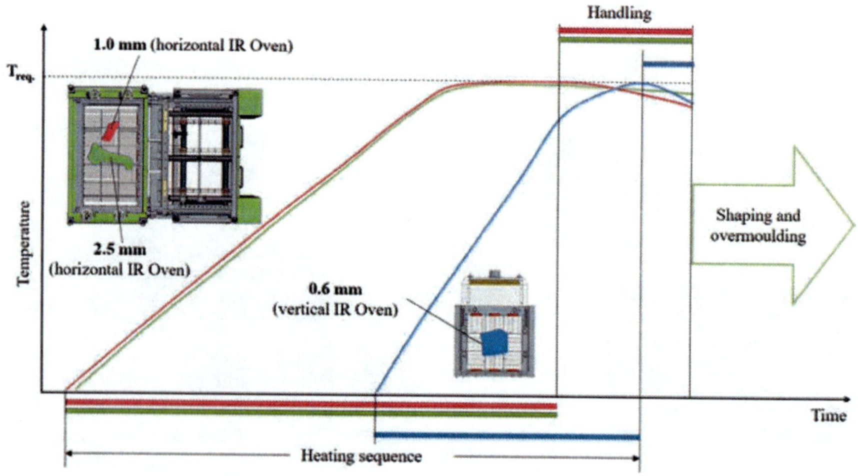

Fig. 4. Heating and handling sequence for the thermoplastic composite sheets with 0.6 mm, 1.0 mm, and 2.5 mm thickness, resp.

improved heating even within the 1.0 mm and the 2.5 mm material, resp. Each controlled region of the blank reaches the desired temperature at the same time whilst the duration of the entire heating sequence stays the same.

With this heating setup, it is possible to have the thermoplastic composite sheets heated within the typical injection moulding cycle time, precisely at the required temperature, with the necessary soaking time considered for the centre region to reach the temperature range for the shaping and overmoulding – and all of that at minimum material impairment, since the time at elevated temperatures is kept as short as possible.

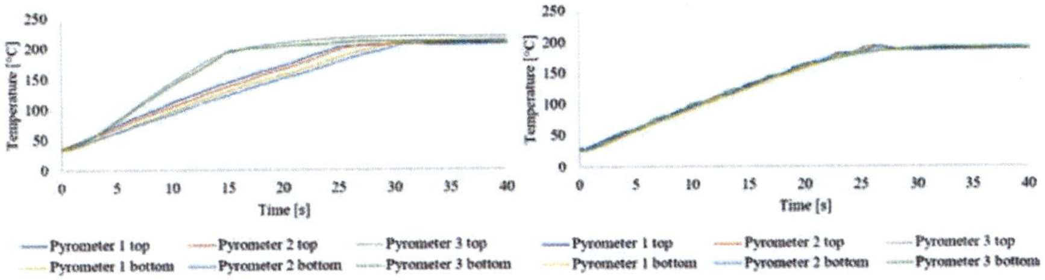

Fig. 5. Heating behaviour of the 2.5 and 1.0 mm material at conventional IR oven control (left) and with utilization of advanced IR oven control (right). Pyrometers 1 and 2 (top and bottom) face the 2.5 mm sheet. Pyrometers 3 (top and bottom) face the 1.0 mm sheet

3.5 Machinery, Mould, and Automation

The shaping of the heated thermoplastic composite blanks and the injection moulding were carried out with an 800 tons two-platen injection moulding machine (ENGEL duo 800/3660). Besides of that, two IR ovens and three articulated robots were necessary for the fully automatic production of the structural door part (Fig. 6). Since the 2.5 mm material was intended to be overmoulded full face on the visible side (inner) and to be equipped with stiffening ribs on the backside (outer), it was necessary to carry out a pre-shaping and re-consolidation of that blank in a separate preforming cavity (Fig. 7). The preforming cavity was operated at elevated mould temperature of 120 °C.

This allowed, on the one hand, already to have a solid and stiff composite component for the next step, which, on the other hand, is still hot enough to end up with

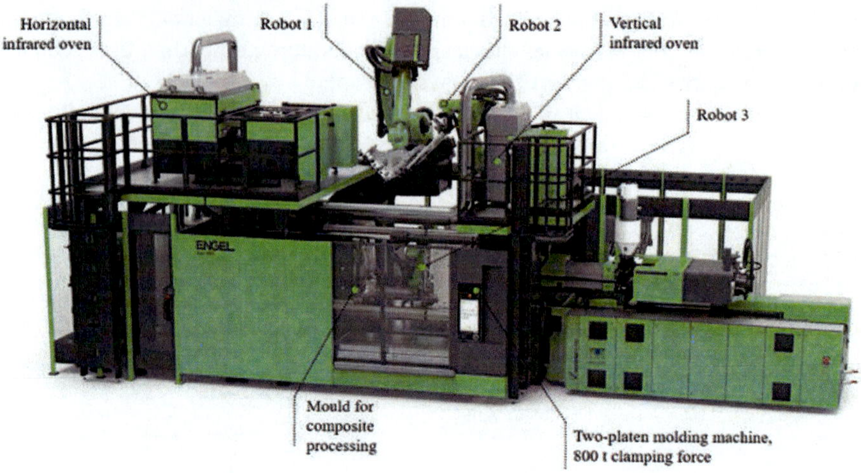

Fig. 6. Injection moulding machine with IR ovens and automation

appropriate adhesive bonding to the thinner composite blank as well as to the injection moulding fraction.

The operations of the three articulated robots of the production cell are denoted in Fig. 8. Robot 1 is picking up all three blanks from a magazine and places two of

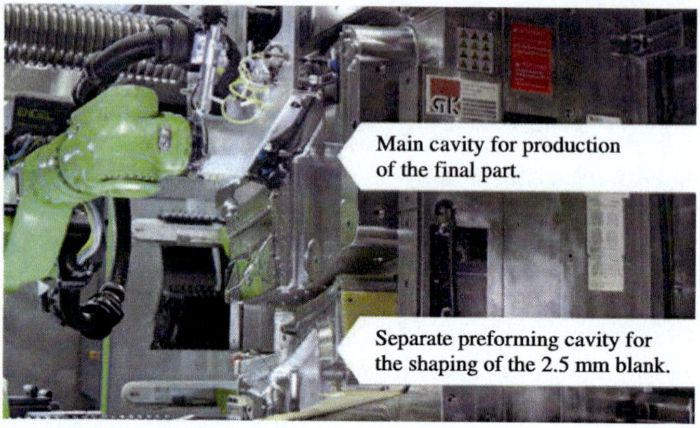

Fig. 7. Mould for the thermoplastic composite processing with a main cavity and a preforming cavity for shaping and reconsolidation of the 2.5 mm blank

them, the 1.0 mm and the 2.5 mm blank, on the tray of the horizontal IR oven. The thin 0.6 mm blank is transferred by means of a robot handshake operation to robot 2. The 0.6 mm blank stays clamped in the end-of-arm-tool of robot 2 during the heating in the vertical IR oven.

Once the mould is opened, robot 3 picks the finished part to demould it. Furthermore, it picks the pre-shaped composite component (2.5 mm) to transfer it into

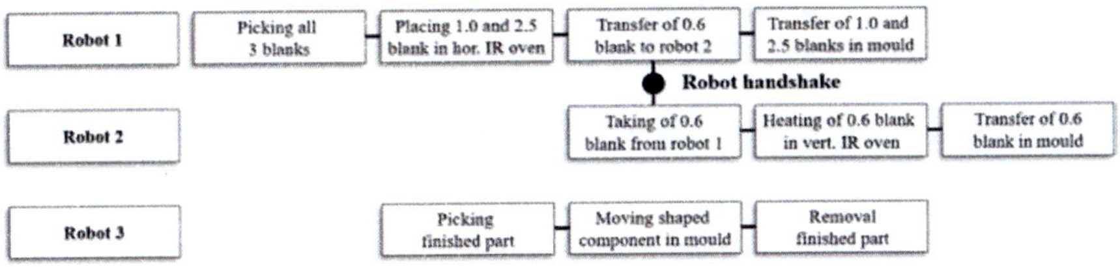

Fig. 8. Operations of the three articulated robots of the production cell

the now vacated main cavity. When this is done, robot 3 brings the finished structural door component to a conveyor belt, Fig. 9. At the start of the next cycle, robot 1 and 2 place the heated blanks in the mould.

The entire productions cell enabled a fully automated production at an overall cycle time of 70 s. All control and data acquisition tasks are performed by the injection moulding machine's control. This means that not only the injection moulding itself in controlled by the machine, but also the two IR ovens as well as the three articulated robots of the process automation. Furthermore, any relevant processing parameters can be shown on a separate control page. These are, for example, robot positions, level of magazines, heating time, current sheet temperature, and number of good parts produced.

Fig. 9. Removal of the finished structural door part

3.6 Recyclability of Thermoplastic Composite Materials

The entire part consists of just one type of fibres, which is continuous glass fibres in the composite fraction and chopped glass fibres in the injection moulding fraction. The only thermoplastic material contained in the component is polypropylene. However, the polypropylene grade in the composite fraction differs from the grade in the injection moulding fraction.

Recycling of continuous fibre reinforced waste was already investigated extensively with carbon-PPS, where subsequent compression moulding was employed to produce the parts from the recycled material [5]. It was found that a major concern is low shear mixing in order to retain the fibre length. Similar correlations are expected when a glass-fibre-PP material is recycled in order to end up with an injection moulding grade.

Consequently, the cut-off from the thermoplastic composite sheets as well as the entire part might be grinded after end of use and recycled to an injection moulding grade of glass fibres and polypropylene. A compounding after grinding is expected to be necessary for several reasons. First, the fibre content needs to be adjusted to a specific value. Second, the material might be equipped with stabilization for the next processing sequence and service period. When the recycling including the compounding is carried out adequately, the material might even be used for moulding of the injection moulding fraction of a similar component as it was before.

Also, during the production of an actual thermoplastic door inner, some waste will appear. On the one hand, cutting the thermoplastic composite sheet leads to a certain amount of scrap. On the other hand, a small fraction of reject in the running production, as well as some waste parts from start-up and shut-down arise. These material fractions and waste parts might also be utilised as components of the recycling PP-GF grade. Therefore, the technology allows a 100% re-use of both, the thermoplastic composite material and the injection moulding grade.

4 Conclusions and Outlook

For a widespread use of thermoplastic composite components, especially in automotive applications, the processing needs to be highly cost efficient. Furthermore, to end up with a high lightweight effect, the composite fraction of the final part should be adjusted precisely to the final product's demands. This often leads to a part design where thickness variations in the composite fraction may occur.

A promising way, to cover such designed demand's is, to assemble several thermoplastic composite blanks with different thicknesses directly in the mould. Within the scope of the project, which was set up as a proof-of-concept, it was proven that a fully automated production of structural components, which consist of thermoplastic composite blanks of different thickness, can be established. Even the production of components with a wide range of composite thickness is possible, for example from 0.6 to 2.5 mm as it was demonstrated within the project.

However, specific characteristics of the materials with respect to their heating behaviour in an IR oven and the cooling during the transfer to the mould require specialized technical solutions. Especially very thin thermoplastic composite materials need to be treated separately. With an adequate combination of sophisticated IR ovens, automation, and end-of-arm-tools for the hot handling, a process setup is possible where the efficiency comes close to the typical conditions that are well known from injection moulding. This is an important key to success: When the production of thermoplastic composite component is performed in the same cycle time as a comparable injection moulding process would be done and the entire production is operating fully automatically, then a competitive economic condition is reached.

The processing setup aimed on a proof-of-concept for structural parts that consists of thermoplastic composite components with three different thicknesses as well as of an injection moulded fraction. It was found that, similar as with a one-shot process for a component that contains only one composite sheet with uniform thickness, a robust and fully-automated production is possible within a typical cycle time of the injection moulding of just 70 s. Whilst the proof-of-concept was finalizes successfully, the outlook focuses on extending the technology to larger parts in order to cover typical dimensions of structural components for automotive applications.

Acknowledgements. The joined R&D project was performed in close cooperation with Brose Fahrzeugteile GmbH & Co. KG, Hallstadt (D), HRSflow, San Polo di Piave (IT), Kingfa Science & Tech, Guangzhou (CN), and Georg Kaufmann Formenbau GmbH, Remetschwil (CH). The authors gratefully acknowledge the contributions of the partners.

References

1. Müller, N.: Converging technologies – Manufacturing of thermoplastic composites and injection molding of structural parts. In: 18th Automotive Composites Conference & Exhibition (SPE ACCE), Conference Proceedings, Society of Plastics Engineers, Novi, USA (2018)
2. Müller, N.: Manufacturing of load optimized structural parts from thermoplastic tapes with focus on flexibility and efficiency. In: 4th International Conference & Exhibition on Thermoplastic Composites (ITHEC), Conference Proceedings, Messe Bremen, Bremen (2018)
3. Kropka, M., Mühlbacher, M., Neumeyer, T., Altstädt, V.: From UD-tape to final part – a comprehensive approach towards thermoplastic composites. In: 1st Cirp Conference on Composite Materials Part Manufacturing, pp. 96–100 (2017)
4. Thienel, M.: Organobleche – Die Zukunft für Leichtbautüren in der Großserie. In: NMB TechDays – Vom UD-Tape zum thermoplastischen Faserverbundbauteil im Minutentakt, Fachtagung, Neue Materialien Bayreuth, 16.–17.03.2016, Bayreuth (2016)
5. de Bruin, T.A., Vincent, G., van Hattum, F.W.J.: Recycling C/PPS laminates into long fibre thermoplastic composites by low shear mixing. In: 21st International Conference on Composite Materials, Xian, China (2017)
6. Akkerman, R., Bouwman, M., Wijskamp, S.: Analysis of the thermoplastic composite overmolding process: interface strength. Front. Mater. **7**, 27 (2020). https://doi.org/10.3389/fmats.2020.00027

Generative Manufacturing

Evaluation of Technologies for the Fabrication of Continuous Fiber Reinforced Thermoplastic Parts by Fused Layer Modeling

Daniel Pezold[1](✉), T. Rosnitschek[2], A. Kleuderlein[1], F. Döpper[3], and B. Alber-Laukant[2]

[1] Manufacturing and Remanufacturing Technology, University of Bayreuth, Universitaetsstr. 30, 95447 Bayreuth, Germany
{daniel.pezold,andre.kleuderlein}@uni-bayreuth.de
[2] Engineering Design and CAD, University of Bayreuth, Universitaetsstr. 30, 95447 Bayreuth, Germany
{tobias.rosnitschek,bettina.alber}@uni-bayreuth.de
[3] Fraunhofer Institute for Manufacturing Engineering and Automation IPA – Bayreuth, Universitaetsstr. 9, 95447 Bayreuth, Germany
frank.doepper@ipa.fraunhofer.de

Abstract. In this research technologies for the production of continuous fiber-reinforced thermoplastics using additive manufacturing are investigated and evaluated. The focus is on the "Fused Layer Modeling" (FLM) process, which is based on an additive, thermoplastic extrusion process. The possibility of combining the plastic filament with continuous fibers allows a specific fiber reinforcement to be introduced into the part to increase the mechanical properties. First, an overview of the technologies for processing continuous fibers is presented. These strategies differ in the design of the machine (hardware) and the possibilities for the constructive insertion of the continuous fibers in slicing (software). The differences of the technologies are the processing method of the fiber as well as the fiber roving used in the extrusion process. The maximal fiber volume content and the interlaminar fiber-matrix adhesion are investigated in various commercial technologies by means of tensile and bending tests. In conclusion, the different technologies are evaluated with regarding the maximal fiber volume content and quality of interlaminar fiber-matrix adhesion.

Keywords: Additive manufacturing · Continuous fiber · Carbon fiber · Fiber-plastic composites

K. Dröder and T. Vietor (Eds.): *Technologies for economic and functional lightweight design*, Zukunftstechnologien für den multifunktionalen Leichtbau, pp. 125–141, 2021. https://doi.org/10.1007/978-3-662-62924-6_11

1 Introduction

In recent decades, fiber-reinforced plastics have become more and more important in technical applications. This has led to an increase in the number of high-performance lightweight structures in various industrial sectors [1]. Compared to most alternative materials, carbon fiber reinforced plastics (CFRP) offer significant weight reductions possibilities in terms of strength and stiffness as well as thermal properties [1, 8]. Additive Manufacturing (AM) represents a promising technology for the production of prototypes and functional parts with complex structures. Although the AM of plastics has increased significantly in recent years, most of the parts manufactured by additive manufacturing are used in the field of prototype development because the material properties are not yet sufficient compared to conventional injection molded parts [1–8].

Although it is state of the art in injection moulding to use different fibers (short or long fibers) to extend the range of application. The AM of fiber-reinforced plastics has a great potential to extend the area of application of new fiber materials. Fiber reinforcement in AM in particular reinforcement with continuous carbon fibers (CCFRAM), can be used to significantly improve the mechanical properties of plastic parts and to optimize their load-bearing capacity [2–4, 8, 10, 12–15]. In addition, as [1] emphasizes, the elastic properties of continuous fiber reinforced parts in particular are significantly higher than those of short fiber reinforced parts. Thus increase the potential for applications that go beyond prototype production.

The aim of this research work is to evaluate test specimens with the maximal fiber volume content using commercially available technologies for CCFRAM by FLM. Furthermore to investigate and compare them with respect to the achievable mechanical properties for tensile and bending strength.

2 Technologies for the Production of Continuous Fiber Reinforced Thermoplastics Using FLM

Beyond its initial role as a prototype manufacturing process, the FLM has begun to evolve into a manufacturing process for series parts. Previously achieved mechanical properties of the manufactured parts using FLM have been further enhanced by optimizing the process parameters and by adding fillers such as nanomaterials, particles or fibers. These composite materials are characterized by high performance and very good functionality. However, the mechanical properties of the additive manufactured composites are still relatively low compared to composites manufactured with conventional methods. FLM promises to be an alternative to conventional process chains for the production of continuous fiber reinforced thermoplastic parts, since no cost-intensive equipment such as tools or autoclaves is required [20, 21]. There are two strategies for the insertion of the continuous fibers into the plastic filament. On the one side, the co-extrusion process for combining the fiber with the matrix can be carried out in the extruder unit. On the other side, the co-extrusion process can be upstream to external machines. In general, the dimensions of the continuous filament must be taken into account when introducing the continuous filament as it cannot be processed like the plastic due to its stiffness.

Markforged uses the latter strategy of the upstream co-extrusion process to insert the continuous filament in the machine. In Markforged's **Continuous Filament Fabrication (CFF)**, the extruder unit consists of the composite extruder and is extended by an additional extruder for processing a pure plastic filament. For the processing of continuous filaments, a cutting unit is necessary for cutting the fibers (Fig. 1). With the dual extrusion method different fiber volume contents can be realized in the parts. The fiber volume content of Markforged composite filaments is 34.5% [5, 20].

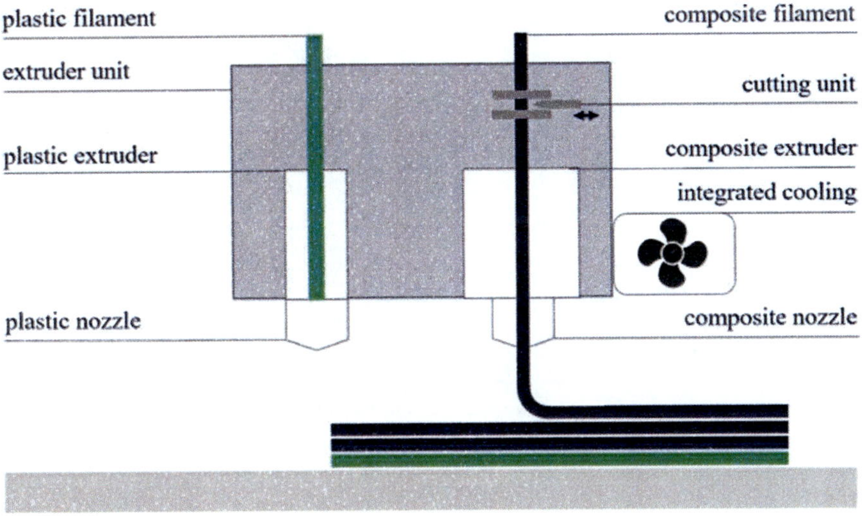

Fig. 1. Continous Filament Fabrication (CFF) – FLM machine Markforged

The definition of the fiber arrangement is done in the web-based Eiger software. For the orientation of the fibers in the part, the arrangement of the part in the installation space is initially decisive, as the fibers can only be inserted in the XY plane. The continuous fiber reinforcement can be carried out layer by layer in the defined patterns for the surface (concentric/isotropic) and number of fiber paths for wall areas. It is not possible to define manually the course of the continuous fiber or a cutting point of the continuous fiber. The processing temperatures during the manufacturing process are predefined.. A wall and a top/bottom layer of plastic must be foreseen for the manufacturing of parts with maximal fiber volume content on the Markforged machine.

Anisoprint's **Composite Filament Co-Extrusion (CFC)** process has integrated the co-extrusion process into the extruder unit. In the manufacturing process, the coated fiber is fed perpendicular to the building platform and the plastic filament is fed laterally into the composite extruder. The coated fiber can be combined with any thermoplastic material. The fiber volume content of Anisoprint's composite filaments has the highest fiber volume content of the technologies considered at 60% [18] It is significantly reduced by the integrated co-extrusion process. Anisoprint's machine works with the dual extrusion method described above. In addition, the feed rate of the plastic filament can be changed during the depositing process of the composite filament. This gives Anisoprint a second possibility to modify the fiber volume content. Due to the co-extrusion process in the machine, three filaments (two plastic, one continuous fiber) are necessary for the operation of the machine (Fig. 2).

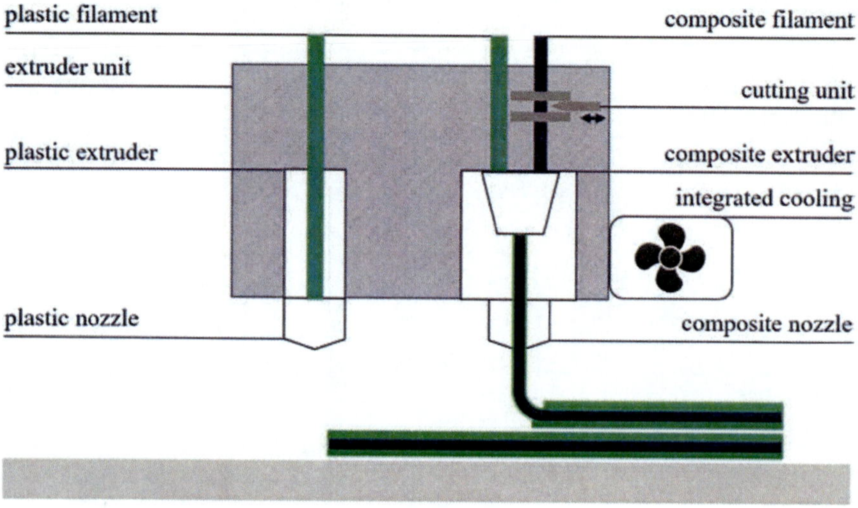

Fig. 2. Composite Filament Co-Extrusion (CFC) – FLM machine Anisoprint

Anisoprint's software Aura is used for the arrangement of the part and thus the fiber. Similar to Markforged only defined patterns can be selected for fiber placement. Due to the open material system the processing parameters for the plastic filament must be entered into the database before the manufacturing process starts. Anisoprint does not require a top or bottom layer for the manufacturing of parts with maximal fiber volume content. However, a wall layer must be foreseen.

The **Additive Fusion Process (AFP)** from 9T Labs uses the strategy of the upstream co-extrusion process and the dual extrusion method. The fiber volume content of the composite filament of 9T Labs has the second highest fiber volume content of the presented FLM machines at 50% [19]. Another difference to Markforged is the external cooling of the extruder unit. To prevent oozing of the plastic the extruders have to be cooled and that leeds to a high downtime during the manufacturing process when changing to the second material (Fig. 3). The processing temperatures can be set in the same way as it can be set for Anisoprint.

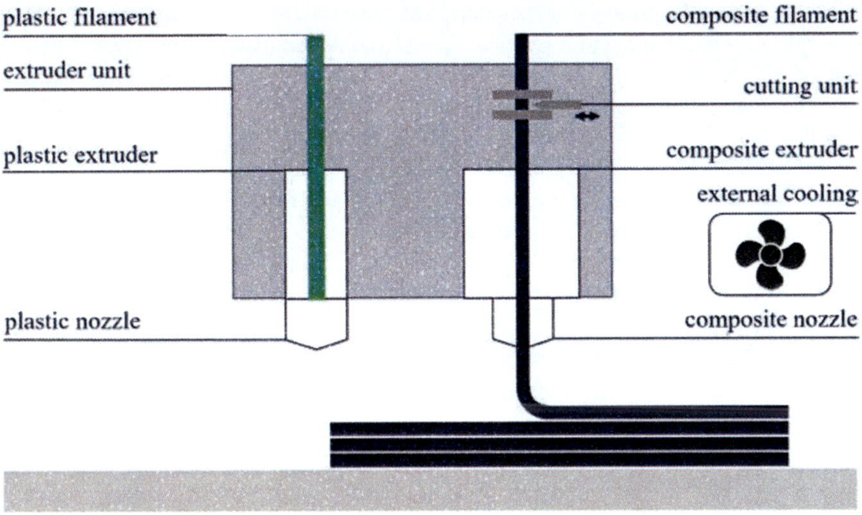

Fig. 3. Additive Fusion Technology (AFT) – FLM machine 9T Labs

9T Labs also works also with its own software. A file for continuous fiber reinforcement and a file for the plastic must be loaded into the Fibrify software. Subsequently, the corresponding materials and parameters are assigned to the parts. At 9T Labs, the arrangement options are determined primarily by the design. In addition, known patterns for fiber placement can be defined in the software. The positioning of the parts is determined by the transfer of the coordinate system from the design file. In contrast to Markforged and Anisoprint, no restrictions are specified for wall, top or bottom layers.

On the commercial market, Desktop Metal also offers a further strategy for the production of CCFRAM with **Micro Automated Fiber Placement (μAFP)**. This is a combination of the FLM process and the tape laying process [22]. The strategy of **Continuous Fiber 3D Printing (CF3D)** of Continuous Composites is also available on the commercial market. In CF3D, fast-curing thermosetting resins are deposited together with the continuous fiber [23]. Due to the very different processing of the continuous fibers in μAFP and the use of a thermoset matrix in CF3D, these two strategies are not considered in this research paper.

3 Experimental Methodology

3.1 Process Parameters

To ensure the comparability of the different technologies, carbon fiber was defined as the fiber material and polyamide as the matrix material for the manufacturing of test specimens. The Markforged MarkTwo with Markforged's composite filament Carbon and Markforged's plastic filament Nylon-White was used to manufacture the

Markforged test specimens. For Anisoprint test specimens, the Composer A4 with the composite filament Composite Carbon Fiber from Anisoprint and the plastic filament Luna PA 12 (polyamide 12) from 9T Labs were used. For the test specimen manufacturing, the CarbonKit (beta test phase) from 9T Labs was used as a conversion kit for the Ultimaker S3 Extended with the composite filament Kepler CFRP and the plastic filament Luna PA 12 from 9T Labs.

The layer heights of the individual technologies differ significantly (Table 1). Markforged has the lowest layer height of 125 µm and requires a plastic layer (wall/top/bottom layer). The deviation in the layer height from the standard value of 360 µm for anisoprint is due to the increase in the plastic feed rate. The standard value was increased by 25%. In addition, a wall layer is required for Anisoprint. 9T Labs works with a layer height of 270 µm and does not specify a plastic layer.

Table 1. Overview of material and layer data

	Plastic	Continuous fiber	Layer height (µm)	Wall layer	Bottom/Top layer
Markforged	Nylon	C-Roving, preimpregnated	125	Yes	Yes
Anisoprint	PA 12	1,5-K C-Roving	500	Yes	No
9T Labs	PA 12	24-K C-Roving, preimpregnated	270	No	No

For the manufacturing of the Markforged test specimens, the specified temperature settings were used. For the manufacturing of the test specimens using Anisoprint and 9T Labs, the same temperature was set on all extruders due to the same material. The building platform temperature corresponds to room temperature for Markforged and the recommended parameters were adopted for Anisoprint and 9T Labs (Table 2).

Table 2. Overview of process temperature

	Plastic extruder-temperature	Composite extruder-temperature (°C)	Building platform temperature
Markforged	232 °C	273	RT
Anisoprint	250 °C	250	40 °C
9T Labs	–	200	90 °C

Table 3 lists the specific material properties from the manufacturers[1] data sheets.

[1] As no explicit data on the fibre characteristics for Markforged are available, the values for the linear mixing rule were chosen based on the data for standard C-fibres in [16] and nylon matrix.

Table 3. Material data and fiber volume of test specimens

	Tensile strengh plastic (MPa)	Tensile strengh fiber (MPa)	E-Modulus plastic (GPa)	E-Modulus fiber (GPa)
Markforged [16, 17]	51	2258	1.7	230
Anisoprint [18]	60	2206	1.4	149
9T Labs [19]	60	4300	1.4	240

3.2 Tensile Test

The tensile test is one of the most frequently used test methods for determining mechanical material properties. In the following tensile test specimens are manufactured with the described technologies and analyzed. Force-extension diagrams are recorded and stress–strain diagrams are determined. The material-characteristic strength values are derived from them and compared [9].

The focus of the test specimen design for the tensile test is to achieve the maximal possible fiber volume content in the standardized geometry. The technology-specific layer height results in the variation of the fiber-reinforced layers between the manufacturing technologies. For example, while two fiber-reinforced layers are required for Anisoprint, Markforged manufactures test specimens with six fiber-reinforced layers. 9T Labs uses only the composite filament, which means that the fiber volume content corresponds to that of the composite filament. Figure 4 shows the test specimen structures resulting from the restrictions of the three technologies explained.

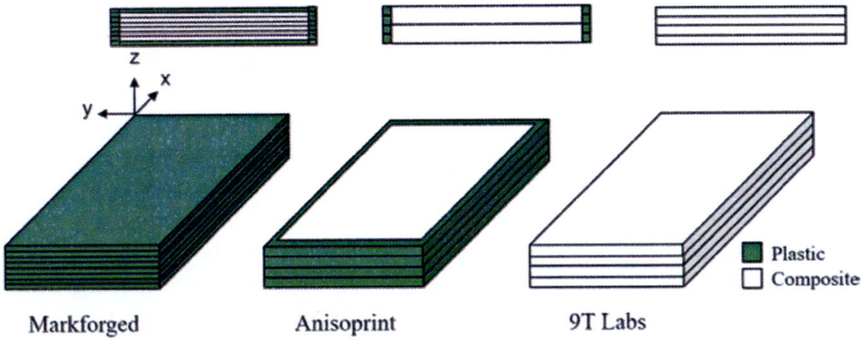

Fig. 4. Restriction of FLM machine for design of the test specimens with maximal fiber volume content

The tensile test specimens were manufactured in the XY plane in unidirectional direction and were manufactured flat so that the layer lines and fibers are oriented in the direction of loading. Just because interlayer porosity is a problem with FLM, this can prove to be a limiting factor for comparison. The tensile test specimens are oriented so that the plastic and composite filaments run parallel to the direction of the load and therefore the interlayer porosity is as low as possible.

The dimensions of the test specimens are in accordance with ISO 525-5 Type A and are listed in Table 4. The force was applied using glass-fiber reinforced glues, which were cut out of a+/– 45° oriented plate with a length of 50 mm and the edges in the middle of the body were bevelled at an angle of 45° using a grinding machine. Then they were applied to the test specimens with an industrial adhesive and cured for 16 h under permanent pressure. The fiber volume contents of the tensile and flexure test specimens were calculated from the ratio of the weight percentages of plastic and composite.

Table 4. Tensile test specimen's parameter

	Dimensions (mm^3)	Layers fiber/plastic	Fiber volume content (%)
Markforged	250 × 15 × 1	6/2	22.13
Anisoprint	250 × 15 × 1	2/0	24.40
9T Labs	250 × 15 × 1	4/0	50.00

The tensile test was performed at a constant test speed of 2 mm/min on a Zwick universal testing machine. The tensile test specimens were tested to failure. The specimens are evaluated and compared with regard to modulus of elasticity (E), tensile strength (σ_{Rm}) and elongation at break (ε_b), which are linked by Hooke's law (Eq. 1) [9]:

$$\sigma = F / A = E * \varepsilon = E * \Delta L / L_0 \tag{1}$$

$$E = (\sigma_2 - \sigma_1) / (\varepsilon_2 - \varepsilon_1) \tag{2}$$

A is the initial cross section, F the force and L_0 the initial gauge length. The E-modulus is determined via the secant modulus in the range from 0.05 to 0.25% elongation (Eq. 2).

3.3 Bending Test

As already mentioned, other material properties than tensile behavior are hardly considered by other studies [1]. Therefore, the three compared technologies are also evaluated on the basis of the bending behavior of test specimens. Since bending stress is one of the most frequently occurring types of stress, it is of considerable importance for determining the properties of plastics and fiber composites [9] and should also be taken into account for manufactured parts using CCFRAM.

Since studies are concerned with the mechanical behavior of tensile properties or fatigue [1, 3–5, 8], only a few of them investigate the bending behavior of manufactured parts using CCFRAM [10, 12–14]. The aim of this research work is to further reduce the research gap in the mechanical properties of additive manufactured composites. This leads to an increase in the potential for use in parts that go beyond prototype production. To ensure that the results obtained with the different technologies are comparable, the test specimen design is such that it has five unidirectional fiber-reinforced layers.

The recommended settings of the FLM machine manufacturer were also used to manufacture the bending test specimens (Fig. 4). In contrast to the test specimens, the Markforged test specimens were manufactured with four bottom or top layers and two wall layers, as this setting is intended to achieve the best possible production quality. These process-specific settings have a significant influence on the effective fiber volume content. Therefore, five unidirectional fiber reinforced layers are manufactured for the bending test with all technologies and the specified manufacturing parameters. The difference in the fiber volume content between tensile and flexure test specimens can also be described as a discrepancy between the recommended and maximal achievable fiber volume content.

The ratio of length to height (*l/h*) of the flexure test specimens has the greatest influence on the test results. Therefore, the support distance on the base of the required test specimen height is adjusted to obtain a constant *l/h* ratio of 25. This minimizes the influence of the disturbing shear stress and achieves a one-dimensional stress state within the measuring length of the test specimen. This corresponds to the recommendations in [9] for the testing of fiber-reinforced thermoplastics and is advantageous to avoid undesired effects of delamination or fiber pull-out. Due to the specification of five unidirectional fiber-reinforced layers, the dimensions with widely varying fiber volume contents, depending on the production technology, are very different; the test geometries are listed in Table 5. Due to the restrictions of the wall/top and bottom layer, the fiber volume content is significantly reduced for small test geometries in the CFF.

Table 5. Bending test specimen's parameter

	Height (mm)	Bearing distance (mm)	Layers fiber/plastic	Fiber volume content (%)
Markforged	1.625	40.625	5/8	8.26
Anisoprint	2.500	62.500	5/0	21.10
9T Labs	1.350	30.750	5/0	50.00

The tests were performed by a three-point bending test according to ASTM D790 Method A [11] and evaluated for flexural strength (σ_f) and flexural modulus (E_f), which are defined by the following three equations [9], where ε_f describes the flexural strain:

$$\sigma_f = 3Fl/2wh^2 \tag{3}$$

$$\varepsilon_f = 6fh/2wl^2 \tag{4}$$

$$E_f = (\sigma_{f2} - \sigma_{f1}) / (\varepsilon_{f2} - \varepsilon_{f1}) \tag{5}$$

Where F is the actual bending force of the load fin, l, h and w are the length, height and width of the test specimens and f is the actual deflection of the test specimens.

The flexural modulus is determined analogous to the tensile test with the secant modulus in the range of 0.05–0.25% outer fiber strain. A testing machine from Hegewald und Peschke Meß- und Prüftechnik GmbH with the corresponding software was used for the bending test. For each technology, seven test specimens were tested without further processing.

4 Results

Considering the results it can be noted that due to the different strategies regarding fiber embedding, the fiber volume contents and thus the elastic properties of the three technologies vary considerably. To enable a comparison between the technologies, the linear mixing rule (Eqs. 6 and 7) is applied to determine the expected moduli and strengths for fiber and matrix (Table 6) based on the material data (Table 3).

$$E_c = \Phi E_f + (1-\Phi)\,E_m \tag{6}$$

$$\sigma_c = \Phi\sigma_f + (1-\Phi)\sigma_m \tag{7}$$

Table 6. Calculated E-Modulus and strength for tensile and bending test specimens

	Calculated e-modulus tensile test Sp (GPa)	Calculated flexural modulus bending test Sp (GPa)	Calculated tensile strength tensile test Sp (MPa)	Calculated flexural strength bending test Sp (MPa)
Markforged	52.2	20.4	539.4	56.1
Anisoprint	37.4	31.0	583.6	70.6
9T Labs	120.7	114.7	2180.0	137.5

These calculated parameters are used as reference for the test results. The determined deviation from the linear mixing rule serves as a comparison value in the linear-elastic range to have an estimation of the potential of the composite.

4.1 Result Tensile Test

Seven test specimens were tested for each technology. The test specimens that failed outside the measuring range were not evaluated. The qualitative progression of the test series is shown in Fig. 5 and the measurement results in Table 7.

Table 7. Result tensile test

	E-Modulus (SD)	Tensile strength (SD)	Elongation at break (%)
Markforged	43.70 GPa (3.91)	557.00 MPa (86.10)	1.23
Anisoprint	29.40 GPa (1.06)	290.00 MPa (32.60)	0.94
9T Labs	58.30 GPa (3.12)	698.00 MPa (36.00)	1.24

Contrary to the assumption when considering the fiber volume contents of the individual test specimens, the 9T Labs test specimens achieve about 50% of the tensile strength of the other test specimens. Anisoprint and Markforged differ slightly in terms of fiber volume content, but are otherwise very similar in structure, which is also reflected in the results. The rovings used by 9T Labs have the highest number of individual filaments and are therefore significantly stiffer than, for example, the rovings from Markforged. This leads to more voids within the part when the pre-impregnated composite filament is deflected.

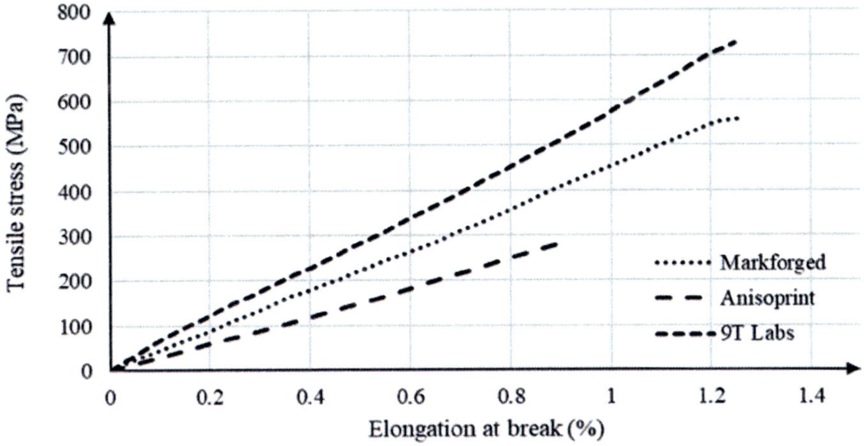

Fig. 5. Qualitative result of the tensile test

In general, a brittle failure of the test specimens is expected during the tensile test in the fiber direction, which was also confirmed in the test. The failure manifests itself in characteristic fracture patterns which can be described as smooth or brush-shaped. A comparison of the fracture patterns from Markforged to Anisoprint and 9T Labs (Fig. 6) shows that the former manufactured a smooth fracture pattern. The localization of the failure indicates good fiber-matrix adhesion. On the other hand, a brush-shaped fracture is observed in the test specimens from Anisoprint and 9T Labs, which indicates weaker fiber-matrix adhesion. The better fiber-matrix adhesion at Markforged is due to the different fiber embedding technologies, as the composite filament is covered by wall, top and bottom layers, thus improving the adhesion of the individual layers. The top and bottom layers thus level out the unevenness caused by the process and at the same time partially close the voids on the surface of the composite caused by deflection.

Fig. 6. Fracture pattern of tensile test specimens

If the results are set in reference to the linear mixing rule, Fig. 7, it can be seen that the E-modulus and tensile strength in the fiber direction can be approximated for Markforged and Anisoprint.

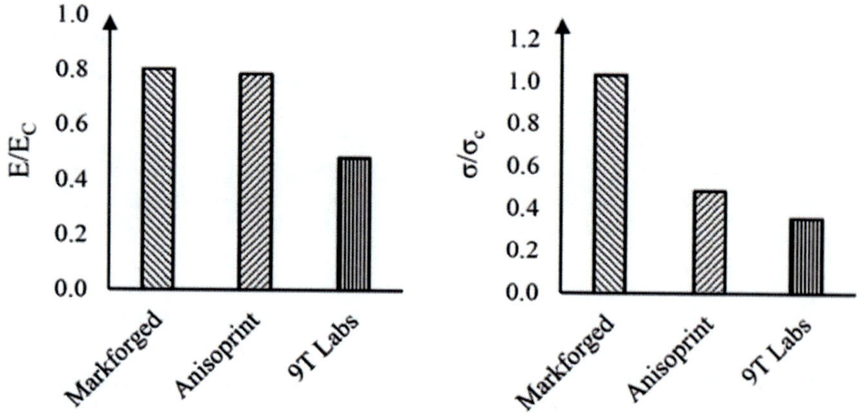

Fig. 7. Normalized E-modulus (left) and tensile strength (right) of the tensile test specimen in relation to the results of the linear mixing rule calculated values

The 9T Labs test specimens also show a clear discrepancy to the calculated value of the mixing rule. Together with the fracture patterns, this behavior can be attributed to the low bonding force between the fiber-reinforced layers and the voids within the part caused by the deflection of the composite filament. Markforged recommends at least two wall layers and several top and bottom layers for good fiber-matrix adhesion. This recommendation has not been taken into account in order to maximize the fiber volume content of the test specimens and thus better observe the effect of the fiber-reinforced layers.

4.2 Result Bending Test

Under bending loads, areas of tensile and compressive stress occur in the test specimen, which can lead to fiber buckling if the fiber-matrix adhesion is poor. This can be observed with the Anisoprint test specimens and partially with the 9T Labs test specimens (Fig. 8).

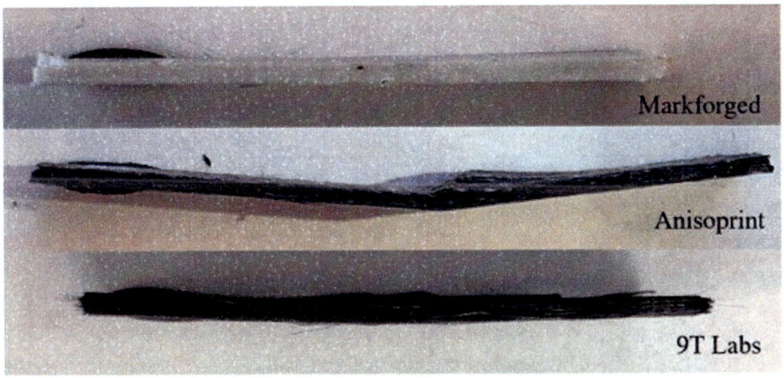

Fig. 8. Fracture pattern of bending test specimens

When manufacturing the test specimens with Anisoprint, the top and bottom layers are omitted, as already described; only one wall layer is required. When the test specimens are manufactured using 9T Labs, only the compound filament is processed. This results in a significantly higher fiber volume content, which is also reflected in the measured values obtained for strength and flexural modulus (Table 8).

Table 8. Result bending test

	E-Modulus (SD)	Tensile strength (SD)
Markforged	109.29 MPa (10.83)	3.04 GPa (0.22)
Anisoprint	110.60 MPa (11.62)	15.72 GPa (2.55)
9T Labs	133.50 MPa (30.77)	22.01 GPa (10.40)

In contrast to the results of the tensile test, the flexural modulus determined behaves as expected and increases with increasing fiber volume content of the test specimens. It is noticeable that the bending strength in comparison to the calculated value is smaller than in the tensile test. One reason for this can be seen in the fiber embedding. Although Markforged describes fiber-matrix adhesion as a deficiency in [10], in comparison with the other technologies, greater emphasis is placed on the adhesion of the individual layers in order to prevent delamination and also to improve haptics. The CFF technology is disadvantageous with regard to achievable fiber volume contents, as the stiffness of the parts is behind the test specimens of Anisoprint and 9T Labs at recommended system parameters.

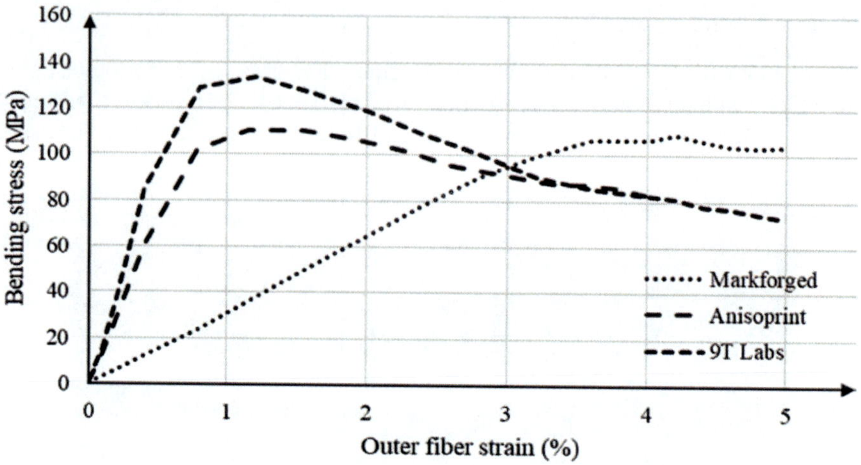

Fig. 9. Qualitative result of the bending test

The qualitative progression of the bending test in Fig. 9 corresponds to the characteristic progression of the composite materials. The Anisoprint test specimens buckled in the range of 4% outer fiber strain and thus failed before the end of the test. The maximal bending stress of the Anisoprint and 9T Labs test specimens both reach this value at about 1% outer fiber strain, whereas the Markforged test specimens only reached this value at about 4% outer fiber strain due to their lower stiffness. The comparison with the values calculated according to the mixing rule, Fig. 10, has a significantly higher discrepancy in the bending test. This is due to the fact that the linear mixing rule cannot be transferred to the bending stress range without restriction. Despite the limited informative value of the mixing rule, the results are representative, with the Anisoprint test specimens clearly standing out despite the buckling of individual layers. This is due to the fact that only wall layers were made of plastic and top and bottom layers were not used. At Markforged, there is a risk that the additional layers required could have a negative effect on the elastic properties due to the aforementioned interlayer porosity, but these are on a par with the test specimens from 9T Labs. With regard to the strengths achieved, it is apparent that the Markforged test specimens show the greatest deviation from the calculated values. This can be attributed to the better fiber-matrix adhesion compared to other technologies for manufacturing parts using CCFRAM and ultimately results in the fact that, despite the significantly lower fiber volume content and the associated more elastic behavior, similar values can be achieved for the bending strength. Regarding the reproducibility of the technologies, it can be stated that the standard deviation is higher at 9T Labs, which can be attributed to the poor fiber-matrix adhesion and the voids described. The influence of the fiber embedding in the matrix can also be confirmed by the Markforged test specimens, where the matrix volume content is significantly increased in the flexure test and the standard deviation is clearly lower than in the tensile test.

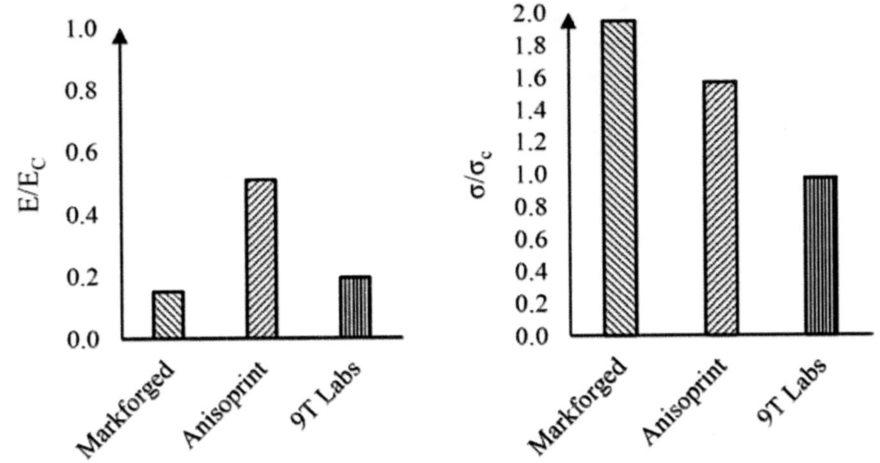

Fig. 10. Average flexural modulus (left) and flexural strength (right) of the bending test specimens in relation to the results of the linear mixing rule calculated values

5 Conclusions

The three investigated commercial technologies (CFF, CFC, AFT) use different strategies for processing the continuous fibers and embedding them in the matrix. At Markforged and 9T Labs, for example, the co-extrusion process is upstream and at Anisoprint it is integrated into the machine. Markforged attaches great importance to fiber-matrix adhesion by using wall, top and bottom layers on which the composite filament is laid and by using pre-impregnated fiber rovings. This has a corresponding negative effect on the maximal achievable fiber volume content, which is the lowest with the CFF technology. CFC technology from Anisoprint offers the advantage that any matrix material can be used due to the open material system. Also the fiber volume content can be adjusted within certain limits due to the integrated co-extrusion process. In comparison, it has been shown that the interlaminar fiber-matrix adhesion is slightly lower than that of the CFF from Markforged, which is compensated in the measurements by the higher fiber volume content. The technology of 9T Labs offers the possibility to realize a fiber volume content of up to 50% whereas tensile tests showed that the high fiber volume content is not equal to maximum mechanical properties. The relatively hardly deformable roving with 24-K single filaments had less surrounding matrix due to the processing of only the composite filament, which could accordingly compensate the unevenness during the melting process. In addition, the deflection of the stiff composite filament led to voids, resulting in poorer fiber-matrix adhesion and thus lower mechanical properties than the results of other technologies. One approach to solve this problem is to compress the parts under pressure and temperature after the manufacturing of the parts. As a result, voids can be reduced and the fiber-matrix adhesion can also be improved. The AFT technology of 9T Labs offers many approaches to better use the high fiber volume content by optimizing the production strategy or post-processing, not least because of the design freedom for the insertion of the reinforcement fibers.

Acknowledgements. This research was supported by the European Fund for Regional Development (EFRE) and the Oberfrankenstiftung.

References

1. van de Werken, N., Tekinalp, H., Khanbolouki, P., Ozcan, S., Williams, A., Tehrani, M.: Additively manufactured carbon fiber-reinforced composites: state of the art and perspective. Addit. Manuf. **31**, 100962 (2020). https://doi.org/10.1016/j.addma.2019.100962
2. Suzuki, T., Fukushige, S., Tsunori, M.: Load path visualization and fiber trajectory optimization for additive manufacturing of compo-sites. Addit. Manuf. **31**, 100942 (2020). https://doi.org/10.1016/j.addma.2019.100942
3. Dutra, T.A., Ferreira, R.T.L., Resende, H.B., Guimarães, A.: Mechanical characterization and asymptotic homogenization of 3D-printed continuous carbon fiber-reinforced thermoplastic. J. Braz. Soc. Mech. Sci. Eng. **41**(3), 133 (2019). https://doi.org/10.1007/s40430-019-1630-1
4. Mohammadizadeh, M., Imeri, A., Fidan, I., Elkelany, M.: 3D printed fiber reinforced polymer composites – structural analysis. Compos. B Eng. **175**, 107112 (2019). https://doi.org/10.1016/j.compositesb.2019.107112
5. Der Klift, F.V., Koga, Y., Todoroki, A., Ueda, M., Hirano, Y., Matsuzaki, R.: 3D Printing of Continuous Carbon Fibre Reinforced Thermo-Plastic (CFRTP) tensile test specimens. OJCM **06**(01), 18–27 (2016). https://doi.org/10.4236/ojcm.2016.61003
6. Ming, Y., Duan, Y., Wang, B., Xiao, H., Zhang, X.: A novel route to fabricate high-performance 3D printed continuous fiber-reinforced thermosetting polymer composites. Materials **12**(9), 1369 (2019). https://doi.org/10.3390/ma12091369
7. Li, N., Li, Y., Liu, S.: Rapid prototyping of continuous carbon fiber reinforced polylactic acid composites by 3D printing. J. Mater. Process. Technol. **238**, 218–225 (2016). https://doi.org/10.1016/j.jmatprotec.2016.07.025
8. Fidan, I., et al.: The trends and challenges of fiber reinforced additive manufacturing. Int. J. Adv. Manuf. Technol. **102**(5–8), 1801–1818 (2019). https://doi.org/10.1007/s00170-018-03269-7
9. Grellmann, W., Seidler, S., Altstädt, V. (eds.): Kunststoffprüfung, 3., [updated] Ed. München: Hanser (2015)
10. Chacón, J.M., Caminero, M.A., Núñez, P.J., García-Plaza, E., García-Moreno, I., Reverte, J.M.: Additive manufacturing of continuous fibre reinforced thermoplastic composites using fused deposition modelling: Effect of process parameters on mechanical properties. Compos. Sci. Technol. **181**, 107688 (2019). https://doi.org/10.1016/j.compscitech.2019.107688
11. ASTM D790-17, Standard test methods for flexural properties of unreinforced and reinforced plastics and electrical insulating materials. ASTM International, West Conshohocken, PA, 2017, www.astm.org
12. Dickson, A.N., Barry, J.N., McDonnell, K.A., Dowling, D.P.: Fabrication of continuous carbon, glass and Kevlar fibre reinforced polymer composites using additive manufacturing. Addit. Manuf. **16**, 146–152 (2017). https://doi.org/10.1016/j.addma.2017.06.004

13. Czasny, M., Goerke, O., Kaba, O., Koerber, S., Schmidt, F., Gurlo, A.: Influence of composition on mechanical properties of additively manufactured composites reinforced with endless carbon fibers. KEM **809**, 335–340 (2019). https://doi.org/10.4028/www.scientific.net/KEM.809.335
14. Tian, X., Liu, T., Yang, C., Wang, Q., Li, D.: Interface and performance of 3D printed continuous carbon fiber reinforced PLA composites. Compos. A Appl. Sci. Manuf. **88**, 198–205 (2016). https://doi.org/10.1016/j.compositesa.2016.05.032
15. Domm, M., Schlimbach, J., Mitschang, P.: Optimizing mechanical properties of additively manufactured FRPC. In: 21st International Conference on Composite Materials, Xi'an, p. 12 (2017)
16. Johnson, D.J.: Structure-property relationships in carbon fibres. J. Phys. D. Appl. Phys. **20**(3), 286. https://doi.org/10.1088/0022-3727/20/3/007.
17. Markforged Inc.: Material specifications composites, www.markforged.com. Accessed 11 Feb 2020
18. Anisioprint LCC: Products CCF&CBF, www.anisoprint.com. Accessed 11 Feb 2020
19. 9T Labs AG, Material Datasheet CF/PA12, 2019
20. Chacóna, J.M., Caminerob, M.A., Núñezb, P.J., García-Plazab, E., García-Morenob, I., Revertea, J.M.: Additive manufacturing of continuous fibre reinforced thermoplastic composites using fused deposition modelling: effect of process parameters on mechanical properties. Compos. Sci. Technol. **181**, 107688. https://doi.org/10.1016/j.compscitech.2019.107688
21. Wang, X., Jiang, M., Zhou, Z., Gou, J., Hui, D.: 3D printing of polymer matrix: a review and prospective. Compos. Part B **110**, 442–458 (2017)
22. Desktop Metal: Products FiberTM, www.desktopmetal.com. Accessed 11 Feb 2020
23. Continuous Composites: Continuous fibre 3D-Printing, www.continuouscomposites.com. Accessed 11 Feb 2020

Design of Additively Manufactured Heat-Generating Structures

Karl Hilbig(✉), Hagen Watschke, and Thomas Vietor

Institute for Engineering Design, Technische Universität Braunschweig, Braunschweig, Germany
{k.hilbig,h.watschke,t.vietor}@tu-braunschweig.de

Abstract. Multi-material additive manufacturing provides new design freedom for integration of functions and opens new possibilities in innovative product design due to local material variations. In particular, material extrusion (MEX) allows for combination of different industrial-grade thermoplastic materials to enhance the functionality of a product by integration of functions. Thus, for instance, electrically conductive structures or heat-generating surfaces can be incorporated in a part by using conductive polymers filled by carbon black (CB), carbon nanotubes (CNT) or copper nanowires (CNW).

The resultant properties of additively manufactured parts are mainly influenced by the choice of process parameters. In addition to mechanical properties (e.g. stiffness and strength), electrical properties are also like resistivity and volumetric power density influenced. In order to design heat-generating structures in a targeted manner, the dependencies between process parameters and electrical performance must be determined. Thus, in this article the dependencies between the process parameters extrusion temperature, raster angle orientation and extrusion speed are investigated experimentally. In order to adjust the resistivity of an additively manufactured part and surface temperature by resistive heating, these dependencies are transferred into mathematical descriptions. The setup of design of experiment is based on model selection for analytical description of material-specific characterization.

In order to demonstrate the potential of additively manufactured heating structures by material extrusion a garnish mold with incorporated heating panels is built as multi-material design. Finally, the heating of the prototypical panel is analyzed by thermographic analyses. Thus, the approach for achieving certain surface temperatures by varying process parameters and part geometry based on the mathematical description is validated.

Keywords: Additive manufacturing · Multi-material · Material extrusion · Design integration of functions · Electric conductivity · Heat-generation

K. Dröder and T. Vietor (Eds.): *Technologies for economic and functional lightweight design*, Zukunftstechnologien für den multifunktionalen Leichtbau, pp. 142–155, 2021. https://doi.org/10.1007/978-3-662-62924-6_12

1 Introduction

Multi-material parts manufactured by additive manufacturing (AM) provides new design freedom for the integration of material-specific functions. This functions, e.g. electrical conductivity, can be integrated locally by using multi-material designs and, thus, minimizing additional mounting or joining processes and needed installation space.

The application area of electrical conductive structures by MEX can be categorized by the conducting paths (e.g. [5, 6]), embedded sensors (e.g. [1, 7, 15]) and heat generating structures (e.g. [2, 3, 14]). Thus, these new opportunities for the integration of functions provides a high potential for more innovative and efficient designs e.g. regarding material consumption. To integrate locally additively manufactured electrical conductive structures the specific knowledge related to AM's design limitations in form of design rules or specific knowledge regarding the performance characteristic is needed. In this contribution, a model for the prediction of the electrical resistance for additively manufactured structures is developed for different materials and by varying selective process parameters. This approach is validated by the manufacturing of a garnish mold with integrated heating panels that differ in geometry and resulting surface temperatures due to an adjustment of the resistance by process parameter changes.

2 State of Research

Current research show the increasing impact of additively manufactured heating panels, by using composite materials with different fillers, e.g. carbon nanotubes, copper nanowires, carbon black or graphene. Dul et al. (2018) [4] compare different weight percentages of carbon nanotubes in a matrix of acrylonitrile butadiene styrene (ABS) and raster angle orientations to the heating up behavior and the surface temperature. They achieved at 12 V a surface temperature of approx. 36 °C and at 24 V a surface temperature of approx. 135 °C by using 8 wt% of CNT and assume that the conductivity of additively manufactured structures correlate with surface temperature due to dissipation of thermal energy [4].

Zhuang et al. [16] develop and evaluate a MEX-based manufacturing process that allows mixing of different materials in order to modify continuously the filler content during the manufacturing process. They find out that a higher content of the filler in the structure leads to a lower resistance, which results in a lower surface temperature. Due to a variation of the filler content in the heating structure, an anisotropic heat distribution is achieved. The maximum temperature difference is about 60 °C at an input voltage of 25 V [16]. Watschke et al. [14] verify the dependency of surface temperature on process and geometry-specific influencing factors as well as on the material composition. The examined process parameters are raster angle orientation, extrusion temperature, flow rate, and extrusion speed [14].

The existing research approaches show a high demand of energy input for heat generation so that solely low surface temperatures at limited surface areas are reached

[4, 16]. For application purposes, the energy demand has to be reduced with simultaneously increasing the maximum surface temperature.

Hampel et al. (2017) [5] develop a rudimentary analytical model for the determination of the electrical resistance of additively manufactured parts by MEX. This model contains only the geometric dimension of the conductive path. In addition, Watschke et al. (2019) [14] prove the dependency of the resistivity by a wide variation of process parameters for different materials. Consequently, specific influencing factors are identified and quantified in order to estimate the electrical resistance of additively manufactured parts by using MEX. Apart from this, in this research an analytic cubic regression model is developed by experimental data set in order to design additively manufactured structures with a specific electrical resistance based on a definition of a process parameter set and a certain geometry.

3 Experimental Set-Up for Additive Manufacturing of Test Specimens

The following section presents the experimental set-up that is used for manufacturing of test specimens for the characterization of the electrical resistance by using X400® (German RepRap GmbH). MEX offers a broad range of commercially available electrically conductive filaments, in this research the following materials are examined: Proto-Pasta Conductive PLA (PPC) [12], 3dkonductive PLA (3dk) [8], Functionalize F-Electric™ PLA (FFE) [11], MULTI3D Electrifi Conductive Filament (Multi3d) [10], and BlackMagic3d PLA (BM3d) [9].

Table 1 presents the design of experiments for determining material and process specific dependencies of resistivity. For each material, a full factorial parameter variation is chosen. As influencing factors, the process parameters of infill orientation by raster angle, nozzle temperature, flow rate, and printing speed are investigated.

Table 1. Design of experiment of full factorial parameter variation

Name	Filler/matrix material	Raster angle (°)	Temperature (°C)	Flow rate (%)	Speed (mm/s)
		(Δ=45)	(Δ=10)	(Δ=5)	(Δ=20)
PPC	CB/PLA	0; ± 45; 90	210–240	95–105	20–60
3dk	CB/PLA				
FFE	CNT/PLA				
BM3d	Graphene/PLA				
Multi3d	CNW/PCL		170–180		

The design of experiment led to a total number of 459 batches of process parameter settings, consequently 1377 specimens are additively manufactured and measured (according to [14]). The high number of parameter settings requires a customized specimen shape in which the process parameters are independently modified, see Fig. 1. The electrical bonding is realized with the silver paste (EMS 12,640) [13].

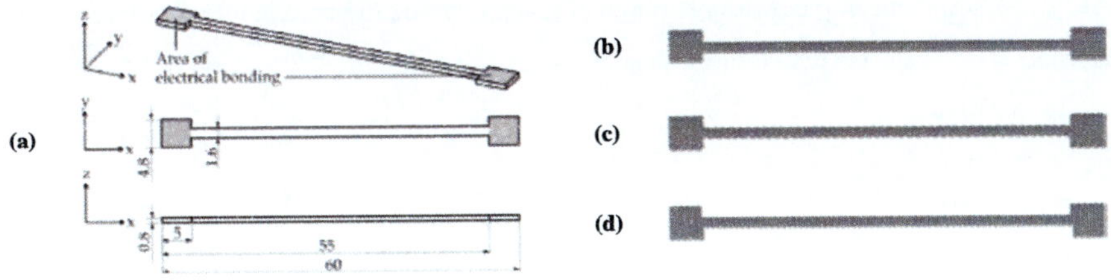

Fig. 1. Specimens for the determination of the influencing factors on resistivity: a) dimensions and isometric view; investigated raster angle orientations, b) 0°, c) ±45°, and d) 90°

Figure 1 shows geometry of used test specimens for the electrical characterization for investigation of the influence of different process parameter sets. Resistance is measured between the electrical contacts. Resistivity was calculated using Eq. (1). Measurement of the electrical resistance of the specimen ($R_{specimen}$) is executed by multimeter (FLUKE 87 V, Fluke Corporation). The cross-section (A) of the specimen is 1.28 mm^2 and the length (L) between the areas of electrical contacting is 50 mm.

$$\rho = R_{specimen} \cdot \frac{A}{L} \tag{1}$$

For analysis of surface temperature generated by resistive heating the VarioCAM® HD head 800 thermographic camera (InfraTec GmbH) with the software IRBIS® 3 is used.

4 Results of Specific Regression Model of Resistivity

The analysis of experimental data of the resistivity is conducted with Cornerstone® 7.1.2.1 (camLine, Germany) in order to develop a material-specific regression model that contains the main influencing process parameters on resistivity. Table 2 illustrates the limits in which resistivity of a specific material can be adjusted by variation of selected process parameters within the boundaries shown in Table 1. Hence, these indicate the main influence factors related to behavior of resistivity of additively manufactured structures by using MEX.

Table 2. Material-specific adjustment range of resistivity due to manufacturing process

Material		Resistivity [Ωm]	Raster Angle [°]	Temperature [°C]	Flow rate [%]	Speed [mm/s]
PPC	min	6.7925333E−02	0	240	105	20
	max	23.3472E−02	90	210	95	40
3dk	min	12.586667E−02	0	230	100	20
	max	32.7424E−02	90	210	105	60
FFE	min	0.6852267E−02	0	220	100	40
	max	0.3219627E−02	90	220	95	40
Multi3d	min	0.0287573E−02	0	170	105	20
	max	0.2587307E−02	45	180	95	20
BM3D	min	1.2484267E−02	0	220	100	20
	max	6.6807467E−02	90	210	100	40

In order to analyze design freedom of the adjustability of resistivity by a variation of material and process parameters a boundary value analysis is carried out. The main factor influencing electrical performance is the raster angle orientation. It determines the electrical resistance along the direction of the current flow in the additively manufactured specimen or part respectively. An orientation of the extruded strands parallel to current flow (raster angle of 0°) a direct connection of electrical bonding areas is achieved. Therefore, a minimal number of transfer resistance between the extruded strands have to be overcome. The major increase of the transfer resistance at a raster angle orientation of 90° results from the orientation of the extruded stands orthogonal to the current flow. This leads to the assumption that a high resistance between the strands generate a broad range of adjustability for electrical conductivity, while a close performance window results from a low transfer resistance between the strands.

The material limits, see Fig. 2, display a filler and matrix material-specific behavior of resistivity. PPC and 3dk have a high intersection of the adjustment range; this can be explained by the same polymer matrix (PLA) and the same filler (CB). Both FFE and BM3d show a deficit in adjustability of absolute resistivity but is suited for applications with requirements of a lower resistivity. Both materials have a more conductive carbon allotrope as filler compared to PPC and 3dk. Multi3d has the lowest determined resistivity and the most precise adjustability. The behavior of precise adjustability is caused by the small process window of Multi3D, whereas the high conductivity results from the CNW filler.

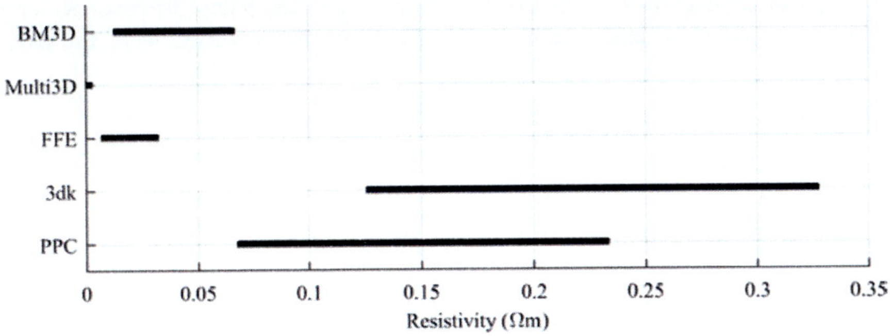

Fig. 2. Limit analysis of the resistivity for an application area based on material selection

4.1 Material-Specific Regression Windows

For analysis of the material models, a cubic regression is used. This assumption of the material behavior enables adaptations of a regression function curve of third order by the experimental data. The trail function enables investigation of linear, curved and quadratic behavior or interaction effects of process-specific influencing factors. Basic assumption for the development of regression models for resistivity of additively manufactured structures, is a normal distribution of experimental data, see Fig. 3.

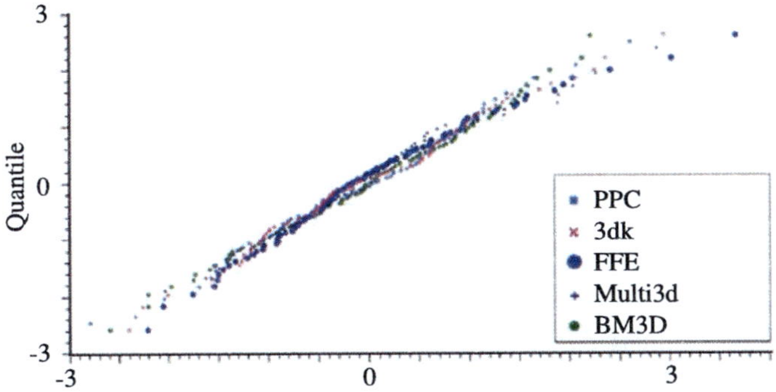

Fig. 3. Material-specific residuals probability plot of experimental data

The material-specific residuals probability plot of experimental data points shows a good fit for a normal distribution. The standard deviation of the specific material models is described by the width of the normal distribution. All characterized materials of the experimental data show that the resistivity of an additively manufactured structure has an interval of deviation $\pm\sigma$ of the estimated value. In order to design additively manufacture electrically conductive structures a scattering of the performance can occur by the estimated value. The adjusted coefficient of determination of the material models represents the fitting quality of the regression model for the experimental data (see Table 3).

Table 3. Regression window of regression models of resistivity.

Term	Material				
	PPC	3dk	FFE	Multi3d	BM3d
R-Square [−]	0.943	0.940	0.988	0.976	0.980
Adj R-Square [−]	0.935	0.930	0.985	0.966	0.977
RMS Error [Ωm]	0.01	0.013	9e−04	0.002	0.067

The summary shown in Table 3 confirms the assumption of a cubic material model behind the behavior of the resistivity. All models have adjusted coefficients of determination higher than 90%. As a result, the terms of the regression models depend on the investigated process-specific factors.

For the integration of electrically conductive structures in real applications, a small root-mean-square (RMS) error is necessary at a specific performance characteristic. The RMS error state the quality of the regression function for the experimental data. For PPC, 3dk, Multi3D and BM3d the investigated RMS errors show a high deviation of the regression function. In comparison, the material FFE has a lower RMS error of the resistivity. Therefore, FFE should be used for the integration of electrically conductive structures for real applications. Based on the high deviation the resistivity of PPC, 3dk, Muli3d, and BM3d solely FFE is utilized for the investigations regarding heat generation and the integration in a garnish mold in order to realize conformal temperature control in a vehicle interior.

4.2 Coefficient Table of the Average Resistivity

The optimization of influencing factors for a specific target value of resistvity is not necessary for development of a functional structure. Table 4 displays the estimated values of the coefficient in the regression model. The terms present the investigated process parameters with the coefficient of the cubic regession model.

Table 4. Coefficient table of the average resistivity of FFE

Term	Coefficient	Std error
c	−0.56678777	0.186672491
f_1	0.021229078	0.003078474
f_2	0.00501736	0.001623703
f_3	1.59895E−05	4.33308E−05
f_4	0.000358699	7.09896E−05
f_1^2	1.38948E−05	2.49996E−06
$f_2\ f_3$	−0.00018702	2.67042E−05
$f_3\ f_1$	1.15949E−05	2.35333E−06
$f_1\ f_4$	−8.9894E−07	1.30258E−06
f_2^2	−1.1033E−05	3.52821E−06
f_4^2	−4.7053E−06	8.82385E−07
$f_3\ f_2^2$	−1.3436E−07	2.48162E−08
$f_4^2\ f_1$	−1.8796E−08	6.41784E−09
$f_1\ f_2^2$	3.92347E−07	5.80369E−08
$f_1\ f_4^2$	3.22362E−08	1.45554E−08

c = constant value; f_1 = raster angle [°]; f_2 = temperature [°C]; f_3 = flow rate [%]; f_4 = speed [mm/s].

By reference to Table 4, the main influences on resistivity are raster angle orientation and nozzle temperature. In addition to the adjustability of the electrical resistance by material and process-specific factors, the adjustment of resistance can be achieved by a variation of the geometry provided by AM's design freedom, e.g. for realizing heat generating structures with customizable surface temperatures.

The design criteria for dimensioning such structures have to be adapted for the area of application. The design criteria for additively manufactured electrically conductive paths is the resultant resistance of the structure. The structure can be seen as a resistor network and the resulting voltage drop can be considered in the circuit diagram. The design criteria of heat-generating structures is the volumetric power density that transfers electrical energy in thermal energy by using resistive heating (Joule effect).

5 Application

The regression model is used for designing and manufacturing of a functional prototype of a garnish mold with integrated heating panels. The integration of these panels is realized by multi-material AM using MEX. On the one hand, the prototype is utilized to demonstrate a local integration of heat generating structures in order to increase comfort by AM. On the other hand, the adjustability of the surface temperature by varying geometry, raster angle orientation, and input voltage is validated.

Based on the boundary conditions of battery electric vehicles the exhaust heat cannot be used for warming vehicle interiors. To realize an electric heating in the interior, a prototype of a garnish mold that is placed inside of the front-seat passengers' dashboard is designed and manufactured by using MEX (see Fig. 4). The setup of the prototype is composed of two separately additively manufactured parts – a carrier structure that captures the shape and connecting requirements of the dashboard and the heating panels. For the carrier structure a silver-metallic PLA of Ultimaker® is used. The three heating panels with a grid-type isolation have various geometries and raster angle orientations and are manufactured by multi-material MEX without any additional mounting or joining process. FFE is used for the heating structure and surrounded by a thermochromic PLA (DAS FILAMENT) that changes the color from red to white depending on temperature (≈40 °C) for an incorporation of an additional design feature. The electrical contacting of the heat panels is realized by a combination of silver paste and flat connectors at the back of the garnish mold. The contacting module of screwed flat connectors serves as a strain relief; this enables a detachable industrial-like connection of this application.

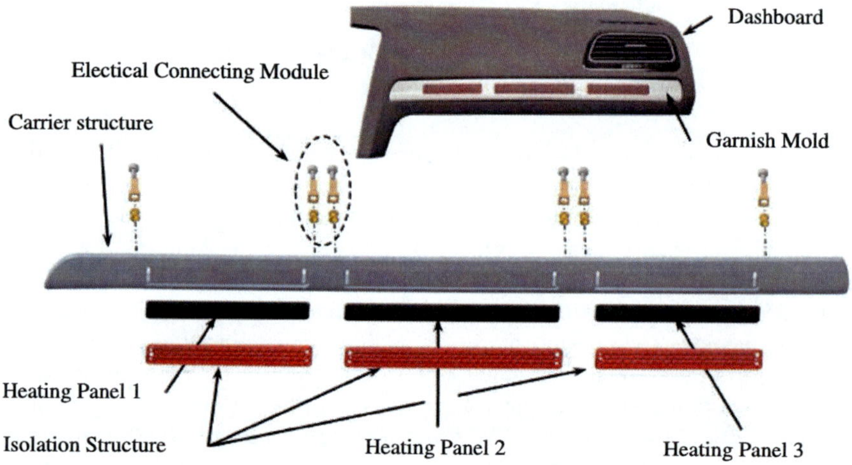

Fig. 4. Functional prototype of near shape heat areas in a garnish molding

For the dimensioning of the three heating panels two design criteria are formulated. First, the left and the middle heating panel should have a similar surface temperature compared to the room temperature. Second, the surface temperature of the right heating panel is about 75% of the left heating panel referred to the room temperature. This leads to the following condition:

$$\Delta T_1 = \Delta T_2 = 0.75 \cdot \Delta T_3 \quad (2)$$

Based on the assumption that the behaviour of the surface temperature is proportional to volumetric power density the process parameter setting and geometrical layout of each heat panel is based on this design criterion:

$$\phi_1 = \phi_2 = 0.75 \cdot \phi_3 \tag{3}$$

Consequently, the parameter sets shown in Table 5 are used to manufacture the individualized heat panels. The selection of the parameter set is based on the regression models.

Table 5. Parameters variation for the heating panels*

Panel no.:	Raster angle [°]	Extrusion temperature [°C]	Extrusion speed [mm/s]	Flow rate [%]	Cross section (A) [mm^2]	Length (l) [mm]
1	90	240	20	105	10.2	92
2	45					92
3	0					122

In order to reduce an undue influence of input voltage by the heating panels a parallel electrical connection is realized. This enables the characterization of each specific heating structures (see Fig. 5). The heat distribution of the surfaces is recorded by the VarioCAM® HD head 800 thermographic camera (InfraTec GmbH) and electrical properties are measured by a multimeter (FLUKE 87 V, Fluke Corporation).

The experimental investigations regarding surface temperatures are carried out at standard climate environmental conditions for polymers (ISO 291-1). The room temperature is set to 23 °C with a relative humidity of 50%. The allowed deviation for the temperature is ±2 °C and for the relative humidity ±10%.

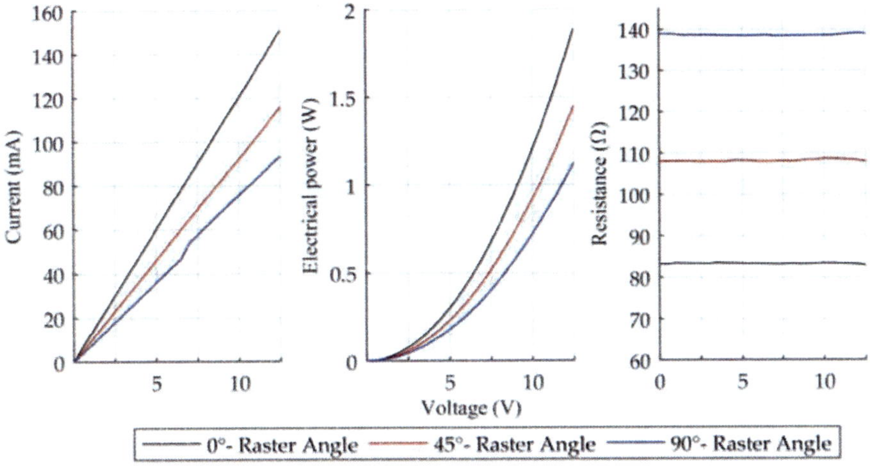

Fig. 5. Performance characteristic of current, electrical power and electrical resistance depending on voltage of the three heating panels: left (±45°), middle (0°), and right (90°)

The current–voltage characteristics show a linear behavior for all raster angle orientations due to temperature independent electrical resistance. The current–voltage characteristic illustrates the varying electrical power (P) in a constant time interval (Δt) for each heating panel ($P_{0°} > P_{\pm 45°} > P_{90°}$). The behavior of electrical power in relation to input voltage has a progressive curve shape. The heating panel with a raster angle orientation of 0° has the highest power consumption because of the lowest electrical resistance. The resistance profiles of the heating panels are constant across the measuring range.

An important performance indicator of the panels is the heating curve. The maximum input voltage is limited to 12 V since the heat deflection temperature (HDT) of the matrix polymer (PLA) is exceeded at this voltage. This operating criterion ensures reversible operating conditions. For measurement of heat distribution, three points per heating panel are set and averaged to determine surface temperature. The heating curves of the panels are represented in Fig. 6.

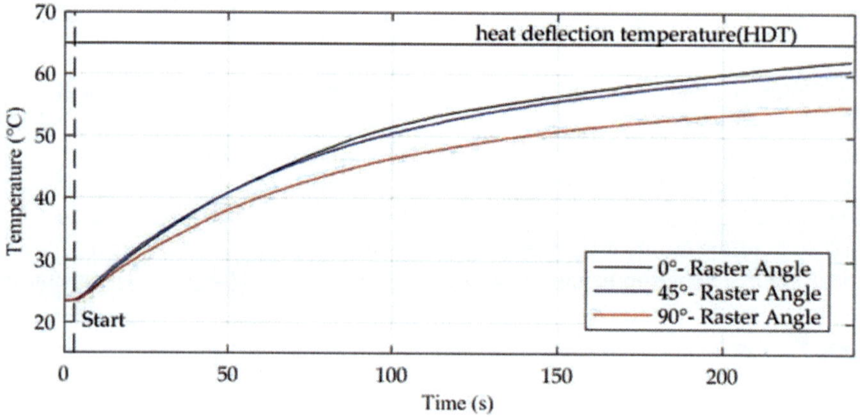

Fig. 6. Representation of the heating curves of the heating panels at an input voltage of 12 V

The heating curve shows a degressive temperature behavior for all heating panels. Each profile has the trend of an asymptotic heating behavior up to a maximal surface temperature at a defined input voltage. The maximal reached surface temperatures at an input voltage of 12 V are 62.2 °C (0°), 60.6 °C (±45°), and 54.9 °C (90°). The heating panels with a raster angle orientation of 0° and ±45° show a comparable behavior, whereas the heating curve of the panel with a raster angle orientation of 90° is less steep. A direct comparison between the heating curves of the heat panels can be achieved by a comparison of the volumetric power density (ϕ) with the surface temperature (see Table 6).

Table 6. Comparison of volumetric power density and surface temperature of the heating panels

Panel no.:	Raster angle [°]	Resistance [Ω]	Power (at 12 V) [W]	Volume [mm^3]	Volumetric power density [W/mm^3]	Surface temperature [°C]
1	90	138.88	1.0356	938.4	0.001104	54.9
2	±45	108.03	1.332	938.4	0.001419	60.6
3	0	83.19	1.7928	1244.4	0.001441	62.2

The heat distribution appears on the possible volumetric power density in the heating panels. This behavior confirms the assumption that the surface temperature correlates with the volumetric power density. The used design criteria of the application proves that similar volumetric power densities leads to similar surface temperatures. Therefore, the developed mathematical models are suitable for the design of additively manufactured heating panels with specific surface temperatures (Fig. 7).

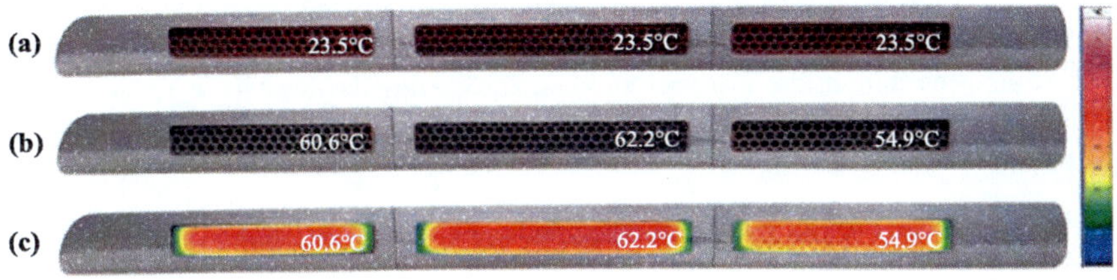

Fig. 7. Illustration of the garnish mold at different states: a) chilled state at room temperature; b) heated state with color change of thermochromic PLA; c) heat distribution at 12 V

The temperature distribution on the heating panels shows a strong decrease in temperature at the border areas due to the electrical bonding. The high temperature gradient can be explained through the higher thermal conductivity of the silver paste and the electrical contact. Within the heating panels a homogeneous temperature distribution is achieved. The results of the thermographic analysis display a similar surface temperature of the left and the middle panel. The right heating panel shows a surface temperature of about 55 °C, this correspond with the defined design criterion. Thus, a feasibility of integrated additively manufactured heating panels is shown by the prototype of the garnish mold. In order to compare the performance of these heating panels to conventional applications further research have to be conducted.

6 Conclusion

This paper presents an investigation of the integration of additively manufactured heating panels by using material extrusion. Five different materials with four different fillers were examined by investigating the adjustability of the electrical conductivity by varying process parameters. By selecting the material with the lowest RMS error a garnish mold with integrated heating structures is manufactured. The utilized material has a linear current–voltage characteristic that allows surface temperature to be controlled via voltage. A specific adjustment of the characteristic curve is achieved by defining the geometry and the raster angle orientation. In this way, a required temperature can be set via geometry and process parameters as a function of a specified input voltage. At 12 V, approximately 45 °C is reached after 60 s and approximately 65 °C after 180 s respectively. Surface textures and a thermochromic material, which undergoes a thermally induced color change from red to white, have been used as customizable design elements of the application. The low mass of the heating panels enables high surface heating rates at short heating times. As a result, additive manufacturing of electrical conductive layers for heat generation allows for customization of the resulting surface temperature and reduces cycle times by simultaneously saving energy requirements. Future research will concentrate on three-dimensional conformal heating structures. As a result, the influencing geometry factors will comprehensively be examined in respect to the design freedom of AM. Another focus is the development of a methodical approach for designing additively manufactured heating structures through use of material extrusion by supporting several steps within the product development process form conceptual until detail design, e.g. by providing design principles and rules. For this reason, the design criteria have to be specified especially for the area of application, the performance characteristics, and the system requirements.

References

1. Alsharari, M., Chen, B., Shu, W.: 3D printing of highly stretchable and sensitive strain sensors using graphene based composites. MDPI **2** (2018). https://doi.org/10.3390/proceedings2130792
2. Dijkshoorn, A., Schouten, M., Wolterink, G., Sanders, R., Krijnen, G.: Characterizing the electrical properties of anisotropic, 3D-printed conductive sheets. In: 2019 IEEE International Conference on Flexible and Printable Sensors and Systems (FLEPS), July 2019. https://doi.org/10.1109/FLEPS.2019.8792279
3. Dorigatoa, A., Morettia, V., Dula, S., Unterberger, S.H., Pegoretti, A.: Electrically conductive nanocomposites for fused deposition modelling. Synth. Metals **226**, 7–14 (2017). https://doi.org/10.1016/j.synthmet.2017.01.009
4. Dul, S., Fambri, L., Pegoretti, A.: Filaments production and fused deposition modelling of ABS/carbon nanotubes composites. Nanomaterial **8**, 49 (2018). https://doi.org/10.3390/nano8010049
5. Hampel, B., Monshausen, S., Schilling, M.: Properties and applications of electrically conductive thermoplastics for additive manufacturing of sensors. Tech. Mess. **84**, 593–599 (2017). https://doi.org/10.1515/teme-2016-0057
6. Gnanasekaran, K., Heijmanns, T., van Bennekom, S., Woldhuis, H., Wijnia, S., de With, G., Friedricht, H.: 3D printing of CNT- and graphene-based conductive polymer nanocomposites by fused deposition modeling. Appl. Mater. Today **9**, 21–28 (2017). https://doi.org/10.1016/j.apmt.2017.04.003

7. Hedges, M.: Additive manufacturing process chains for 3D printed electronics. In: Rapid. Tech- International Trade Show Conference for Additive Manufacturing, part 5, pp. 152–171, June 2017
8. Technical Data—3dkonductive—Elektrisch Leitfähig. https://3dk.berlin/de/spezial/169-3dkonductive.html. Accessed March 2020
9. Technical Data—BlackMagic3d. https://graphene-supermarket.com/Conductive-Graphene-PLA-Filament.html. Accessed March 2020
10. Technical Data—Electrifi Conductive Filament. https://www.multi3dllc.com/product/electrifi. Accessed March 2020
11. Technical Data—Functionalize F-Electric™ PLA. https://functionalize.com/about/functionalize-f-electric-highly-conductive-filament. Accessed March 2020
12. Technical Data—Proto-Pasta Conductive PLA. https://www.proto-pasta.com/pages/conductive-pla. Accessed March 2020
13. Technical Data—Silver Paste EMS #12640. https://www.emsdiasum.com/microscopy/technical/datasheet/12640.aspx. Accessed March 2020
14. Watschke, H., Hilbig, K., Vietor, T.: Design and characterization of electrically conductive structures additively manufactured by material extrusion. Appl. Sci **9**(4) (2019). https://doi.org/10.3390/app9040779
15. Zapciu, A., Constantin, G.: Additive manufacturing integration of thermoplastic conductive material in intelligent robotic end effector systems. Proc. Manuf. Syst. **11**, 201–206 (2016)
16. Zhuang, Y., Song, W., Ning, G., Sun, X., Sun, Z., Xu, G., Zhang, B., Chen, Y., Tao, S.: 3D-printing of materials with anisotropic heat distribution using conductive polylactic acid composites. Mater. Des. **126**, 135–140 (2017). https://doi.org/10.1016/j.matdes.2017.04.047

Process Simulation for Screw Extrusion Additive Manufacturing of Plastic Parts

Johannes Albers[1(✉)], Ulf Hillemann[1], Andreas Retzlaff[1], André Hürkamp[2], and Klaus Dröder[2]

[1] Volkswagen AG, Wolfsburg, Germany
johannes.albers@volkswagen.de
[2] Institute of Machine Tools and Production Technology, Technische Universität Braunschweig, Braunschweig, Germany

Abstract. Additive manufacturing of plastic parts is a widely spread production method for prototyping. In the recent past, additionally series applications evolved from different sectors of industry. Especially plasticizing processes are characterised by high printing speed and a broad range of materials which can be provided both as filament and granulate. The latter offers the advantage of lower material costs compared to filaments but requires a screw extruder to plasticise and homogenise the plastic melt prior to its deposition through a nozzle. Screw extruders have a wide processing range and therefore are capable of producing a huge variety of printing bead shapes. This shape is directly affected by several extrusion settings and has a great impact on manufacturing time as well as mechanical properties and surface quality of fabricated parts. It would take great effort to determine the influence of process parameters like temperatures and velocities in experimental trials.

In this contribution a process simulation is presented which predicts the bead shape for screw extrusion additive manufacturing. A computation of material flow is performed including nozzle outlet and bead shaping in the gap between nozzle and printing platform. The free surface of the plastic melt is tracked using a volume of fluid method. Numerical investigations follow the concept of Design of Experiments in order to identify significant relationships between extrusion settings and bead shape. By this means, processing windows can be estimated virtually and conclusions can be drawn regarding slicing parameters and manufacturing time.

Keywords: Additive manufacturing · Plastics · Process simulation · Screw extrusion

---

The results, opinions and conclusions expressed in this thesis are not necessarily those of Volkswagen Aktiengesellschaft.

K. Dröder and T. Vietor (Eds.): *Technologies for economic and functional lightweight design*, Zukunftstechnologien für den multifunktionalen Leichtbau, pp. 156–169, 2021. https://doi.org/10.1007/978-3-662-62924-6_13

1 Introduction

Over the last years research and industry have intensified the development of additive manufacturing processes for plastic parts. The main goal is to reduce costs and realize complex and functionally integrated geometries. For prototyping, various additive processes are long known to rapidly produce concept parts at affordable costs. At the same time these processes are rarely used for series application because of small machine sizes, low material throughput and insufficient part quality [1, 2].

In terms of thermoplastic materials, the recently developed screw extrusion additive manufacturing (SEAM) challenges the economical limitations. SEAM uses a screw extruder to plasticise granulate and deposit it through a nozzle. The screw extruder has a high material throughput and is capable of processing injection moulding granulate. Therefore it offers significant advantage over conventional filament-based processes [2, 3]. In combination with a jointed-arm robot or tilt table the manufacturing system has a wide processing range to deposit different bead shapes [3, 4]. This setup is illustrated in Fig. 1.

The bead shape has a significant impact on manufacturing time, surface quality and mechanical properties. A trade-off has to be done between sufficient part quality and economical production. For this purpose it is mandatory to know the influence of extrusion settings on bead shape. In practical use it takes inefficiently long time to analyse these relationships in experimental testing due to the great amount of relevant parameters. On the contrary, a virtual prediction of the printing bead shape saves time and costs in the process development stage. This can be accomplished with a numerical simulation of the deposition process. As the bead shaping occurs when the plastic is in melt state, the process simulation has to cover the governing equations of computational fluid dynamics (CFD). These are referred to in the literature [5].

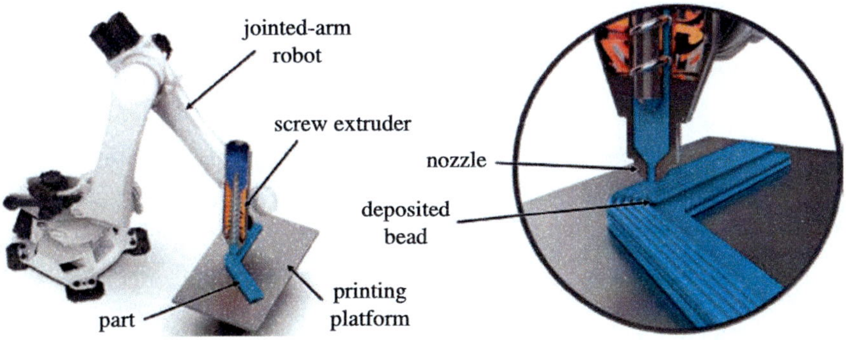

Fig. 1. Potential system setup for SEAM

Several studies exist regarding the simulation of bead deposition. Du et al. [6] as well as Xia et al. [7, 8] and Verma et al. [9] published investigations using temperature-dependent CFD models which can predict the bead shape for filament extrusion processes. However, these models are primarily used to evaluate the reheating area when several beads of material are deposited next to or on top of each other. Therefore they have to calculate the resulting temperature field. These non-isothermal calculations are computationally expensive which is drawback for routine design work.

Instead, Comminal et al. [10] proposed an isothermal CFD model for filament extrusion which was successfully validated by Serdeczny et al. [11]. For the evaluation of their results, the dimensionless ratio of feeding speed to printing speed was introduced. By this means, they described the process dynamics with only one parameter. This procedure is consistent as the non-linear shear-thinning effect of polymer materials is neglected. Serdeczny et al. therefore investigated the influence of shear thinning on bead shape and found a small but – for typical parameters in filament extrusion – negligible difference [12].

In contrast, this work considers the difference in shearing between common filament-based processes and SEAM. Throughputs and velocities in SEAM are at least one order of magnitude higher and significantly higher local shear rates have to be expected. A process simulation for SEAM should consequently account for shear-thinning behaviour. Hence, the process dynamics cannot be described with only one dimensionless parameter. Additionally, the temperature dependence of viscosity may have an impact on bead shape. To investigate this in a simple way, the shear-dependent viscosity function can be determined for different temperature levels. This function is then used as a variable input for the calculation. Therefore, this approach can simulate different temperature settings without temperature field calculation.

Based on these considerations, the process window for SEAM can be estimated in terms of extrusion settings. In the context of this work, also the impact on manufacturing time will be demonstrated exemplarily for the production of a large-scale tension bar.

Following the introduction, Sect. 2 of this contribution will explain the methodology used for the predictions in more detail. The modelling of bead deposition is focussed as well as the trial design in order to determine the impact on manufacturing time. Simulation results are discussed in Sect. 3. Hereafter, Sect. 4 states a conclusion and outlook for further applications.

2 Methodology

In order to gain information on bead shape and manufacturing times, the deposition process is investigated virtually using a Design of Experiments (DoE).

2.1 Modelling of Bead Deposition

The simulation model used in this work covers a small part of the extrusion nozzle with its exit and the gap between nozzle and printing platform. Figure 2 depicts an experimental setup and the corresponding simulation model.

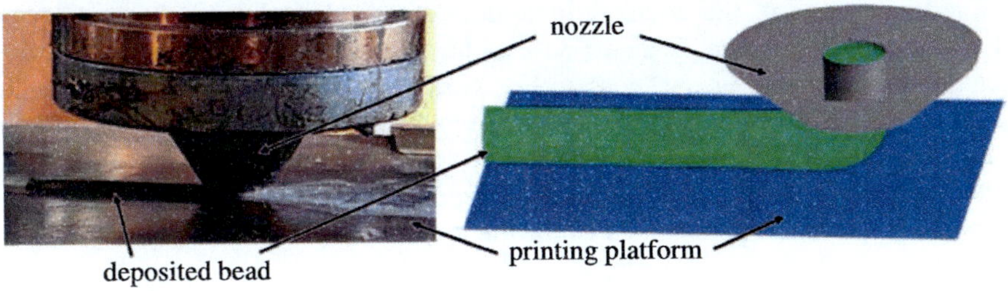

Fig. 2. Photograph (left) and simulation (right) of bead deposition

Figure 3 illustrates the components and dimensions of the simulation model. As the domain is symmetrical, only one half is modelled. The diameter of the nozzle is fixed to 1 mm. The upper face of the nozzle is considered as material inlet and all boundary faces except the nozzle and platform faces determine the possible outlet region. Results are evaluated at a slicing plane near the end of the domain in extrusion direction since evaluating directly at the outlet would contain boundary effects.

The numerical simulations are carried out in the software Ansys Polyflow, Version 19.2, which is based on the finite volume method. Free surface of the material flow is tracked using a volume of fluid method [13]. Tetrahedral elements with an edge length of 0.08 mm are used to discretise the computational domain (Fig. 3). Time step increments are automatically calculated according to the flow front propagation.

Modelling the process dynamics of SEAM can be done in two different ways. A representation with a moving extrusion nozzle requires a sliding mesh which is computationally expensive. To avoid this and consider the motion in a simple manner, a constant velocity is determined for the printing platform. As the movements of nozzle and printing platform are relative to each other and the simulation does not account for accelerations, both approaches are admissible [10].

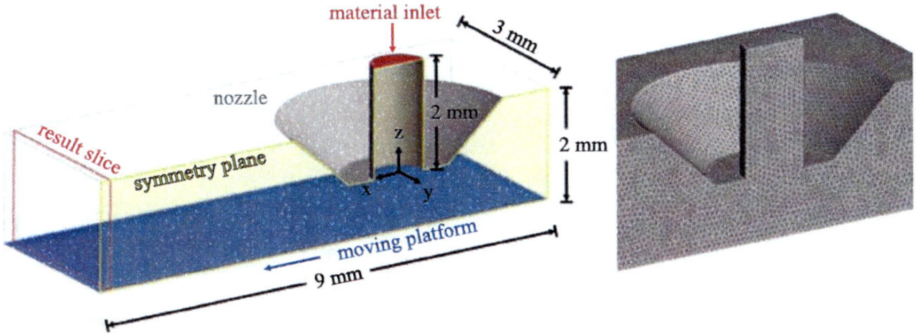

Fig. 3. Model of the computational domain (left) and section of the tetrahedral mesh (right)

The dimensions of the computational domain are chosen according to previous test studies. Following requirements have to be fulfilled for all process settings: representation of the whole bead cross section (Fig. 4b), representation of the backflow against extrusion direction (Fig. 4c) and achieving a converging flow for cross section of the deposited bead (Fig. 4c). For several process settings, flow instabilities may occur when the material detaches from the tapered nozzle wall (Fig. 4d). Therefore, the length in extrusion direction has to be the longest dimension to achieve a converging flow.

The material flow is characterised being isothermal, Newtonian and incompressible. Inertia effects are neglected as the flow is creeping and laminar due to a very small Reynolds number (Re<<1). The velocity field at the material inlet face is determined to be fully developed which means that a parabolic profile is applied. This considers the assumption that a real nozzle is significantly longer than this model. Gravitational force is accounted for. Wall adhesion of plastic melt on surfaces of nozzle and printing platform is assumed.

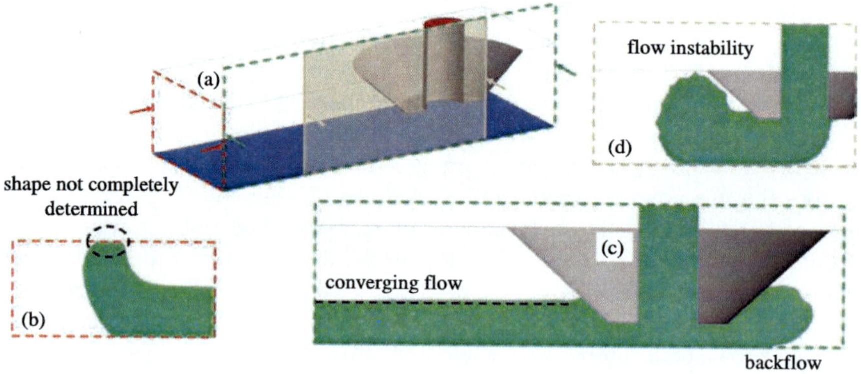

Fig. 4. Aspects to be considered for dimensioning of the computational domain (polymer material in green colour)

Furthermore, the material flow is considered as shear thinning. Previous test runs showed local shear rates much higher than 100/s especially in the area of the nozzle exit (Fig. 5). The influence of shear rate on viscosity of a relevant injection moulding compound was measured via high pressure capillary rheometer. The resulting viscosity function shown in Fig. 6 underlies a significant impact of shear rate for this process. Therefore, the model has to account for shear rate dependency of viscosity. In this contribution, the Cross model

$$\eta(\dot{\gamma}) = \frac{\eta_0}{1 + \left(\frac{\eta_0 \cdot \dot{\gamma}}{\tau *}\right)^{1-n}} \tag{1}$$

is applied [14], where η denotes the viscosity, $\dot{\gamma}$ the shear rate, η_0 the zero shear viscosity, $\tau*$ the critical shear stress and n is the Power Law index. As seen in Fig. 6, the viscosity functions can be obtained for different temperatures. This is done using the approach of Williams, Landel and Ferry (WLF) [14]. The Cross model and WLF approach can be combined to have only one single viscosity function. This single function is dependent on shear rate and temperature and therefore needs a temperature field. Usually this method is used for non-isothermal simulations of injection moulding. Local viscosity changes due to shear heating and forced cooling have a great impact on flow behaviour in these cases.

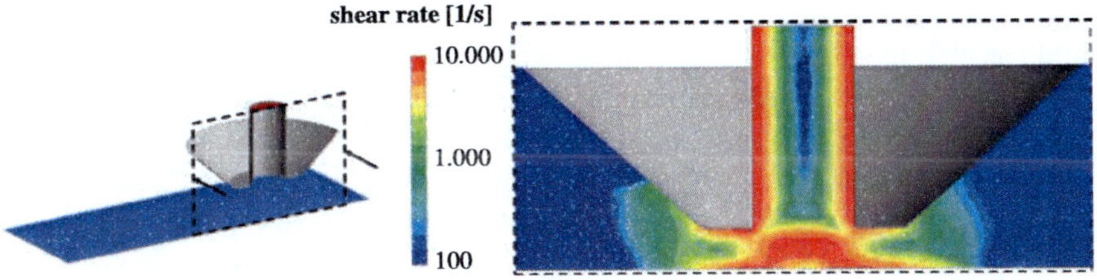

Fig. 5. Exemplary distribution of local shear rates during bead shaping

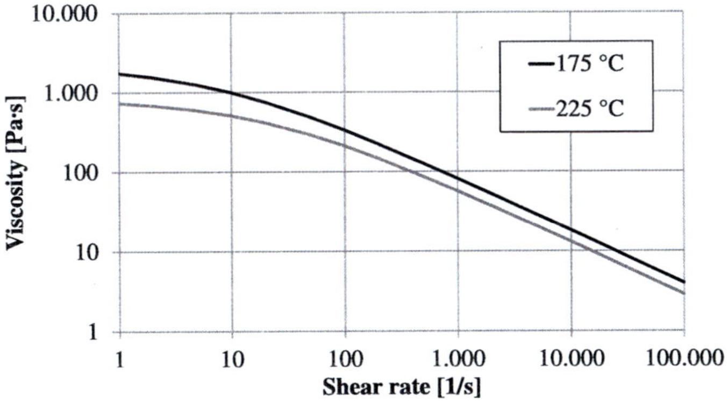

Fig. 6. Viscosity function of the selected injection moulding compound for two different temperatures

For the process simulation of SEAM these aspects are assumed to be negligible. The shear rates are significantly lower than in injection moulding. Additionally, the residence time of the fluid in the model domain (0.01–0.15 s) is too small for significant cooling of the material. Therefore, it is not advisable to include these effects as the computation time would increase.

Nevertheless, the influence of a different initial temperature of the melt is investigated in a simple way for this work. The isothermal Cross function is shifted to different temperature levels using the WLF approach. Hence, the variation of temperature settings is just an imaginary variation here as the temperature is not an input parameter. Indeed, the complete Cross function is the variable input parameter. The global viscosity is changed to the level of a different temperature instead of computing local viscosity changes. However, under given assumptions this method yields a first approximation of flow behaviour for different temperatures at moderate computation time.

2.2 Estimation of Manufacturing Time

SEAM provides several setting options. In order to regulate the process with regard to quality criteria, it is important to know the influence of parameter adjustments. The shape of the printing bead especially affects the manufacturing time, the surface roughness as well as the global density and bonding strength of parts. Therefore, profitability as well as surface quality and mechanical properties are influenced. This calls for a trade-off between those objectives. To accomplish this, a systematic optimisation of process settings is required.

In the recent past, the concept of DoE has gained in importance for those applications. It quantifies the influence of parameters and their interactions while keeping the testing effort as low as possible. The best-known method of this concept is the two-level factorial trial design. Each relevant parameter – a so-called factor – of the process is varied at two different levels. For the use case of this contribution, the four factors extruder throughput ($\dot{m}$), platform velocity (v), gap (h) between nozzle and platform and material temperature (T) are considered. The two different levels for each of these factors are chosen according to practical experience. The values are given in Table 1.

Table 1. Factors and investigated factor levels for DoE analysis

Level	Throughput ($\dot{m}$) [kg/h]	Platform velocity (v) [m/s]	Gap (h) [mm]	Material temperature (T) [°C]
low (−)	0.6	0.2	0.4	175
high (+)	1.2	0.6	1.0	225

For trial planning, the most common approach would change only one factor at a time to determine its influence. Contrary to this, DoE varies several factors at the same time. With a special trial design it is still possible to clearly assign the effects to their inducing factors [15, 16]. The amount of trials

$$n_r = n_l^{n_f} \tag{2}$$

in a factorial design is dependent on the amount of investigated factors n_f and the amount of levels n_l for each factor [15]. For the application of this work an amount of 16 simulation trials results.

Every trial will output a corresponding bead shape. As there is no given optimal bead shape, this is an auxiliary result. The bead shape is required to calculate values, which serve as a quality criterion for a specific part.

In this contribution, the approach will be carried out exemplarily for the manufacturing time of a part which is an economical quality criterion. The selected part is a tension bar based on DIN EN ISO 527–2. The tension bar is scaled by factor 8 in length and width and by factor 4 in thickness. By this means, the impact of bead shape on manufacturing time of large-scale parts is investigated as SEAM is especially suitable to produce these structures. The dimensions of the part are illustrated in Fig. 7. The manufacturing time for the part is estimated using the slicing software Ultimaker Cura, Version 3.4. The input parameters bead width, layer height and printing speed are set according to the simulation results. The two different slicing orientations 0° and 90° (see Fig. 7) are investigated. All other slicing settings are kept constant.

Following, the effect of each factor on manufacturing time of the part is calculated. According to [15], the average manufacturing time of all trials with low level is subtracted from the average manufacturing time of all trials with high level. For example, the effect $\dot{m}$ characterises the average change in manufacturing time if the extrusion throughput is changed from 0.6 kg/h (−) to 1.2 kg/h (+). Therefore, the influences of different settings on manufacturing time can be compared quantitatively by comparing their effects.

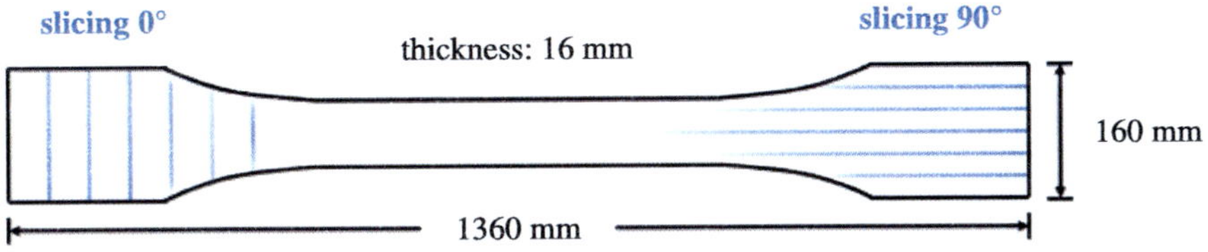

Fig. 7. Dimensions and slicing orientations of the large-scale tension bar

Additionally, factorial design allows its user to calculate the interactions of factors. This would not be possible if only one factor at a time was varied. Interaction means that the effect of a factor depends on the level of other factors, thus it may be illustrated as the influence of boundary conditions. The calculation of interactions follows the same scheme as that of effects [15, 16].

The interactions of factors allow a decision whether process settings should be optimized separately or not. Factors with high interactions have to be adjusted with great caution. This information is a great advantage of factorial trial design. However, the two-level design only provides linear relationships. To record non-linear relationships, at least three levels for each factor are required. This would increase the amount of simulations to an impractical effort. Moreover, in many industrial processes interactions have a greater impact than non-linearities [15]. This justifies the approach to determine many effects in a simple way instead of investigating only one factor and describing its influence in a complex way.

3 Results

In this section the results concerning the bead shape as well as the resulting manufacturing time for a large-scale tension bar are illustrated.

3.1 Prediction of Bead Shape

The simulation results yield the shape of the deposited bead for different extrusion settings of SEAM process. In the following, significant impacts on the bead shape are illustrated using comparisons between two settings at a time. The corresponding figures show the specified result slice (illustrated in Fig. 3) with the polymer material in green colour and the contour line of the nozzle in the background. For this purpose, Fig. 8 illustrates two testing points which only differ in the specified throughput.

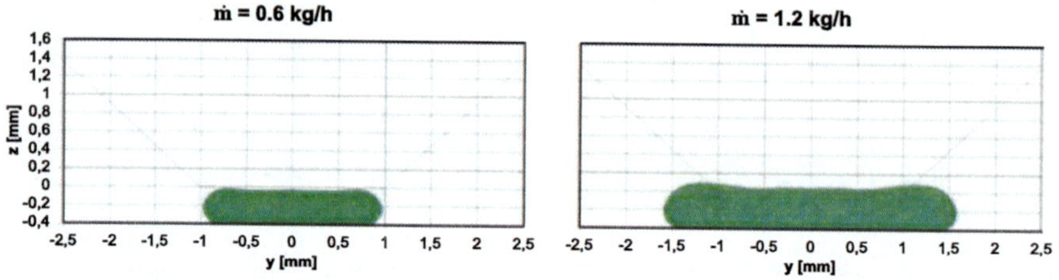

Fig. 8. Exemplary influence of throughput on bead shape. v=0.6 m/s; h=0.4 mm; T=175 °C

The higher throughput leads to a broadening of the bead as the gap between nozzle and printing platform is fixed. However, the increase in throughput also causes an increase in layer height to a value above the gap size. The nozzle forces the molten material to climb its tapered sidewall. When the plastic detaches from the nozzle, a backflow in y-direction to the middle of the bead occurs. This effect may depend on nozzle geometry and melt viscosity. Further increase in throughput or slowing down the platform could result in an unfavourable bead shape. This problem is illustrated in Fig. 9.

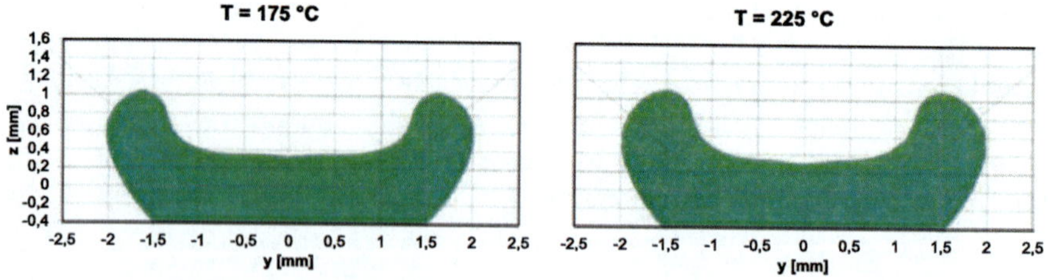

Fig. 9. Unfavourable bead shape due to wall adhesion on tapered nozzle wall. ṁ=1.2 kg/h; v=0.2 m/s; h=0.4 mm

The combination of high throughput, low platform velocity and small gap leads to a bead shape, which is not stackable without large void areas. The left and right result of Fig. 9 compare the setting for different temperatures. An increase in temperature and therefore a decrease in viscosity cannot improve the condition in this case. Figure 10 shows a contrary example for a larger gap. For these process settings an increase in temperature leads to a flattening of the bead due to backflow. However, when considering all simulation trials, only small changes in bead shape are notable due to temperature changes. Differences in temperature and therefore differences in global viscosity seem to be important only in cases where the material adheres to the tapered nozzle wall and a consequent backflow to the middle is possible.

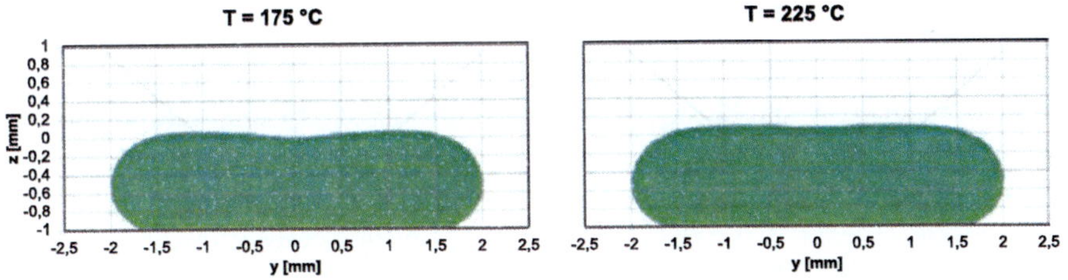

Fig. 10. Material backflow due to higher process temperature and lower viscosity. $\dot{m} = 1.2$ kg/h; $v = 0.2$ m/s; $h = 1.0$ mm

The influence of the gap on bead shape is illustrated exemplarily in Fig. 11. The unfavourable cross section at the small gap significantly improves when the gap is increased.

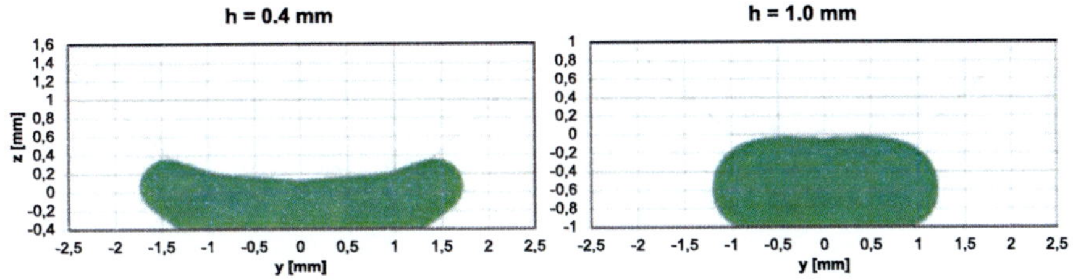

Fig. 11. Exemplary influence of the gap on bead shape. $\dot{m} = 0.6$ kg/h; $v = 0.2$ m/s; $T = 175$ °C

Figure 12 depicts the influence of platform velocity on bead shape for an exemplary configuration. As the velocity increases, the cross section changes from oblong to a filament-like rather circular shape.

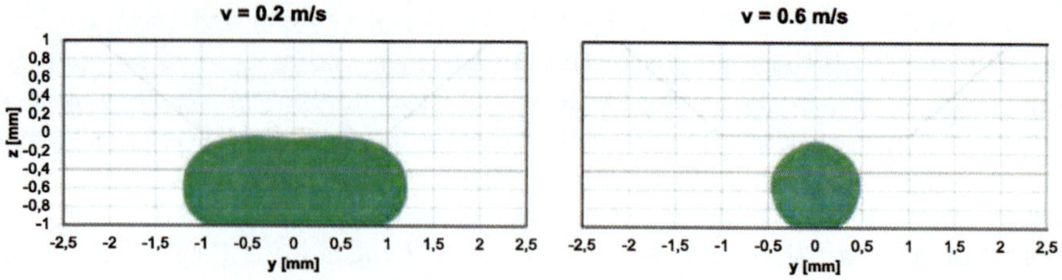

Fig. 12. Exemplary influence of platform velocity on bead shape. $\dot{m} = 0.6$ kg/h; h = 1.0 mm; T = 175 °C

In general, these simulation results are useful to determine the processing window of SEAM according to the part requirements. For example with regard to mechanical properties large bonding areas and voids as small as possible are preferable. The optimal bead shape would be approximately rectangular. Extrusion settings should be chosen in a way that the volume under the flat nozzle surface is filled by the bead as good as possible. On the other hand, overflow adhering to the tapered face of the nozzle should be avoided. With these considerations in mind, extrusion settings can be adopted to fulfil these requirements – for example according to a given material, nozzle geometry and throughput. In addition, the outer nozzle geometry can be designed to meet the requirements of bead shaping.

3.2 Influence on Manufacturing Time

The simulation results of bead deposition are used to estimate the manufacturing time of a large-scale tension bar. The input data for slicing are bead width, layer height and printing speed. Bead width is determined to be the greatest y-dimension and layer height is determined to be the greatest z-dimension of the bead. Printing speed is set to the value of platform velocity. The results of four simulation trials are neglected for this evaluation since their bead shape is out of the reasonable process window and would falsify the results. The DoE evaluation is illustrated in Fig. 13. The effect of factor T (temperature) and its interactions are not considered due to their negligible small value. Interaction vh cannot be determined as the above mentioned trials are not considered.

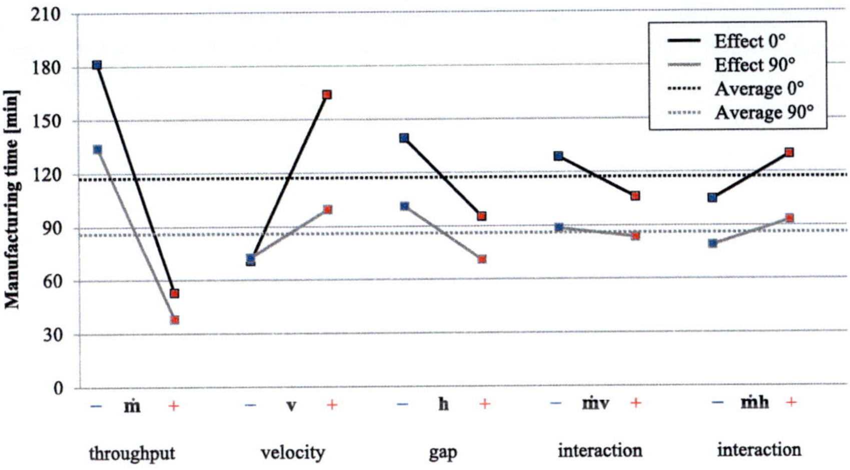

Fig. 13. Influence of factors and interactions on manufacturing time

The primary influence on manufacturing time comes from extruder throughput ($\dot{m}$). The high slope of the corresponding function shows that changing the throughput level from 0.6 kg/h (−) to 1.2 kg/h (+) significantly reduces manufacturing time. This is notable for both of the slicing orientations 0° and 90° with a similar magnitude.

The platform velocity (v) shows a contrary influence on manufacturing time. When increasing the velocity from 0.2 m/s (−) to 0.6 m/s (+), it takes significantly more time to produce the part. This is explained with the simulation results of bead deposition. The predicted bead shapes show that a higher printing speed induces a smaller bead width. Therefore, every layer has more paths that the platform has to travel along. As the motion has to stop at every end of a path and to accelerate again afterwards, the manufacturing time suffers from a high amount of pathways. The slicing process assumes a constant bead width and layer height. Hence, the accelerations are synchronised with the extruder throughput. Consequently, more accelerations lead to a higher percentage of manufacturing time without full extruder throughput. For the same reason, the average manufacturing time for orientation 90° is smaller than for 0°. The paths are oriented in direction of the longest dimension of the tension bar, which results in less acceleration procedures. According to this, the effect of platform velocity becomes significantly smaller if the slicing orientation is switched from 0° to 90°.

Increasing the gap (h) from 0.4 mm (−) to 1 mm (+) results in significantly shorter manufacturing time although the influence is smaller than that of throughput. The slicing orientation has a comparatively small influence on this effect. The shorter manufacturing time will likely come along with more void areas in this case because the larger gap induces a rather circular bead shape.

Furthermore, the diagram shows that interactions of factors $\dot{m}$ and v as well as $\dot{m}$ and h are relevant. This means that boundary conditions influence the effects of those factors. For example, an increase in throughput ($\dot{m}$) at a high platform velocity (v+)

is more effective than an increase in throughput ($\dot{m}$) at a low platform velocity (v−). However, this interaction of throughput and velocity is only significant for slicing orientation 0°. In comparison, interaction $\dot{m}h$ must be considered for both orientations.

4 Conclusion

The simulation model used for this contribution offers the desired option to calculate the bead shape as a function of extrusion settings in SEAM. A qualitative comparison of simulation results justifies the assumption that a change in global viscosity level due to a different process temperature has only a small impact. The simulation results additionally suggest that the outer nozzle geometry may significantly influence the process window because of material adhering and detaching from it. For a given nozzle geometry the processing window can be determined – for example when considering overflow at the nozzle side wall to be undesirable. Additionally, with the help of the introduced simulation model a nozzle can be designed to produce the required bead shape.

Nevertheless, the model does not account for viscoelastic die swelling after the nozzle exit as well as for volume contraction due to material cooling. These factors may likely lead to a scaling of the bead rather than changing its shape. However, they have to be kept in mind for a comparison with real processes. In this context, further development of the simulation model is conceivable to cover these options. Additionally, the investigation of different material types and nozzle geometries and the validation of the simulation model with practical machine trials are reasonable next steps.

Moreover, using calculated bead shapes to estimate the manufacturing time of a simple part could be demonstrated. Via DoE, the complex relationships of several process settings can be evaluated with a minimum effort. The selected trial concept provides comprehensible knowledge of the impact that different settings may have on manufacturing time. Therefore, this approach is suitable to be used for routine design work. Apart from the economical measure of manufacturing time also mechanical quality criteria like global density or bonding area can be optimized. Furthermore, a trade-off between different criteria is feasible.

Although potentially non-linear relationships cannot be quantified with a two-level factorial design, it is still possible to interpret the results and expand the process knowledge. However, in order to state universal conclusions and detect functional dependencies, more extensive studies have to be carried out.

References

1. Dröder, K., Heyn, J.K., Gerbers, R., Wonnenberg, B., Dietrich, F.: Partial additive manufacturing. experiments and prospects with regard to large series production. Proc. CIRP **55**, 122–127. Elsevier, Amsterdam (2016)

2. Hopmann, C., Hellmich, C., Lammert, N., Dormanns, J., Rinderlin, J., Stier, T., Zgrzebski, J.: Schichtweise vom Granulat zum Bauteil. Additive Fertigung von faserverstärkten PA6-Strukturbauteilen in hoher Geschwindigkeit. Kunststoffe **108**(11), 22–25 (2018)
3. Kausch, M., Blasé, J., John, C., Kroll, L., Holzinger, M., Reinhardt, A.: Additive Herstellung von Kunststoffbauteilen im High-Speed-Verfahren unter Einsatz von Standardgranulat. Konstruktion **70**(5), 58–60 (2018)
4. Dröder, K., Reichler, A.-K., Mahlfeld, G., Droß, M., Gerbers, R.: Scalable process chain for flexible production of metal-plastic lightweight structures. Proc. CIRP **85**, 195–200. Elsevier, Amsterdam (2019)
5. Anderson, Jr., J.D.: Governing equations of fluid dynamics. In: Wendt, J. F. (ed.): Computational Fluid Dynamics. An Introduction. Springer, Berlin, Heidelberg (2009).
6. Du, J., Wei, Z., Wang, X., Wang, J., Chen, Z.: An improved fused deposition modeling process for forming large-size thin-walled parts. J. Mater. Process. Technol. **234**, 332–341 (2016)
7. Xia, H., Lu, J., Dabiri, S., Tryggvason, G.: Fully resolved numerical simulations of fused deposition modeling. Part I – fluid flow. Rapid Prototyping J. **24**(2), 463–476 (2018)
8. Xia, H., Lu, J., Tryggvason, G.: Fully resolved numerical simulations of fused deposition modeling. Part II – solidification, residual stresses and modeling of the nozzle. Rapid Prototyping J. **24**(6), 973–987 (2018)
9. Verma, A., Vishnoi, V., Sukhotskiy, V., Furlani, E.P.: Numerical simulation of extrusion additive manufacturing: fused deposition modeling. TechConnect Briefs **4**, 118–121 (2018)
10. Comminal, R., Serdeczny, M.P., Pedersen, D.B., Spangenberg, J.: Numerical modeling of the strand deposition flow in extrusion-based additive manufacturing. Addit. Manuf. **20**, 68–76 (2018)
11. Serdeczny, M.P., Comminal, R., Pedersen, D.B., Spangenberg, J.: Experimental validation of a numerical model for the strand shape in material extrusion additive manufacturing. Addit. Manuf. **24**, 145–153 (2018)
12. Serdeczny, M.P., Comminal, R., Pedersen, D.B., Spangenberg, J.: Numerical study of the impact of shear thinning behaviour on the strand deposition flow in the extrusion-based additive manufacturing. In: Euspen's 18th International Conference & Exhibition. Venice, Italy (2018)
13. Noh, W.F., Woodward, P.: SLIC (Simple Line Interface Calculation). In: Proceedings of Fifth International Conference of Fluid Dynamics, pp. 330–340. Springer, Berlin, Heidelberg (1976)
14. Osswald, T.A., Rudolph, N.: Polymer Rheology. Fundamentals and Applications. Hanser Publishers, Munich (2015)
15. Siebertz, K., van Bebber, D., Hochkirchen, T.: Statistische Versuchsplanung. Design of Experiments (DoE). Springer, Heidelberg (2010)
16. Antony, J.: Design of Experiments for Engineers and Scientists. Elsevier, Amsterdam (2014)

Bio-based Innovations

Biomimetic Soft Robotic Peristaltic Pumping System for Coolant Liquid Transport

Falk J. Tauber[1,2,3](✉), Tom Masselter[1,2], and Thomas Speck[1,2,3]

[1] Plant Biomechanics Group, Faculty of Biology, Botanic Garden University Freiburg, Germany

[2] FMF – Freiburg Materials Research Center, Freiburg im Breisgau, Germany

[3] Cluster of Excellence Living Materials Systems (livMatS), University of Freiburg, Germany

falk.tauber@biologie.uni-freiburg.de

Abstract. In nature and technology, liquids are transported and distributed in a directed manner via pumping systems. Technical pumps often show signs of wear due to abrasion on moving parts, erosion and liquid contamination, which can lead to damage and unwanted noise. Innovative pump systems for electro mobility applications should have particularly low noise emission. In the course of evolution, various solutions have emerged in nature and can serve as a source of inspiration for the development of biomimetic pumping systems. In the last decade, the development of various biomimetic peristaltic systems highlight this pronounced biomimetic potential. These systems are based on principles behind bowel and esophageal peristalsis and incorporated within soft robots and medical devices. The main goal of this study was the biomimetic implementation of peristalsis into a flexible, silent, robust, energy efficient, space-saving and low cost technical application for the usage in combustion engines, electric engines and cooling systems. The biomimetic pump of the present study is based on the esophageal peristalsis and enables an easy, quiet and safe transport of a variety of Newtonian and non-Newtonian fluids with variable viscosities. The characterization of individual actuators as well as the entire peristaltic pump system in terms of closing rate and volume flow proved the influence of the actuator frequency and different peristaltic actuation patterns on the generated flow rate. The results show that the biomimetic flexible and elastic self-priming peristaltic pump based on silicone achieves sufficient flow rates of more than 250 l/h and thus offers an excellent alternative to conventional technical pumps in the field of electro mobility.

Keywords: Peristalsis · Soft robotics · Peristaltic pumping system · Biomimetic · Coolant pump

K. Dröder and T. Vietor (Eds.): *Technologies for economic and functional lightweight design*, Zukunftstechnologien für den multifunktionalen Leichtbau, pp. 173–181, 2021. https://doi.org/10.1007/978-3-662-62924-6_14

1 Introduction

In nature and technics, peristaltic pumping is used to safely transport a wide variety of media. Most peristaltic pumps in technics are roller or finger driven [1, 2]. These pumps consist of a rigid framing and a tube or hose as a flexible member, which is compressed via rollers or fingers (Fig. 1A). The current state-of-the-art coolant pumps are centrifugal, rotary vane pumps and gear pumps with flow rates between 100 and 1000 l/h depending on the engines conditions [2–4]. These technical pumps often show signs of wear due to abrasion on moving parts, erosion and liquid contamination, which can lead to system damage and unwanted noises. Innovative systems, especially for electro mobility applications, should be particularly quiet and wear-resistant. As the sound provided by combustion engines is missing in electric vehicles, a loud pump could possibly be heard inside the vehicle.

In contrast to their technical counterparts, biological pumps are generally highly integrated and specifically adapted systems that can only be considered to a limited extent detached from the complex overall system [4–6]. They consist of flexible, self-healing composite materials, which form in case of peristaltic systems soft hollow organs (e.g. mammalian esophagus and intestines, dorsal vessels of arthropods) [2, 5, 6]. Media transport is generated via contraction by flexible actuators (muscles) integrated into the pump [2, 6, 7] (Fig. 1B). This arrangement does not require any bearings or moving parts to rotate, offering advantages over technical systems in terms of wear [6]. In addition, biological peristaltic pumps can realize a continuous flow of viscous liquids with solids, generally have a long lifetime and are flexible, quiet and adaptive systems [2, 4–6, 8].

Within the last decade various biomimetic peristaltic pumping systems were developed. Esser et al. (2019) gave a comparative overview of biomimetic, biological and technical peristaltic pumping systems in terms of performance, volume flow rate and generated pressure [2]. Most biomimetic peristaltic pumps were based on bowel and esophageal peristalsis. A wide variety of actuation systems were utilized to drive the peristaltic motion e.g. dielectric and magnetic actuators [9–12], shape memory alloy actuators [13, 14] and pneumatic actuators [2, 15–25]. The fields of application of the systems differed widely. Dielectric and magnetic systems are mostly used in microfluidics, whereas shape memory alloy and pneumatic system also include macroscopic applications. For example, a pneumatic peristaltic conveyor was developed for the transport of printer powders [22, 24] and a swallowing robot as a biomimicry in vitro alternative for the investigation of food bolus rheology [26]. In our systems the flexible biomimetic peristaltic pump serves as a proof-of-concept alternative for industrial coolant pumping systems [18–20] (Fig. 1C). The comparison shown in

Table 1 gives an overview about the great variety of possible applications of the peristaltic principle for pump systems from highly precise systems for drop transport in the µl range [10, 27, 28] to systems that can cope with any fluid in most varied viscosities [1, 2, 20, 28–30] with flow rates from ml/h to several 100,000 l/h [2, 29, 30]. These examples show that the transfer of the underlying principles of the biological role models into technology can lead to the development of flexible, silent peristaltic pump systems.

Table 1. Overview of biological, biomimetic and technical peristaltic pumps

Pump system	Inner tubing diameter	Transport velocity/ Volume flow rate	Transported media
Human esophagus	13–25 mm [2, 31]	3–5 mm/s [2]	Water, food bolus, slurries
Peristaltic conveyor	30 mm [22–24]	120 l/h [22–24]	Water, bolus starch thickener
SBPP- flexible biomimetic peristaltic pump	20 mm	Up to 300 l/h	Water, oil cooling liquids, slurries
Microfluidic peristaltic pump	0.12–25 mm [29]	0.1–20 l/min [29]	Infusion media, blood, food slurries, liquids, chemicals
Hose pump	20 mm [30]	300,000 l/h	Food, paints, abrasive slurries, corrosive acids, large solids and gaseous liquids

The silicone based soft robotic biomimetic pumping system (SBPP) developed within the presented project is already capable of transporting various liquid with flow rates up to 298 l/h [20] (Fig. 1C). The developments within last decade confirm the high biomimetic potential of the peristaltic principle for the transfer into to novel technological applications as flexible, silent, robust, energy efficient, space-saving and low cost biomimetic pumping systems for the usage in combustion engines, electric engines and cooling systems.

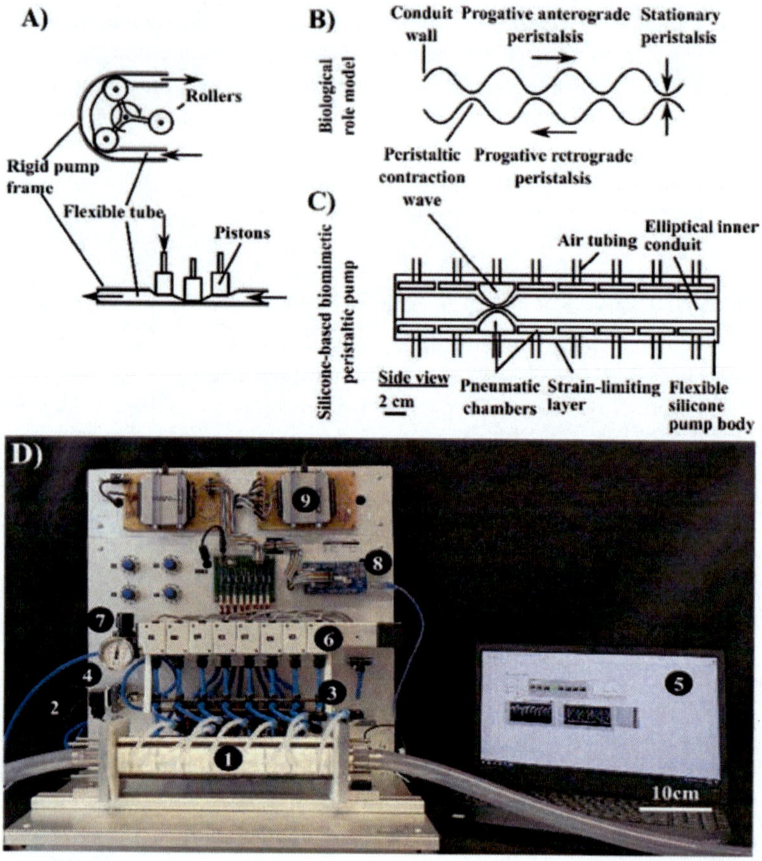

Fig. 1. Overview of peristaltic pumping systems. A: Two main designs of technical peristaltic pumps. Upper: Rotary peristaltic tube/roller pump; Lower: Linear peristaltic pump with piston actuators (adapted from [20]) B: Two modes of peristalsis pumping shown in a biological conduit, distinguishable into propagative anterograde (forwards) peristalsis and propagative retrograde (back-wards) peristalsis. C: Simplified blueprint of a soft robotic silicone-based biomimetic peristalsis pump (adapted from [20]). D) Test bench with biomimetic peristaltic pump. 1) Silicone-based biomimetic peristaltic pump. 2) Gas volume flow sensor (mounted on the rear). 3) Solenoid valves. 4) System air pressure sensor. 5) PC with test bench program. 6) Actuator air pressure sensors. 7) Pressure gauge for setting the system pressure. 8) Microcontroller. 9) Data acquisition devices

2 Characterization of the Silicone Based Biomimetic Peristaltic Pump (SBPP)

In contrast to other biomimetic and technical systems, a fully flexible design was achieved with the SBPP (Fig. 1C, D) [19, 20]. The pump is outfitted with eight pneumatic actuators (each with two pneumatic chambers perpendicular to each other) and a strain-limiting layer on the outside. The pump remains flexible in longitudinal direction but becomes inextensible in the radial direction when pressurized, achieving an entirely flexible system capable of transport without rigid framing. When pressurized the strain-limiting layer on the outside directs all expansion inward and the inner elliptical conduit is occluded by the expanding pneumatic chambers. The system is able to produce sufficient flow rates by consecutive activation of the actuators with a peristaltic activation pattern (see [19, 20]). The whole systems can be bent, pulled and twisted without damaging the system, so the system could be adapted to various build spaces.

The developed self-priming SBPP was characterized with a specially designed pump test bench in terms of produced flow rate and power consumption at different sequential activation durations (AD) of each actuator (75 ms, 100 ms, 125 ms). Within the characterization, four pumps were tested 10 times with 50 actuation cycles during each individual test. Additionally, the activation patterns (pattern 1: Sequential; pattern 2: Including a pause after each cycle (see Table 2) and transport directions (forward and backward) were varied for the transported fluids (water, oil and cooling liquid) (Fig. 1D) [20]. No check valves were incorporated in the system. With the integration of the pause it was possible to investigate if fluid inertia could be utilized to augment the fluid flow rate. In addition, the pause could lower the energy consumption, as no energy (air) is consumed during the pause [20].

Table 2. Peristaltic actuation patterns used in the SBPP characterization, one actuation cycle per pattern is shown, each cycle was repeated 50 times per measurement with 75 ms, 100 ms and 125 ms AD. O: open state, **X**: occluded state

Pattern 1	Pattern 2 incl. pause
XOOOOOOO	**X**OOOOOOO
O**X**OOOOOO	O**X**OOOOOO
OO**X**OOOOO	OO**X**OOOOO
OOO**X**OOOO	OOO**X**OOOO
OOOO**X**OOO	OOOO**X**OOO
OOOOO**X**OO	OOOOO**X**OO
OOOOOO**X**O	OOOOOO**X**O
OOOOOOO**X**	OOOOOOO**X**
	OOOOOOOO

Figure 2 exemplary highlights the different volume flow at 100 ms AD for the investigated fluids. Within this study, a first characterization of the SBPP transporting a commercially available cooling liquid mixture for cars was conducted. In general, the SBPP achieved for the transport of water and cooling fluid volume flow rates of over 200 l/h and for oil of over 120 l/h depending on the AD and actuation pattern. In a direct comparison significantly higher flow rates were achieved with pattern 2 for all transported fluids [20]. The highest flow rate achieved for water was 298 l/h at 75 ms AD, for oil 169 l/h at 100 ms AD and for cooling liquid 215 l/h at 100 ms AD. The reversal of the pumping direction had only marginal effect on the volume flow rate.

This was consistent with the null hypothesis that the deformation of the actuators was not affected by the direction of actuation [20]. The developed systems produced no additional noise during contraction and showed no signs of wear after 50.000 actuation cycles. For a more in detail characterization of the SBPP pump performance and energy consumption, the authors refer to Esser et al. 2019 [20].

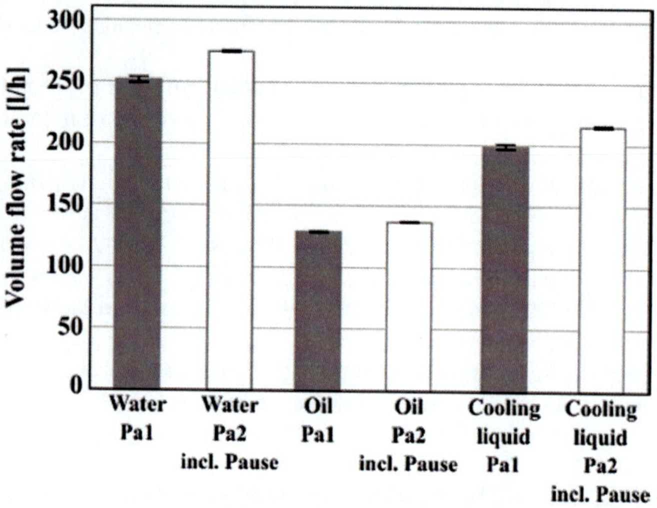

Fig. 2. Exemplary data comparing volume flow rate measurements for water, oil and cooling liquid pumping at 100 ms actuation duration (n = 10 à 50 actuation cycles). The SBPP produced higher flow rates with a pause in the actuation cycle utilizing the fluid inertia. Pa1 = pattern 1 sequential activation; Pa2 = pattern 2 with a pause corresponding to the AD after each cycle

3 Discussion

In previous studies, the influence of different actuation patterns as well as the in-corporation of check valves into the system was investigated in pumps with a tubing diameter of 8 and 20 mm at inlet and outlet of the SBPP (inner pump diameter was in both cases 20 mm) [19, 20]. The flow rate increased with a higher actuation frequency and a higher tubing diameter, but not with the number of simultaneously active actuators. With 8 mm tubing a flow rate of 50 l/h was achieved with the higher tubing diameter flow rate of over 250 l/h for water could be achieved. Additionally, a higher number of simultaneously active actuators had a negative impact on the flow rate [19]. Check valves posed as a high flow hindrance in the 8 mm tubing version, the 20 mm tubing system was self-priming so no check valves were needed. By including a pause in the pump cycle the fluid could flow unhindered through the pump once within each cycle, through this the inertia of the fluid was exploited and the volume flow augmented to reach higher transport values. The inclusion of a pause had not only a positive effect on the flow rate but also on the power consumption, which was significantly lower as no actuator was active and no "energy" (pressured air) was consumed during the pause as seen in [20]. The tests with cooling liquids showed that the system could generate sufficient flow rates to transport liquids with different viscosities and densities. As expected, liquids with higher density and viscosity than water were transported more slowly, and a significantly higher flow rate was achieved for cooling liquid than for oil. The results highlight the potential to use the presented system as an alternative for current coolant pumps. The system is able to produce sufficient flow rates for currently used cooling liquids.

Compared to existing biomimetic pump systems, the SBPP distinguishes itself from other demonstrators [17, 22, 23] by its completely flexible design and twice as high reached volume flow rates (4 l/min SBPP vs. 2 l/min produced by the demonstrator of [22]). In varying the actuation pattern and frequencies, the performance of the pump can be adjusted as required. Through changing the base material and in-corporation of an electrical actuation, the SBPP system can be used for electro mobility. This makes it possible to replace the previous centrifugal pumps and coolant hoses with self-pumping tube systems between the system components, which can be controlled as required. This saves installation space and weight by gaining in addition a noiseless pump system with low wear. In future studies alternative electrically based actuation principles will be developed and utilized to actuate an electrically based biomimetic soft-robotic tube pump.

4 Conclusions

The extreme progress in the producibility of new types of pump systems within the last decade was only made possible by new and now economically feasible manufacturing processes for novel material and actuator systems. This allows the production of previously unimaginable geometries for pump systems in hitherto unconsidered areas of application, as in case of peristaltic pumps as alternative coolant pumps.

The developed biomimetic peristaltic pumping systems presented in literature and within this project show the high potential for peristaltic pumping in the field of electro mobility. Within the scope of this project, a first demonstrator system was realized to demonstrate the technical feasibility of transferring the peristaltic pump principle into a biomimetic peristaltic pump as an alternative for coolant pumps in electric vehicles. The system described in this article achieved sufficient flow rates of fluids with different viscosities by peristaltic activation patterns. A significant positive impact of a short pause of actuation on fluid flow rate and power consumption, as well as an influence of the actuation duration on the flow rate could be identified. Future projects will focus on the development of purely electrically driven low-energy pumping systems. As well as the further investigation of influencing factors of viscous fluid transport.

Acknowledgements. We thank various colleagues within the framework of the cluster of excellence 'Living Materials Systems (*liv*MatS)', funded by the Deutsche Forschungsgemeinschaft (DFG, German Research Foundation) under Germany's Excellence Strategy – EXC-2193/1 – 390951807, for inspiring scientific discussions.

References

1. Nesbitt, B.: When is a tube not a tube?: when it's a pump! World Pumps **2004**(448), 20–23 (2004)
2. Esser, F., Masselter, T., Speck, T.: Silent pumpers: a comparative topical overview of the peristaltic pumping principle in living nature, engineering, and biomimetics. Adv. Intell. Syst. **1**(2) (2019)
3. Karassik, I.J. (ed.): Pump Handbook, 4th edn. McGraw-Hill, New York (2008)
4. Bach, D., Schmich, F., Masselter, T., Speck, T.: A review of selected pumping systems in nature and engineering—potential biomimetic concepts for improving displacement pumps and pulsation damping. Bioinspir. Biomim. **10**(5), 51001 (2015)
5. Vogel, S.: Nature's pumps. Am. Sci. **82**(5), 464–471 (1994)
6. Vogel, S.: Living in a physical world X. Pumping fluids through conduits. J. Biosci. **32**(2), 207–222 (2007)
7. Cannon, W.B.: The nature of gastric peristalsis. Am. J. Physio. Leg. Content **29**(2), 250–266 (1911)
8. Fratzl, P.: Biomimetic materials research: what can we really learn from nature's structural materials? J. Royal Soc. Interface **4**(15), 637–642 (2007)
9. Lotz, P., Matysek, M., Schlaak, H.F.: Peristaltic pump made of dielectric elastomer actuators. In: Bar-Cohen, Y., Wallmersperger, T. (eds.) Electroactive Polymer Actuators and Devices (EAPAD) 2009, p. 72872D. SPIE, Bellingham (2009)
10. Bowers, A.E., Rossiter, J.M., Walters, P.J., Ieropoulos, I.A.: Dielectric elastomer pump for artificial organisms. In: Bar-Cohen, Y., Carpi, F. (eds.) Electroactive Polymer Actuators and Devices (EAPAD) 2011, p. 797629. SPIE, Bellingham (2011)
11. Fuhrer, R., Schumacher, C.M., Zeltner, M., Stark, W.J.: Soft iron/silicon composite tubes for magnetic peristaltic pumping: frequency-dependent pressure and volume flow. Adv. Func. Mater. **23**(31), 3845–3849 (2013)
12. McCoul, D., Pei, Q.: Tubular dielectric elastomer actuator for active fluidic control. Smart Mater. Struct. **24**(10), 105016 (2015)
13. Miki, H., Okuyama, T., Kodaira, S., Luo, Y., Takagi, T., Yambe, T., Sato, T.: Artificial-esophagus with peristaltic motion using shape memory alloy. Int. J. Appl. Electromagnet. Mech. **33**(1–2), 705–711 (2010)
14. Solano-Arana, S., Klug, F., Mößinger, H., Förster-Zügel, F., Schlaak, H.F.: A novel application of dielectric stack actuators: a pumping micromixer. Smart Mater. Struct. **27**(7), 74008 (2018)
15. Chen, Z., Deng, Z., Dhupia, J.S., Stommel, M., Xu, W.: Modelling of a soft actuator for a planar manipulator table. In: 2018 25th International Conference on Mechatronics and Machine Vision in Practice (M2VIP), pp. 1–6. IEEE, New York, NY, (2018)
16. Dirven, S., Xu, W., Cheng, L.K., Allen, J., Bronlund, J.: Biologically-inspired swallowing robot for investigation of texture modified foods. Int. J. Biomech. Biomed. Robot. **2**(2), 163 (2013)
17. Dirven, S., Allen, J., Xu, W., Cheng, L.K.: Soft-robotic esophageal swallowing as a clinically-inspired bolus rheometry technique. Meas. Sci. Technol. **28**(3), 35701 (2017)
18. Esser, F.J., Steger, T., Bach, D., Masselter, T., Speck, T.: Development of novel foam-based soft robotic ring actuators for a biomimetic peristaltic pumping system. In: Mangan, M., Cutkosky, M., Mura, A. (eds.) Biomimetic and Biohybrid Systems, vol. 10384, pp. 138–147. Springer, Cham (2017)

19. Esser, F.J., Krüger, F., Masselter, T., Speck, T.: Development and characterization of a novel biomimetic peristaltic pumping system with flexible silicone-based soft robotic ring actuators. In: Vouloutsi, V., Halloy, J., Mura, A. (eds.) Biomimetic and Biohybrid Systems, vol. 10928, pp. 157–167. Springer, Cham (2018)
20. Esser, F.J., Krüger, F., Masselter, T., Speck, T.: Characterization of biomimetic peristaltic pumping system based on flexible silicone soft robotic actuators as an alternative for technical pumps. In: Martinez-Hernandez, U., Vouloutsi, V., Mura, A. (eds.) Biomimetic and Biohybrid systems, vol. 11556, pp. 101–113. Springer, Cham (2019)
21. Stommel, M., Xu, W., Lim, P.P.K., Kadmiry, B.: Robotic sorting of ovine offal: discussion of a soft peristaltic approach. Soft Robot. **1**(4), 246–254 (2014)
22. Suzuki, K., Nakamura, T.: Development of a peristaltic pump based on bowel peristalsis using for artificial rubber muscle. In: The IEEE/RSJ 2010 International Conference on Intelligent Robots and Systems (IROS 2010), pp. 3085–3090. IEEE, Piscataway, NJ (2010)
23. Yamada, Y., Ashigaki, K., Yoshihama, S., Negishi, K., Kato, K., Nakamura, T.: Triangular cross-section peristaltic conveyor for transporting powders at high speed in printers. Adv. Robot. **32**(12), 646–658 (2018)
24. Yoshihama, S., Takano, S., Yamada, Y., Nakamura, T., Kato, K.: Powder conveyance experiments with peristaltic conveyor using a pneumatic artificial muscle. In: 2016 IEEE International Conference on Advanced Intelligent Mechatronics (AIM), pp. 1539–1544. IEEE, Piscataway, NJ (2016)
25. Zhu, M., Xie, M., Xu, W., Cheng, L.K.: A nanocomposite-based stretchable deformation sensor matrix for a soft-bodied swallowing robot. IEEE Sens. J. **16**(10), 3848–3855 (2016)
26. Dirven, S., Xu, W., Cheng, L.: Soft robotics for bio-mimicry of esophageal swallowing. In: Verl, A., Albu-Schäffer, A., Brock, O. (eds.) Soft Robotics: Transferring Theory to Application, pp. 282–291. Springer, Heidelberg (2015)
27. Hildebrandt, J.-P.: Circulation in the leech, Hirudo Medicinalis L. J. Exp. Biol. **134**(1), 235–246 (1988)
28. Berg, J.M., Dallas, T.: Peristaltic pumps. In: Li, D. (ed.) Encyclopedia of microfluidics and nanofluidics, Second edition, pp. 2693–2701. Springer, New York (2015)
29. Watson-Marlow, in Watson-Marlow Ind. Overv., Waston-Marlow Fluid Technology Group (2018)
30. LSM Pumper: LSM 200 Technical datasheet. https://www.lsmpumps.com/products/lsm-pumps/lsm-200/ (2020). Accessed 1 Juli 2020
31. Banić, M., Urek, M. C., Prka, L.: Intestinal Function. In: Duvnjak, M., Smirčić-Duvnjak, L. (eds.) Gastrointestinal complications of diabetes. Clinical Gastroenterology, pp 103–113. Humana Press, Cham (2018)

Biomimetic Suction Cups for Energy-Efficient Industrial Applications

Harald Kuolt[1(✉)], Tim Kampowski[2,3], Simon Poppinga[2,3], Thomas Speck[2,3,4], Ralf Tautenhahn[5], Atena Moosavi[5], Jürgen Weber[5], Felix Gabriel[6], Erika Pierri[6], and Klaus Dröder[6]

[1] J. Schmalz GmbH, Glatten, Germany
harald.kuolt@schmalz.de
[2] Plant Biomechanics Group, Botanic Garden, University of Freiburg, Freiburg im Breisgau, Germany
[3] Freiburg Materials Research Center (FMF), University of Freiburg, Freiburg im Breisgau, Germany
[4] Cluster of Excellence livMatS, University of Freiburg, Freiburg im Breisgau, Germany
[5] Technische Universität Dresden, Institute of Mechatronic Engineering, Chair of Fluid Mechatronic Systems Engineering, Dresden, Germany
[6] Institute of Machine Tools and Production Technology (IWF), Technische Universität Braunschweig, Braunschweig, Germany

Abstract. Suctions cups are widely used in industrial applications for handling various goods. In the animal kingdom, a multitude of analogous structures has evolved, e.g. in octopuses and leeches, which allow the respective organisms to securely attach and detach from various substrates. To date, the biological suction cups outperform their technical counterparts in terms of versatility, as they allow attachment also on challenging (e.g. porous, wet and fouled) surfaces.

Keywords: Vacuum handling · Energy efficiency · Bioinspiration · Biomimetics

1 Introduction

Suctions cups are widely used in industrial applications for handling various goods. In the animal kingdom, a multitude of analogous structures has evolved, e.g. in octopuses and leeches, which allow the respective organisms to securely attach and detach from various substrates. To date, the biological suction cups outperform their technical counterparts in terms of versatility, as they allow attachment also on challenging (e.g. porous, wet and fouled) surfaces.

In the "BiVaS" project, which is funded by the German Federal Ministry for Economic Affairs and Energy (BMWi), the consortium partners are working together

K. Dröder and T. Vietor (Eds.): *Technologies for economic and functional lightweight design*, Zukunftstechnologien für den multifunktionalen Leichtbau, pp. 182–188, 2021. https://doi.org/10.1007/978-3-662-62924-6_15

on a new approach to optimize the energy efficiency of vacuum handling technology, based on biomimetic principles. In this context, we investigate the functional and constructional principles of animal suction cups and evaluate their biomimetic potential, especially in the multifunctional leech adhesive system. With the abstraction and technical implementation of working principles found, we develop new biomimetic applications leading to novel and versatile suction cups for effective industrial handling scenarios. Moreover, the authors are further working on the optimization of both the ejector fluid dynamics and the overall operational strategies. Finally, the energy saving potentials of biomimetic suction cups in conjunction with improved vacuum generation technologies and adequate design of the robot-based handling process will be demonstrated.

2 Suction Adhesion in Biological and Biomimetic Structures

Previous studies of biological suction devices have mainly been carried out on aquatic animals, especially molluscs (here especially the octopus), arthropods (e.g. crustaceans, insects), fish (here especially the Northern clingfish and remora) and lampreys. The respective organs of these animals are usually suction discs, which are used, for example, in adhering to hosts for parasitizing for blood or on sexual partners for copulation. Their edges are flexible, often microstructured and characterised by "fringes" (termed villi), which cause contact splitting (better adhesion) and compensate for unevenness of the substrate (leakage compensation) [1]. The microstructured suction disc rim of the octopus (*Octopus vulgaris*), for example, is characterized by a roughness of Ra = 11.3 μm on average, which enables a very good form fit to the substrate [2]. In many animals, such as the clingfish, mucus is released at the sucker, which supports leakage compensation [3]. A good form fit with the substrate is also achieved by the soft edges of the sucker. For example, the edge of the octopus suction disc consists of extremely soft tissue (average modulus of elasticity of E = 7.7 kPa) and can therefore adapt very well to uneven surfaces [2].

Also a smart solution for maintaining negative pressure generated in the suction cup is already present in nature: in order to remain firmly attached, the octopus does not have to use any muscle power. Here, the deformation of a protrusion in the suction cavity ensures that a separate water torus is formed during the suction process. Due to the cohesive force of the water in the torus and due to friction between the surfaces of the protrusion and the walls of the suction cavity, this state is maintained without further muscle contraction and it is balanced by the elastic restoring force of the protrusion [2, 4].

While much research has already been carried out on aquatic organisms, very little is known to date about how (partially) land-based animals with suction discs manage their adhesion and which behaviour-based, morphological and biomechanical adaptations exist in these organisms [1, 5]. These include, for example, terrestrial leeches, which comprise several systematic groups (e.g. Haemadipsidae), and the larvae of the gastropod moth (Limacodidae) [6]. The medicinal leech (*Hirudo verbana*) has a suction system that functions reliably on various surfaces both under and above water [7].

3 Transfer into Suction Cups for Industrial Applications

Conventional technical suction pads are generally made of polymers, which are characterised by material properties different from those of biological tissue. In addition, they typically do not possess any microstructuring and no additional slime or other liquids are used to compensate for leakage. A number of biomimetic approaches have already attempted the transfer of natural principles for improving suction adhesion in technology. For example, it was shown that octopus-inspired, radially running, one to several hundred micrometers wide and deep grooves on the undersides of 3D-printed suction discs made of elastomers increase their adhesive forces under water [8]. In other approaches, the muscular deformation of the octopus suction cavity was abstracted for the generation of suction power and implemented in technical solutions, e.g. by means of dielectric elastomers [9], shape memory alloys [10], or with thermo-responsive hydrogels [11]. Also the principle of adhesion described above for the octopus allowing to adhere onto the substrate over a longer period of time without energy-intensive muscle work was used as model for a permanent biomimetic suction system [12].

When focusing on industrial applications and the components used there, it has to be emphasized that handling of workpieces is not a value-adding activity. Generally, handling means the transport of workpieces from one point to another, including the processes of grasping and releasing. This has to be accomplished rapidly, with high accuracy, efficiency and reproducibility, and without negative influence on the workpiece (no damage or change of properties). When looking on suction adhesion in biology, there are some functional differences and also difficulties in transferring the above mentioned principles into technical solutions. First, most of the studied principles are based on water as the surrounding medium. Furthermore, gripping principles in nature are in many cases not evolved for a smooth and gentle handling of sensitive "workpieces". For example, during the chase for and adhesion on prey, marks on the prey's body may occur.

Having a closer look on the customers' needs, process safety is very important in the first place and, therefore, requires a fail-safe gripper-workpiece-connection. A loss of a workpiece has to be avoided under all circumstances, even during an electrical power failure. A high system availability, no or less downtime and short cycle times are mandatory to achieve low recurring costs. Furthermore, low investment costs are also necessary. Therefore, new biologically inspired suction cups should allow production with common production technologies and in high numbers. Due to the fact that handling is not a value-adding activity, the customer will not pay more for a biological inspired handling solution than necessary, especially when there already exists a proven standard technology to solve his handling task.

At the moment, two different principles are in focus for further development. First, passively working suction cups offer an interesting opportunity for reducing energy consumption. During contact of the workpiece and such a suction cup, the air inside the suction cup is pressed out. When the device lifts up, the volume will increase and a vacuum will concomitantly be applied. When there is no leakage, a handling process with nearly no (external) energy consumption would be possible.

To assure process safety, an additional vacuum generator could be mounted for providing a higher vacuum level or an additional suction flow. This could take effect if the “passive vacuum level” is not high enough or leakage is occurring. Such predominantly passive suction handling would lead to less air and energy consumption during vacuum generation. Another advantage could be the usage of smaller vacuum generators than in standard solutions. In Fig. 1 a special developed suction cup with less inner volume and measures to support passive vacuum generation is presented and first test results are shown in Tables 1 and 2.

Fig. 1. concept of an improved suction cup with less inner volume for active and passive vacuum handling

Table 1. Evacuation time of an improved suction cup with different hardness in comparison with a standard suction cup based on a standard vacuum handling application with a rigid an airtight workpiece with different vacuum generators (vacuum level: 600 mbar)

Nozzle [mm]	Evacuation time [s]		
	Standard Cup	Improved [65 ShA]	Improved [85 ShA]
5	0.36	0.22	0.16
6 l/min/850 mbar			
13 l/min air comsumpt.			
7	0.32	0.16	0.14
12 l/min/850 mbar			
21 l/min air consumpt.			
10	0.14	0.06	0.06
23 l/min/850 mbar			
46 l/min air consumpt.			
Cobot Pump	0.41	0.27	0.28
12 l/min/750 mbar			
14 W			

Table 2. Evacuation time of an improved suction cup with different hardness in comparison with a standard suction cup based on an improved process with passive pre-evacuation in combination with an active vacuum generator to reach the same vacuum end level of 600 mbar with a rigid an airtight workpiece with different vacuum generators

Nozzle [mm]	Evacutation time [s]		
	Standard Cup	Improved [65 ShA]	Improved [85 ShA]
5 6 l/min/850 mbar 13 l/min air comsumpt.	0.06	0.06	0.05
7 12 l/min/850 mbar 21 l/min air consumpt.	0.07	0.06	0.06
10 23 l/min/850 mbar 46 l/min air consumpt.	0.03	0.03	0.03
Cobot Pump 12 l/min/750 mbar 14 W	0.24	0.22	0.22

Enhancing the sealing and friction behavior between suction cup and workpiece would greatly optimize the efficiency of the handling process. In our biomimetic approach, suction cups based on the gripping principle of the northern clingfish were developed. As mentioned in [13], the combination of a rigid suction cup body and a soft sealing "tissue" with an integrated microstructure can improve both, the sealing and friction behavior. First suction cups were realized and tested already, as mentioned in Fig. 3.

	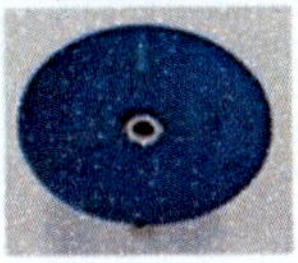		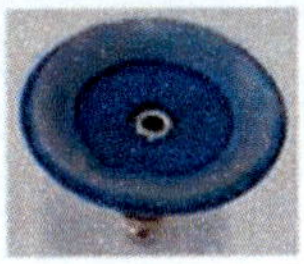
	450 mbar	820 mbar	820 mbar
	310 mbar	670 mbar	760 mbar
	0 mbar	820 mbar	820 mbar
	0 mbar	820 mbar	820 mbar

Fig. 3. Achievable vacuum levels on different structures with standard suction cup (left), special suction cup with EPDM-Sealing (middle) and standard suction cup with sealing based on the Northern Clingfish (done in Cooperation with Petra Ditsche)

Another promising approach to inspire new suction cups are the suction discs of leeches, especially of the medicinal leech (*Hirudo verbana*). As shown in [7, 14], it is possible for leeches to adhere on multiple surfaces with different structures, even on very porous structures with holes. Whereas suction is the main attachment mechanism in the medicinal leech, other principles like clamping are additionally employed. Hence, the suction discs are very versatile and fail-safe. Some first prototypes of biomimetic suction pads based on *Hirudo verbana* will be realized and tested in the next upcoming months.

4 Conclusion and Outlook

We present first results gained during the ongoing BiVaS project regarding the biomimetic potential of selected biological sucker systems. Based on these results, we show how biomimetically advanced technical suction cups could work and anticipate a reduction of the energy consumption in vacuum handling of up to 20%.

Acknowledgements. The authors acknowledge funding by the by the German Federal Ministry for Economic Affairs and Energy (BMWi) under the project number: 03ET1559 ("Increased energy efficiency in vacuum handling by bionic principles and reduction of dead volumes (BiVaS)".

References

1. Ditsche, P., Summers, A.P.: Aquatic versus terrestrial attachment: water makes a difference. Beilstein J. Nanotechnol. **5**, 2424–2439 (2014)
2. Tramacere, F., Kovalev, A., Kleinteich, T., Gorb, S.: Structure and mechanical properties of Octopus vulgaris suckers. J. R Soc. Interface (2014)
3. Wainwright, D.K., Kleinteich, T., Kleinteich, A., Gorb, S., Summers, A.P.: Stick tight: suction adhesion on irregular surfaces in the northern clingfish. Biol. Lett. **9**(3), 20130234 (2013)
4. Tramacere, F., Beccai, L., Kuba, M., Gozzi, A., Bifone, A., Mazzolai, B.: The morphology and adhesion mechanism of *Octopus vulgaris* suckers. PLoS One **8**(6), e6507 (2013)
5. Wolff, J.O., Gorb, S.: „Suction cups“ in Attachment structures and adhesive secretions in arachnids. Biol. Syst. (2017)
6. Epstein, M.E.: Evolution of locomotion in slug caterpillars (Lepidoptera: Zygaenoidea: Limacodid group). J. Res. Lepidoptera **34**, 1–13 (1995)
7. Kampowski, T., Eberhard, L., Gallenmüller, F., Speck T., Poppinga, S.: Functional morphology of suction discs and attachment performance of the Mediterranean medicinal leech (*Hirudo verbana* CARENA). J. R Soc. Interface (2017)
8. Tramacere, F., Follador, M., Pugno, N.M., Mazzolai, B.: Octopus-like suction cups: from natural to artificial solutions. Bioinspir. Biomim. **10**(3), 035004 (2015)
9. Follador, M., Tramacere, F., Mazzolai, B.: Dielectric elastomer actuators for octopus inspired suction cups. Bioinspir. Biomim. (2014)
10. Bing-Shan, H., Li-Wen, W., Zhuang, F., Yan-Zheng, Z.: Bio-inspired miniature suction cups actuated by shape memory alloy. Int. J. Adv. Robot. Syst. **6**(3), (2009). https://dx.doi.org/10.5772/7228
11. Lee, H., Um, D.-D., Lee, Y., Lim, S., Kim, H., Ko, H.: Octopus-inspired smart adhesive pads for transfer printing of semiconducting nanomembranes. Adv. Mater. **28**(34), 7457 (2017)
12. Tramacere, F., Beccai, L., Mazzolai, B.: What can we learn from the Octopus? Biol Biomim. Adhesiv.: Challenges Opportunities, pp. 89–102 (2013)
13. Ditsche, P., Summers, A.P.: Learning from Northern clingfish (*Gobiesox maeandricus*): bioinspired suction cups attach to rough surfaces. Philos. Trans. R Soc. B **374**, 20190204 (2019). https://doi.org/10.1098/rstb.2019.0204
14. Kampowski, T., Thiemann, L.-L., Kürner, L., Speck, T., Poppinga, S.: Exploring the attachment of the Mediterranean medicinal leech (*Hirudo verbana*) to porous substrates. J. R. Soc. Interface **17**, 20200300 (2020). https://doi.org/10.1098/rsif.2020.0300

Bio-Sourced Artificial Leather for Interior Automotive Applications

Stefan Friebel(✉) and Steffen Sydow

Fraunhofer Institut Für Holzforschung, Bienroder Weg 54 E, 38108 Braunschweig, Germany
{stefan.friebel,steffen.sydow}@wki.fraunhofer.de

Abstract. Flexible interior artificial leather is today produced on the basis of soft PVC and PUR and to a small extent of polyolefins, whereby the polyurethanes are primarily suitable for high-quality materials.

PU-based coating systems have several advantages: The mechanical properties of the coating can be adjusted in a wide range and they are thermally stable. Furthermore, they are considered as a substitute for PVC because they are free from plasticizers.

Fraunhofer WKI has developed a series of bio-sourced polyurethane leathers with a bio-sourced proportion of >85 wt.-%. The polymer matrix consists of sugar and vegetable oil derivatives and it is waterborne. The polymer material is UV-curable, providing an improved mechanical performance for the leather as well as an excellent chemical resistance. The elongation at break could be up to 175%, while the bio-source artificial leather is resistant to water and common household chemicals, such as red wine, 48% alcohol, mustard and water. For applications where there is no requirement for highly flexible materials e.g. dashboard lamination or gear stick, we also developed a finish that is resistant to sunscreen. Furthermore, a flexible head rest was designed by memory-wood foam. The final properties of the wood foam are accomplished by a combination of wood and a bio-soured latex to reach a pliable head rest. After the construction of head rest, the material was covered with the bio-sourced polyurethane leather smoothing the surface of the memory-wood foam.

Keywords: Artificial leather · Bio-sourced · Sustainability · UV-curing · Wood foam · Interior

1 Introduction

The Fraunhofer WKI has developed a wood foam and a polymeric material that can be used as a vegan, bio-sourced imitation leather for interior design. Vegetable oils in the formulation create the water-repellent properties and are simultaneously the prerequisite for the pleasant feel of the material. The sugar components enable the cross-linking of the polymer chains, responsible for the good mechanical properties

K. Dröder and T. Vietor (Eds.): *Technologies for economic and functional lightweight design*, Zukunftstechnologien für den multifunktionalen Leichtbau, pp. 189–198, 2021. https://doi.org/10.1007/978-3-662-62924-6_16

and the chemical resistance of the leather surface. The wood foam is used as an alternative to the PU-foams. It can be used in combination with the biosourced imitation leather for fitting e.g. the headrest, the steering wheel and the dashboard.

Through variation of the oil and sugar content, it is possible to produce highly elastic or very strong and resilient imitation leather. The vegan imitation leather is low in emissions and becomes odourless after a very short time. The use of renewable raw materials in the vegan imitation leather significantly increases the proportion of materials with a small CO_2 footprint; the product performance, however, thereby remains unchanged.

Partially biosourced PU-artificial leather are available. The product line Impranil eco provides a waterborne PU-dispersion with upto 65% biosourced materials. There are also niche products on the markets. Ultrafabrics launched a new product: Ultraleather | Volar Bio. The company has incorporated renewable plant-based materials into some layers of the product. They use corn-based chemicals to produce polyols for polycarbonate polyurethane resin. And wood pulp-based materials that are incorporated into the twill backcloth. In the US BioPreferred program, Volar Bio is labelled 29% biobased. Ultrafabrics aims to include biobased ingredients and/or recycled content in 50% of new product introductions by 2025 and in 100% of new products by 2030.

Modern Meadow, a producer of 'biologically advanced materials' has developed sustainable biofabricated materials inspired by leather. They partner with Evonik, the major company of speciality chemicals, to bring its production to commercial scale. Modern Meadow's technology produces animal-free collagen, a protein naturally found in animal hides, via a fermentation process using yeast cells. The start-up will be based in Nutley, New Jersey, USA. The material, called ZoaTM will be produced in a variety of shapes, sizes, textures and colours.

2 Materials and Methods

2.1 Vegan Biosourced Imitation Leather

Imitation leather is usually a composite material that is built from several layers. A typical example is sketched in Fig. 1. A fabric such as a weave or a fleece provides the tensile strength of the imitation leather. Than a thicker polymer layer, that can be foamed, for weight purposes, or consist of fibers or other filling materials builds the largest part of the leather follows. Another layer that undergoes a roll stamping gives the surface a leather like look. Finally, a protective coating can be applied that provides improved scratch and mar resistance as well as resistance against chemicals.

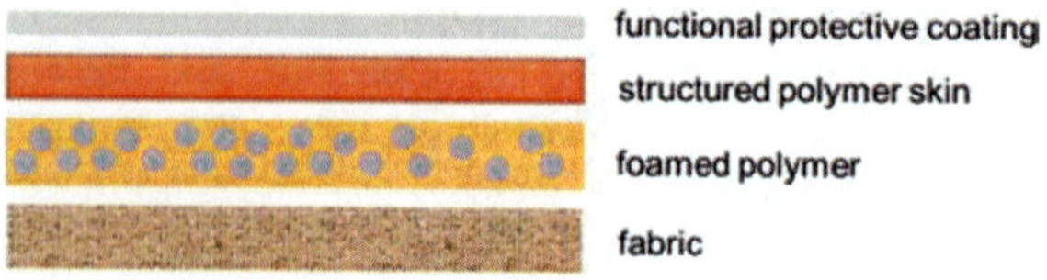

Fig. 1. Schematic composition of artificial leather.

Conventional imitation leather is mainly made from polyvinylchloride, polyurethanes or to a small amount from polyolefines. For highest specification polyurethanes is the first choice. Although there are many actions to avoid the harmful solvent dimethylformamide (DMF), a majority of imitation leather uses this solvent in their production process. Our materials do not require DMF.

Polyester polyurethanes can be made of 100% biosourced materials in theory. In practice >80% provides better mechanical and chemical performance. Typical biosourced building blocks for the polyester synthesis are depicted in Fig. 2. A biosourced isocyanate, pentamethylene diisocyanate, is now also available from Evonic.

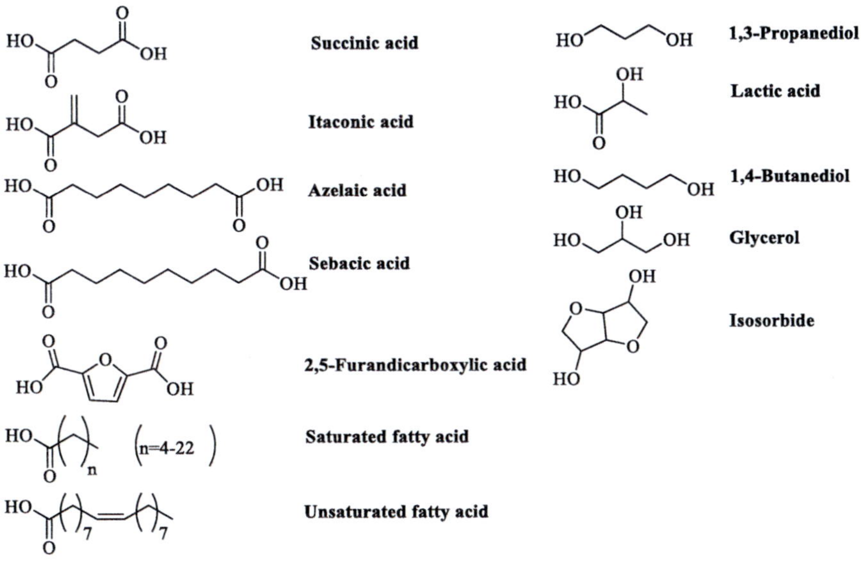

Fig. 2. Some examples of biosourced building blocks for polymers.

2.2 Waterborne Polyurethane Imitation Leather

Polyurethane based imitation leather provides best performance in high-end application, such as automotive interior. Conventional polyester polyurethanes often use 1,6-hexandiol for elasticity reasons. Our work uses the sugar sourced 1,3-propanediol (1,3-PDO) as an underestimated interesting bio-sourced building block for polyester polyurethane coating resins. The company DuPont-Genencor already produces 1,3-PDO from corn sugar by means of a biotechnology process [1]. Other work deals with the conversion of glycerol to 1,3-PDO, which should represent a more cost-effective process [2]. The effect on the mechanical and physical properties of the polyurethane coating was investigated by the synthesis of polyurethane dispersions (PUD) which were prepared either with 1,6-hexanediol (1,6-HDO) or alternatively with 1,3-PDO. The polyurethane dispersions (PUD) were produced according to the acetone process, followed by a chain extension with a diamine. The reaction scheme is outlined in Fig. 3.

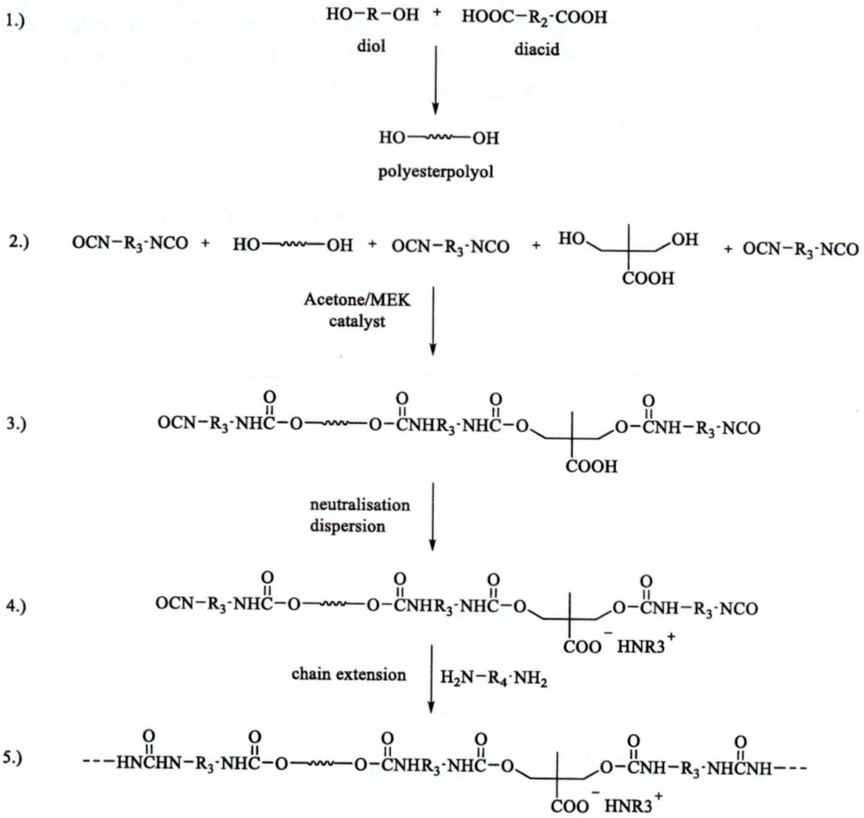

Fig. 3. General synthesis route of polyurethane dispersions.

A fairly new technology regarding imitation leather is the application of UV-curing materials. Our newly developed materials involve sugar sourced itaconic acid. The crosslinking of the prepolymer is accomplished by UV-induced radical polymerization of the itaconic acid double bonds [3]. The monomer reactivity ratio for copolymerisation of hydroxyethyl acrylate (HEA) with itaconic acid reveals that reactivity of the itaconic acid double bonds during radical polymerisation in water is smaller ($r_1 = 0.44$) than that of HEA ($r_2 = 0.52$) [4]. Conventional PU-acrylates have their active double-bonds within the hard-block of the polyurethane. The double bonds of the PU-itaconates are located in the soft and more mobile regions of the polyurethane. This important difference has some major influence of the properties of the crosslinked polyurethane. While the conventional polyurethane prepolymers are crosslinked only at the chain ends, the PU-itaconates crosslink at the whole polymer backbone, leading to a denser and stronger cross-linked network in theory. In Fig. 4 the two polyurethane types are illustrated.

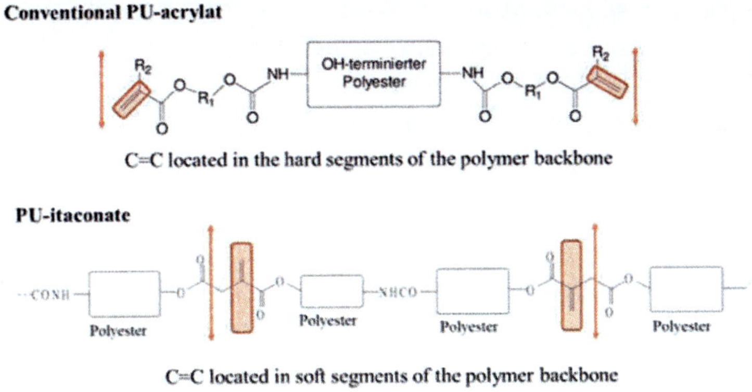

Fig. 4. Comparison of the schematic structures of conventional PU-acrylates and PU-itaconates

2.3 Woodfoam

Foams are usually made of petrochemical-based polymers. For wood foam 100% renewable resources are possible. For production wooden materials as beech or spruce can be used. Furthermore, other lingo cellulose can be processed into wood foam as well. No synthetic adhesives have been used.

The production of wood foam is a similar process to pulp and paper industry. Wood chips will be converted into fiber in a refiner by a thermos-mechanical pulp process (TMP). After this step wood foam has to be refined further by adding water to get a foamable wood/water suspension. Physical (Air, N_2 or CO_2) or chemical (carbonate, peroxide) foaming agents have to be added to the suspension to get wet foaming suspension. After final drying in a vacuum cabinet the wood foam can be processed and sawn to size [5] (Fig. 5).

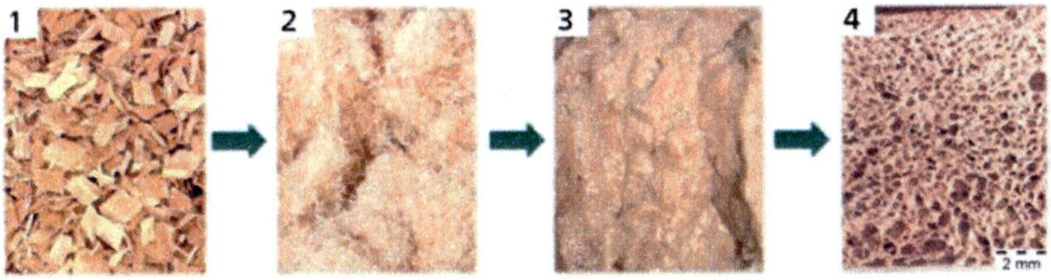

Fig. 5. Production of wood foam (1) wood chips (2) disintegration (3) suspension and additives (4) drying.

3 Results and Specification

3.1 Waterborne PU-Material

The characteristics of the PUD and the free films are summarized in Table 1. The proportion of 1,3-PDO and 1,6-HDO was reduced successively. The composition of the investigated polyesters shows no influence on the particle size and the zeta potential of the corresponding PUD. As expected the glass transition temperature of the polyester precursors and polyurethane films decreases with increasing amount of 1,6-HDO because of the reduced ester density. Surprisingly the viscosity of the polyesterpolyols is independent of their composition. It was also expected that with higher amounts of 1,6-HDO in the composition, the viscosity drops due to the lower ester-bounding density and therefore lower dipole interaction.

In the model polymers which include 1,3-PDO the "odd–even" effect from the C3-building block seems to be responsible for the high values of the elongation at break, comparable to the mechanical properties of the polymers made with 1,6-HDO. The sample dimensions are 80 mm × 0,07 (±0,02 mm). A mean value of 10 samples were determined.

Table 1. Physical properties of the PUDs and the free polymer films

1,3-PDO: 1,6-HDO	particle size [nm]	zeta-potential [mV]	T_g [°C]	max. tensile strength σ_{max} [N/mm²]	Elongation at break [%]
100:0	185	−42	−11	17±2	241±18
80:20	158	−41	−21	14±1	257±7
60:40	93	−47	−24	14±2	264±18
40:60	124	−50	−28	15±3	258±29
20:80	180	−40	−33	14±2	281±34
0:100	139	−35	−36	15±3	323±35

UV-curing polymer films are usually predestined for hard and less elastic materials. However, tt is possible to generate very elastic and flexible films with a high elongation at break value by varying the chemical composition of the polyurethane. Parameters that have been changes are the polyurethane content, the urea content and the crosslinking density of the polymer via the amount of UV-curable double bonds (see Table 2). The results demonstrate that a lower crosslinking density results in a higher elongation at break value. A higher urethane portion results in lower elongation at break and also in higher max. tensile strength due to the hydrogen bonds. PUD-IA-4 includes long chain carboxylic acid and therefore has a large elongation at break, although the cross-linking density is high.

Table 2. Physical properties of the UV-PUDs and the free polymer films

UV-PU-itaconate	Urea [Gew.-%]	Urethane [Gew.-%]	C=C-bonds [mmol/g]	max. tensile strength σ_{max} [N/mm^2]	Elongation at break [%]
PUD-IA-1	6.2	24.8	1.23	17±2	174±16
PUD-IA-2	7.2	28.7	2.11	27±2	51±3
PUD-IA-3	6.7	27.0	1.94	24±1	70±7
PUD-IA-4	6.7	27.0	1.94	13±1	112±9
PUD-IA-5	6.4	25.6	1.79	19±1	64±3

In Fig. 6 illustrates an example of vegan imitate leather for a tractor seat. The leather has been applied with a vacuum laminating process. The vegan imitation leather also resists 24 h of sunscreen application without any marks.

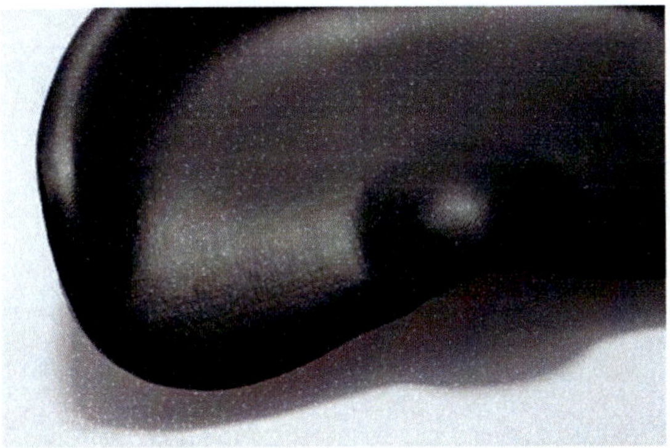

Fig. 6. Tractor seat with UV-cured biosourced imitation leather. The leather was applied by a vacuum laminating process.

3.2 Wood Foam

The received wood foam after foaming and drying steps consists of 100% wood and required no additive like binder or adhesives, because the strength of the foam caused by wood's own bonding forces. Thus, any possible health risks due to emissions from adhesive are excluded. Furthermore, any lignocellulosic material can be used as starting material regardless of quality. Hence, materials can be used which are available in high amounts. The result of the processing is a lightweight base material with a porous, open-cell structure and low bulk density [6] (Fig. 7).

Fig. 7. Foaming suspension (left)and open-cell structure of the wood foam (right).

Foams can be manufactured in a density range of 40–200 kg/m^3. The tensile strength depends on the adjusted density of the material as shown in Fig. 8. The range is between 20 and 200 kPa. Furthermore, compressive strength is also dependent on the density and is up to 600 kPa as shown in Fig. 9.

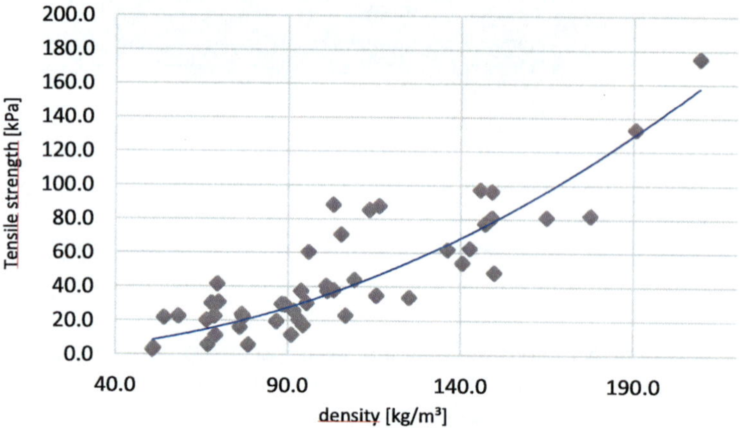

Fig. 8. Tensile strength [kPa] as function of the density [kg/m^3].

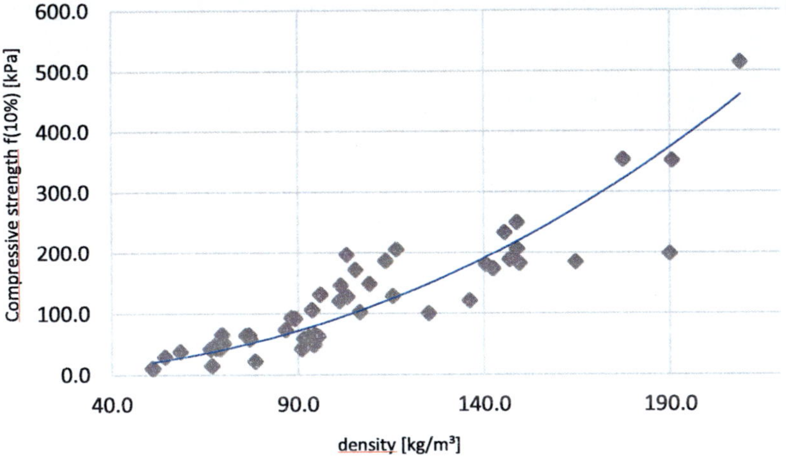

Fig. 9. Compressive strength [kPa] as function of the density [kg/m^3].

By combination of the wood foam with soft materials, e.g. bio-soured latex, a flexible material can be obtained. The application possibilities are versatile. By combination a pliable head rest, flexible armrests or the interior of a car seat could be imagined.

4 Conclusion and Outlook

It has been shown that the production bio-sourced polyurethane leathers with a bio-sourced proportion of >85 wt.-% is possible by using a combination of vegetable fatty acids and sugar sourced building blocks. This is more that the know commercial products. The polymer material could be conventionally drying or UV-crosslinked, providing an improved mechanical performance for the leather as well as an excellent chemical resistance, meeting high standards of the automotive industry for interior application. New is the use of UV-curing polymers for biosourced artificial leathers. Due to the UV-crosslinking of the polymer and appropriate combination of biosourced modified sugar and fatty acidy building blocks very chemical resistant surfaces can achieved. Compared to conventional UV-curing binder, in which the crosslinking double bonds are at terminated polymer backbone, the biosourced UV-curing binder have their double bonds within the polymer backbone. This leads to a more homogen network, resulting in excellent chemical and physical properties. Furthermore, the production of wood foam as an adhesive-free material was possible with a wide range of density. The result is a lightweight base material with a porous, open-cell structure. Adding soft segments like bio-soured latex flexible materials can be created. By combination of this type of bio-soured leather as coating materials and pliable wood foam a wide field of applications can be addressed, e.g. pliable head rests or seats.

References

1. White, G.M., Bulthius, B., Trimbur, D.E., Gatenby, A.A.: Patent WO 9910,356 1999
2. Nakamura, C.E., Whited, G.M.: Curr. Opin. Biotechnol. **14**, 454 (2003)
3. Robert, T., Friebel, S.: Itaconic acid – a versatile building block for renewable polyesters with enhanced functionality. Green Chem. **18**, 2922–2934 (2016)
4. Cowie, J.M.G., McEwen, I.J., Yule, D.J.: Eur. Polym. J. **36**(9), 1795–1800 (2000)
5. Scholtyssek, J., Thole, V.: Wood Foam – innovative, natural and versatile. In: Proceedings of the 5th International Conference on Biofoams, Sorrento, Italy (2015)
6. Bunzel, F., Ritter, N., Scholtyssek, J.: Holzschaum – eine Alternative zu petrochemischen Schaumstoffen. Holztechnologie **58**(1), 31–35 (2017)

Functional Structures

Continuous Profile Production with Hybrid Materials by Pultrusion

Marcus Knobloch, David Löpitz(✉), David Wagner, and Welf-Guntram Drossel

Fraunhofer Institute for Machine Tools and Forming Technology IWU, Reichenhainer Str. 88, 09126 Chemnitz, Germany
david.loepitz@iwu.fraunhofer.de

Abstract. An important factor for the success of electromobility is lightweight design – the lower the mass of a vehicle, the less energy is required for acceleration and movement, and the smaller, lighter and cheaper battery and drive technology can be designed. In close cooperation with partners from science and industry, the Fraunhofer Institute for Machine Tools and Forming Technology (IWU) develops solutions suitable for series production for a sustainable German automotive industry. Lightweight design in economical hybrid design is the focus of numerous development projects. In the research project "Hybrid Pultrusion", the research partners Fraunhofer IWU and the Leibniz-Institut für Polymerforschung Dresden e. V. (IPF) develop new methods for the reliable and reproducible production of hybrid components made of metal and fiber-reinforced plastics (FRP) by using the pultrusion process. A central focus in this project is the bonding mechanism between metal and FRP at the molecular level and the transfer of these phenomena to the pultrusion process.

Keywords: Pultrusion · Hybrid · Lightweight design · Process · Multi-material

1 Introduction

Hybrid materials are able to increase the potential of lightweight design of countless applications and thus make an essential contribution to energy-efficient, resource-saving and CO_2-minimized mobility [1]. However, the combination of two classes of materials (e.g. metal and FRP) to create a hybrid material is a major challenge for engineers, especially for the use of hybrid materials in large-scale production like automotive industry and transportation. The main disadvantages for series production are insufficient joining qualities, time- and cost-intensive pre- or post-treatment processes, long cycle times or the insufficient fatigue strength at changing climate conditions. Due to the different thermomechanical properties of metal and FRP, an excellent adhesion strength at the interface is necessary [2]. Gluing is one of the few processes that has been established on a large-scale production for joining FRP with

K. Dröder and T. Vietor (Eds.): *Technologies for economic and functional lightweight design*, Zukunftstechnologien für den multifunktionalen Leichtbau, pp. 201–210, 2021. https://doi.org/10.1007/978-3-662-62924-6_17

metals in such a way that high safety standards can be maintained with regard to the fatigue strength in automotive applications [3, 4]. Gluing is a process of applying a non-metallic material between two parts to create a material bonded through adhesion and internal strength [5]. As needed pre-treatment, application and curing of the glue are usually associated with additional time and costs, current research has a significant claim to reduce or even to avoid the use of glue by appropriate methods. The difficulties in joining metals and FRP without using glue are mainly caused by large differences in the thermal expansion coefficients and a poor molecular adhesion of both materials. In order to achieve a high molecular adhesion between metal and FRP, current research shows promising concepts and results. For example, adhesion promoter layers can be used, which take over the role of the glue in the hybrid composite and allow significantly faster cycle times [6]. The surface functionalization of the metallic joining partner, for example, offers a further innovative solution for completely dispensing with an adhesion promoter. By applying a microstructure with a laser, the surface is enlarged in such a way that the liquid thermoplastic matrix enters the undercuted microstructure during the joining process and becomes firmly attached to it [7]. Other recently studied pretreatment methods deal with plasma pretreatment on the bonding surface [8, 9]. For the production of metal-FRP-hybrid components with thermoset matrix, approaches with functional powder coatings are particularly promising. As thermosetting reaction resins are initially in liquid form, the chemical processes taking place during curing can be used for the joining process without the need of further processing steps [2]. For the use of polyurethanes, it has already been demonstrated that reactive groups of the polymer forms a solid covalent bond with the powder coating [10, 11].

The objective of the research project "Hybrid Pultrusion" is to modify the powder coating for a covalent bonding to epoxy resins, which are used as standard thermoset for the pultrusion process. Those powder-precoated metals are added to the pultrusion process and their adhesion is tested.

2 Materials

In the following, the examined materials are described, which are used in this project. Furthermore, the methods and processes used to manufacture hybrid specimen are presented.

2.1 Components of the Fiber-Reinforced Plastics

For the trials, materials are used as they are usually applied in the pultrusion process. The fiber material is a glass fiber roving of the type PulStrand 4100 with 4800 tex from the manufacturer Owens Corning. Table 1 shows the components of the anhydride epoxy resin system.

Table 1. Resin system

Component	Component name	Supplier
Resin	Araldite LY 3585 CH	Huntsman
Curing agent	Aradur 917-1 CH	Huntsman
Catalyst	Accelerator DY080	Huntsman
Internal Mold Release	IC25	ChemTrend
Filler	ASP 600 (0,6 μm)	BASF

2.2 Metallic Component and Powder Coating

As the basic metallic component a cold rolled unalloyed steel sheet of the quality DC01/S235JR has been selected. The steel sheet is sandblasted with aluminum oxide (high class corundum) and pre-treated with zinc phosphate leading to a fine crystalline layer. The metal gets a good corrosion protection. As previous investigations have shown this treatment is also necessary in order to achieve the highest possible adhesion in the later hybrid structure [11]. The powder coating materials consist of a two-step curable powder coating based on commercially available uretdione (internally blocked isocyanates) cross linkers, OH-functionalized polyester resins and a specific catalyst system which was developed at the IPF in the past. The curing mechanism is shown in Fig. 1. The primary focus of the coating formulation is the variation of the OH-group number of the polymer resins to investigate their influence on the adhesive strength to the epoxy-matrix of the joint. Additionally, the post formability of the coating system, as well as the surface appearance should be maintained at a high level.

The powder coating is applied to the metal sheets in the form of dry powder by using corona discharge. This technology uses a spray gun with an integrated high voltage cascade which charges an electrode and the powder and also generate an electric field between the gun and the grounded steel sheet. The powder particles moves along this field lines to the metal sheet. At last the powder coated sheets are cured at temperatures of 150 °C. This process step completes the first reaction step of the powder coating material resulting in a polyallophanate network and simultaneously creates a strong bonding to the metal.

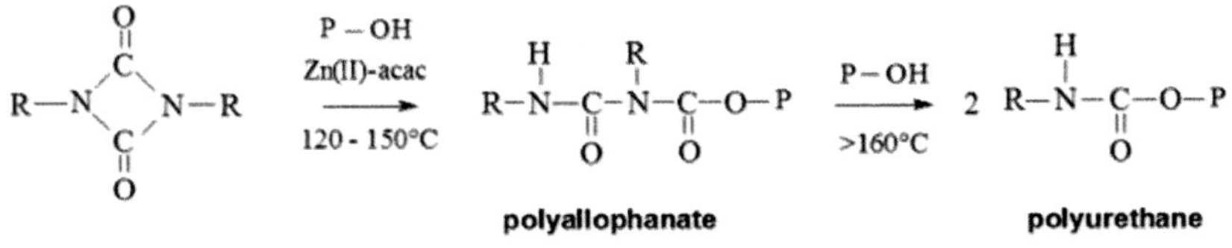

Fig. 1. Two step curing mechanism of the used uretdione powder-coating

3 Specimen Manufacturing

The test specimens were first manufactured under laboratory conditions at the Leibniz IPF in order to identify general process and material parameters. For the basic tests in the laboratory, the manufacturing process Resin Transfer Molding (RTM) was used. Subsequently, the results of the laboratory tests were adapted to the pultrusion process at the Fraunhofer IWU.

3.1 The Pultrusion Process

The pultrusion process, which is focused for the later application, is one of the few manufacturing processes for continuous fiber-reinforced plastic profiles that is suitable for large-scale production. Figure 2 shows the basic features of the process: Semi-finished fiber products (1) are pulled from bobbins by alternately moving pulling devices (4) and pass through a resin bath (2). Afterwards, the impregnated fibers are pulled through a heated die (3), in which the liquid thermoset plastic cures completely within seconds. A saw (5) cuts the profiles to the desired length [13].

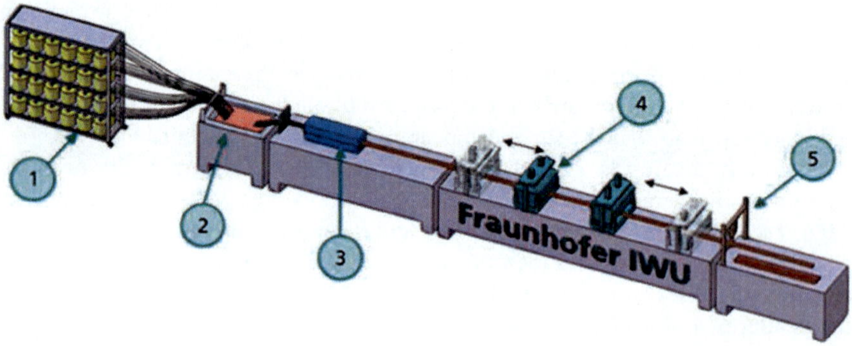

Fig. 2. Pultrusion line

3.2 Manufacturing of Specimen by Pultrusion

In order to be able to produce hybrid test specimens by pultrusion, researchers of the Fraunhofer IWU have developed suitable delivery solutions for the metallic component. For manufacturing representative test specimens, it is necessary that the powder coated metallic sheets can be supplied into the process exactly in the direction of the fibers and that slippage of these is avoided. As the metallic semi-finished products coated by Leibniz IPF have a maximum length of 475 mm, a supply system was developed as shown in Fig. 3. Metal sheets with an optimized powder coating were used for the trials on the pultrusion line. As reference, some sheet metals without coating were also pultruded to hybrid specimens.

Fig. 3. Supply system for the metallic component (① – coated metal sheet; ② – delivery sheet)

In order to avoid skips in fiber volume content (FVC) during the pultrusion process, the metallic component must be supplied continuously, as every change of the FVC results in differences in the curing behavior of the resin system and associates problems such as cracks or material deposits. This is ensured by means of a metal sheet (see Fig. 3, ②) which is fed into the die during the entire process. The metal sheet has cut-outs at defined intervals into which the coated sheets (see Fig. 3, ①) are inserted. Due to the high thermal capacity of the added metallic sheet, it is necessary to adjust the process parameters to ensure the curing of the epoxy resin as well as to make sure that the reactive groups of the resin can react with the powder coated metal to form a strong bond.

Table 2 shows the adjusted process parameters of the standard FRP-pultrusion and the hybrid pultrusion with FRP and metal. Best results for hybrid pultrusion could be achieved by increasing the temperature in heating zone 2 and reducing the process speed to one third.

Table 2. Process parameters

Parameter	Value standard pultrusion	Value hybrid pultrusion
Length of die [mm]	1000	1000
Number of heating zones [–]	5	5
Process speed [mm/min]	600	200
Temperatures in heating zone 1–5 [°C]	150-150-190-190-190	150-190-190-190-190

4 Test, Results and Discussion

4.1 Optical and Non-Destructive Analysis of the Complete Composite System

Figure 4 shows microscopic images of grinding patterns of the hybrid specimens manufactured by pultrusion. The left picture shows a direct bonding of metal to powder coating and powder coating to FRP. The right picture shows a hybrid specimen without powder coating. It can be clearly seen that there is no connection between metal and FRP as embedding material has deposited in the space between the two materials.

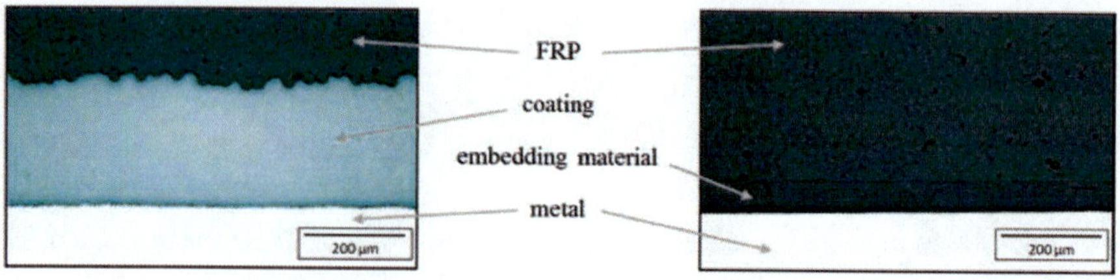

Fig. 4. Microscopic images of boundary layers (left: hybrids with powder coating; right: hybrids without powder coating)

4.2 Specimen Preparation and Lap Shear Test

To determine the bond strength, a lap shear test was carried out in accordance to DIN 65148 [14]. For this purpose, the hybrid parts manufactured by RTM and pultrusion were cut to standardized tensile test specimen (dimension 250 mm × 25 mm × 4 mm) on a water-cooled separating device. Then, two notches are put into the test specimen, resulting in the shearing area of 25 mm × 12.5 mm recommended in DIN 65148. The standard prescribes the use of a supporting device to prevent deformation in the shear zone. Figure 5 shows the final specimen for the lap shear test.

Fig. 5. Specimen preparation and geometry (left); lap shear test with supporting device (right)

The standard prescribes a path-regulated test in which failure occurs within one minute after start. In a preliminary test with a test specimen, in which average adhesion values are expected, the test speed is set at 0.2 mm/min.

4.3 Results of the Lap Shear Test and Discussion

Figure 6 shows the results of the lap shear tests of the RTM and pultruded specimens with the optimized powder coating before and after tempering. For some specimens a subsequent thermal treatment was carried out at 160 °C for 2 h with the aim of a further enhancement of the interlayer bonding between the powder coating layer and the epoxy matrix. In each test series at least 6 specimens were tested.

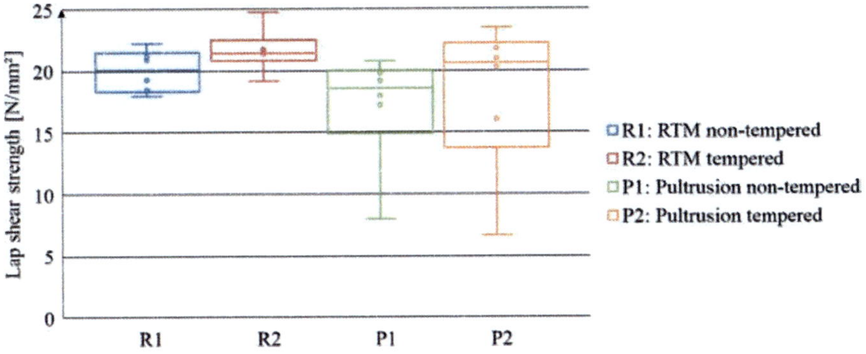

Fig. 6. Results of the lap shear test

As expected, the tempered specimens R2 and P2 show a slightly higher trend for the average lap shear strength than the non-tempered ones (R1 and P1). It is assumed that additional covalent bonds are formed by diffusion during the thermal treatment together with the relaxation of internal stress which resulted from the uneven cooling down after the fabrication in the RTM tool. The trend for the average lap shear strength of the specimens R1 and R2, which were manufactured under laboratory conditions by RTM, is slightly higher than those (P1 and P2) manufactured by pultrusion. The reason for this could be the much faster production speed of the pultrusion process. Due to the rapid heating, curing and cooling, there is a different shrinkage behavior of metal and FRP, resulting in stresses in the boundary layer. These stresses, together with a constant tensile stress on the glass fibers, prevent an optimum bond between metal and FRP. The residual stresses that form in the boundary layer of specimens P1 and P2 result in significantly higher standard deviations of the lap shear strength compared to specimens R1 and R2.

Due to the large standard deviation of the pultruded specimens, no clear statement can be made at this time about the comparison of laboratory and pultrusion specimens. Further investigations with a larger number of samples are planned in order to prove the trend shown in Fig. 6 with statistical certainty.

In addition, for comparison purposes, specimens with uncoated metal inserts were also produced using the pultrusion process. The aim was to quantify the influence of the powder coating on the bond strength. However, when the specimens were cut to size, it could already be clearly established that there is nearly no bond between the metal and the FRP. This is mainly due to the different thermal behaviour and the additional chemical shrinkage of the plastic component, which leads to corresponding internal stresses in the hybrid composite.

5 Pultrusion of a Hybrid Side Sill

The improvement in properties with adhesive integrated metal parts in pultrusion profiles are tested using a demonstrator component. For this purpose, a hybrid side sill is manufactured using a modified pultrusion process. Subsequent crash tests are intended to demonstrate increased energy absorption compared to the pure FRP-variant of the side sill.

Figure 7 shows the cross section of the side sill as well as the tool system, which consists of the pultrusion die, three heated cores and four guide plates. The guide plates are required for the orientation of the fibers and the metal. They also served to hold the cores.

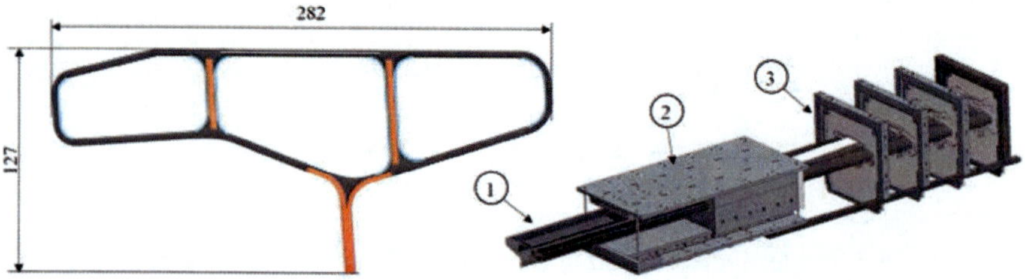

Fig. 7. Cross section of the side sill and tool system (① – hybrid side sill; ① – pultrusion die and heat zones; – guide plates and cores)

The side sill is a complex 3-chamber hollow profile with different fibre orientations in different cross section parts. For this, mostly unidirectional (grey) 0°/90° (orange) – and triaxial (blue) fabrics are used. The metal is integrated in the top center part. Figure 8 shows the manufacturing of the pultruded side sill.

Fig. 8. Pultrusion of a hybrid side sill

The fabrics and rovings pass through a forming section made of various guide plates and a soaking bath. The metal is fed into the process between the top fabric layers. All materials pass through a heated die. The profile geometry is formed in the course of the curing reaction of the pultrusion resin. In addition, the powder coating of the metal component forms a covalent bond with the polymer.

6 Conclusion and Outlook

The results of the tests indicate that powder coating has a high potential for the manufacturing of hybrid components made of metal and FRP. It has also been shown that the laboratory results can be transferred to the pultrusion process without any significant deterioration of the lap shear strength.

The current limit of the hybrid pultrusion process is that only strips with a maximum length of 475 mm could be coated. For a series process, the coating technology must be upscaled so that metallic, coated components can be supplied endlessly from a coil to the pultrusion process. Depending on profile cross section the pultrusion process is only economical at production speeds of more than 500 mm/min. The project aims to significantly increase the pultrusion speed of currently 200 mm/min by optimizing process parameters without reducing the bond strength.

With the production of a hybrid side sill, it was shown that the results can be transferred to complex components. The next objective of the project is to carry out a crash test to show that the energy absorption of pultruded components can be significantly increased by integrated metal elements.

Acknowledgements. This work was conducted within the project "Hybrid Pultrusion", which was funded by the Sächsisches Staatsministerium für Wissenschaft, Kultur und Tourismus (SMWK).

European Union

Europe funds Saxony. EFRE European Regional Development Fund

Diese Maßnahme wird mitfinanziert durch Steuermittel auf Grundlage des von den Abgeordneten des Sächsischen Landtags beschlossenen Haushaltes.

References

1. Vogt, M., Malanoski, N., Glitz, R., Stahl-Rolf, S.: Bestandsaufnahme Leichtbau in Deutschland. Projekt IC 4 (2015). https://www.bmwi.de/Redaktion/DE/Publikationen/Studien/bestandsaufnahme-leichtbau-in-deutschland.pdf
2. Scheik, S., Schleser, M., Reisgen, U.: Thermisches Direktfügen von Metall und Kunststoff– Eine Alternative zur Klebtechnik? In: Leichtbau-Technologien im Automobilbau, pp. 89–94. Springer (2014)
3. Habenicht, G.: Kleben. Grundlagen, Technologien, Anwendungen. VDI-Buch (2006)
4. Brockmann, W., Dorn, L., Käufer, H.: Kleben von Kunststoff mit Metall. Springer, Berlin (1989)
5. DIN EN 923:2016-03, Klebstoffe – Benennungen und Definitionen; Deutsche Fassung EN_923:2015
6. Haider, D., Krahl, M., Koshukow, W., Wolf, M., Liebsch, A., Kupfer, R., Gude, M.: Adhesion studies of thermoplastic fibre-plastic composite hybrid components part 2: thermoplastic-metal-composites (2018)
7. Gebauer, J., Fischer, M., Lasagni, A. F., Kühnert, I., Klotzbach, A.: Laser structured surfaces for metal-plastic hybrid joined by injection molding. J. Laser Appl. **30**(3), 32021 (2018)

8. Gallant, D., Savard, V.: New adhesive bonding surface treatment technologies for lightweight aluminum-polypropylene hybrid joints in semi-structural applications. SAE Int. J. Mater. Manuf. **4**(1), 314–327 (2011)
9. Hopmann, C., Wunderle, J., Neuß, A., Ochotta, P., Bobzin, K., Schulz, C.: Influence of surface treatment on the bond strength of plastics/metal hybrids. Z. Kunststofftechnik **1**, 227–255 (2015)
10. Kühnert, I., Gedan-Smolka, M., Fischer, M., Scholz, P., Landgrebe, D., Garray, D.: Prefinished metal polymer hybrid parts. Technol. Lightweight Struct. **1**(2), 89–97 (2017)
11. Gedan-Smolka, M., Lehmann, F., Lehmann, D. (Hrsg.): Catalysis in uretdione powder coatings enables innovative processing lines. AMER CHEMICAL SOC 1155 16TH ST, NW, WASHINGTON, DC 20036 USA 2003
12. Hopmann, C., Michaeli, W.: Einführung in die Kunststoffverarbeitung. Hanser, München (2017)
13. Starr, T.F. (ed.): Pultrusion for Engineers. Woodhead Publishing Series in Composites Science and Engineering. CRC Press, Boca Raton (2000)
14. DIN 65148:1986-11, Luft- und Raumfahrt; Prüfung von faserverstärkten Kunststoffen; Bestimmung der interlaminaren Scherfestigkeit im Zugversuch

Comparison of the Mechanical Properties of Adhesively Bonded and Mechanically Interlocked Steel/Fibre-Reinforced Thermoplastic Hybrids Produced Using One-Step Forming Process

David Trudel-Boucher[1], Philipp Kabala[2(✉)], Jan Beuscher[2], Michel Champagne[1], and Klaus Dröder[2]

[1] Automotive and Surface Transportation, National Research Council Canada, 75 de Mortagne, Boucherville, Québec J4B 6Y4, Canada
{david.trudel-boucher,michel.champagne}@cnrc-nrc.gc.ca

[2] Institute of Machine Tools and Production Technology (IWF), Technische Universität Braunschweig, Langer Kamp 19B, 38106 Braunschweig, Deutschland
{p.kabala,j.beuscher,k.droeder}@tu-braunschweig.de

Abstract. The continuous increase in the weight of newly registered vehicles is in conflict with the targeted reduction in greenhouse gases emission and fuel consumption imposed by several environmental regulations. To meet these stringent environmental regulations, the fabrication of hybrid structural parts is a particularly suitable alternative for mass-produced vehicles. Through the combination of fibre-reinforced thermoplastics (FRTP's) and steels, an optimum of component costs and weight savings can potentially be achieved. However, the production of hybrid components traditionally involves additional manufacturing steps and a subsequent joining processes. In order to increase the efficiency of the hybrid production process, both materials can be formed and joined together in one forming step. In this paper, two different approaches to join FRTP and steel sheet in a single forming step are compared in respect of their mechanical strength. For this, structured steel sheets designed to obtain mechanical interlocking and an adhesive film are used. Furthermore, process parameters such as forming force, blank holder force and tool temperature are varied during the manufacturing process to determine their influence on the mechanical performance.

Keywords: Hybrid structures · Hybrid forming · FRTP · Joining method

K. Dröder and T. Vietor (Eds.): *Technologies for economic and functional lightweight design*, Zukunftstechnologien für den multifunktionalen Leichtbau, pp. 211–219, 2021. https://doi.org/10.1007/978-3-662-62924-6_18

1 Introduction

In the wake of climate change, environmental protection is becoming an important issue in the automotive industry. This can be seen on the one hand by strict regulations on the emission of greenhouse gases [1] and on the other hand by the growing environmental awareness of potential buyer groups for large scale produced vehicles [2]. One approach for car manufacturers to meet these requirements is lightweighting [3]. By reducing the moving mass, the weight spiral can be reversed and more components, such as the drive train and the chassis, can be dimensioned smaller to achieve further weight reduction [4].

Endless fibre reinforced polymers (FRP's) are known to have high specific mechanical properties, making these materials appealing for lightweight design [5]. However, due to their high material costs and time intensive manufacturing processes, the substitution of conventional materials by FRP's is often uneconomical in large scale production. In order to counteract these disadvantages, fast-processing fibre-reinforced thermoplastics (FRTP's) can be combined with steel sheets to produce hybrid components. Such hybrid components allow a compromise between low material and manufacturing costs, high mechanical properties and mass reduction [6]. According to current state of the art, hybrid components are generally manufactured in two successive production steps [7, 8] (see Fig. 1a). In the first step, the steel sheet is formed. A widely used forming process for sheet metals is the deep drawing process. In the second step, the flat semi-finished FRTP is thermoformed onto the already formed steel sheet. In addition to forming of the FRTP, joining of the dissimilar materials is also obtained in the second forming step. However, both forming steps lead to high cycle time along the process chain and high effort for handling and logistics. In order to reduce cycle time and increase the cost efficiency of the production of hybrid components, both materials can also be formed in a one-shot process (see Fig. 1b).

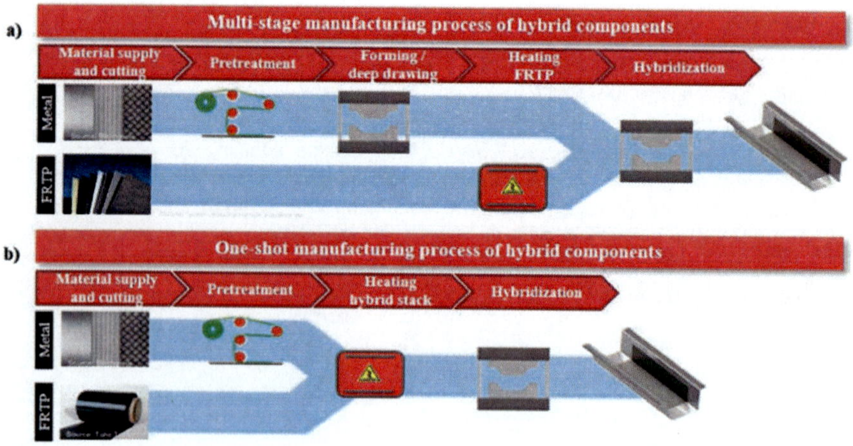

Fig. 1. Reduction of process steps through the one-shot process compared to the multi-stage process.

Behrens et al. showed that FRTP's and metal sheets in sandwich composite structures can be formed using an integrated deep-drawing and thermoforming process. Furthermore, it was shown in their study that the forming process can be carried out with variothermal temperature control [9, 10]. *Gresham* et al. also dealt with the forming behaviour of hybrid sandwich structures. They showed that the blank-holder force and preheating temperature have significant influence on the failure mode [11]. While many authors such as *Gresham* et al. [11] focused on adhesive bonding to join FRTP's and sheet metals, bonding of dissimilar materials can also be realized using structured sheet metals. *Brand* et al. and *Kühn* were able to prove that promising composite strengths can be achieved with structured sheets [12, 13]. *Demes* et al. considered the thermal distortion of hybrid specimens manufactured with structured sheets or adhesive bonding. They showed that the hybrid samples manufactured with structured sheets led to lower distortion [14]. However, structured sheets have not yet been used in an integrated forming and joining process to produce hybrid components. In addition, it is unclear how the mechanical performance of such hybrid structures will compare to those produced using adhesive bonding.

In this study, U-shaped hybrid test specimens are manufactured with one-shot forming process using adhesive bonding or mechanical interlocking to join steel and FRTP sheets. During forming trials, several forming parameters are varied. Then, mechanical tests are carried out to determine the influence of these studied process parameters and to compare both investigated joining approaches.

2 Experimental Procedure

In this study, U-shaped hybrid test specimens manufactured in a matched-die tool are used to compare the mechanical performance of hybrid structures produced using two different joining approaches, i.e. adhesive bonding and mechanical interlocking. For all manufactured specimens, the thickness of steel and FRTP sheets are of 1 mm and 2 mm, respectively. The length and width of semi-finished products used for forming are 361 mm and 80 mm. FRTP material considered in this study is an unidirectional glass fibre/PA6 tape produced by Celanese (Celstran®CFR-TP PA6GF60-01) with a glass fibre content of 60 wt%. To achieve mechanical interlocking, GRIPMetal™ sheets made from AISI 1010 steel are used. The height of the hooks for the selected GRIPMetal™ product is of approximately 1.3 mm. During forming, the hooks penetrate the molten FRTP composite to provide mechanical interlocking. The penetration into the FRTP stabilizes the hooks during the forming process so that they are not destroyed. For the adhesive joining approach, adhesive film NOLAX 45.200 and DC01 steel are used. Both steels used in this investigations have similar mechanical characteristics [15, 16], which ensures that the results of both joining methods are comparable. The geometry of the U-shaped profile as well as the stacking sequence used for both joining approaches is shown in Fig. 2.

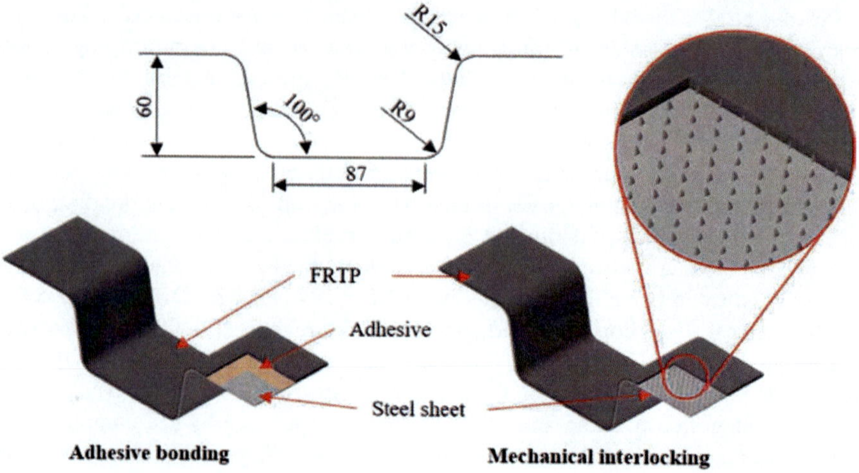

Fig. 2. Geometry of the U-profile and stacking sequence for both joining approaches.

To form U-profiles, a pre-consolidated tape plate with unidirectional fibre orientation is first placed on a steel sheet according to the layer structure shown in Fig. 2. In all cases, the fibres run along the contour of the U-profile. Subsequently, the superposed materials are heated in an oven at 250 °C for 7 min to ensure uniform melting of the polymer matrix material. The heated semi-finished product is then transferred manually into a tool developed for the integrated deep drawing and thermoforming process. Positioning pins in the tool ensure reproducible positioning of the semi-finished product. In case of mechanical interlocking, the hooks penetrate the melted FRTP during the forming process. The thickness of the FRTP is greater than the hook length, which ensures that the hooks are encased by the FRTP and cannot be damaged by the forming force. During the forming process, the tool is kept at a constant temperature. Temperature curves measured in preliminary screening tests using thermocouples inserted in the boundary layer between the FRTP and steel sheet are shown in Fig. 3 for tool temperatures of 80 °C and 120 °C. As shown in this Figure, no significant temperature change occurs in the component after a period of 20 s in the matched-die tool. Therefore, a dwell time of 20 s is selected in this study for the forming trials.

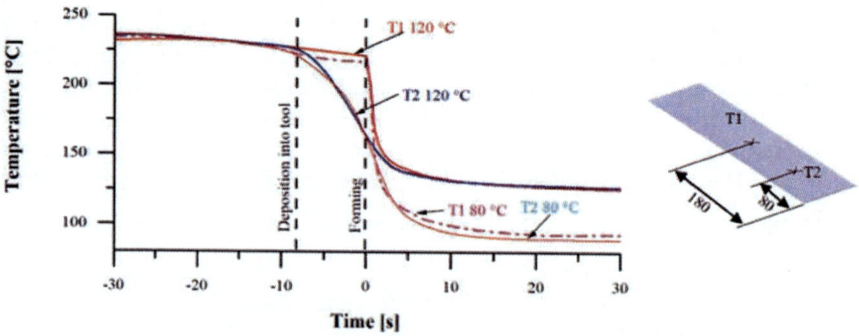

Fig. 3. Temperature profile in the semi-finished product during forming.

During forming trials, the tool temperature, forming force and blank holder force are varied. Table 1 shows the experimental plan. The influence of these process parameters on the mechanical performance is later investigated in order to compare both joining approaches considered in this study and to provide recommendations for the design of the forming process. Test used to evaluate the mechanical performance is based on a 3-point bending test. For this purpose, a clamping device was developed (see Fig. 4). The hybrid U-profiles are clamped at the flange in order to avoid a spreading of the flanks during loading. The tests are performed on a quasi-static testing machine at a constant crosshead speed of 10 mm/s. As soon as a failure of the specimen occurs, the test is stopped. The force at failure as well as the energy absorption up to the maximum force are then examined.

Table 1. Experimental plan of the forming tests.

No	Tool temperature (°C)	Forming force (kN)	Blank holder force (kN)
1	120	1600	10
2	120	160	10
3	80	1600	10
4	80	160	10
5	80	160	30
6	120	160	30

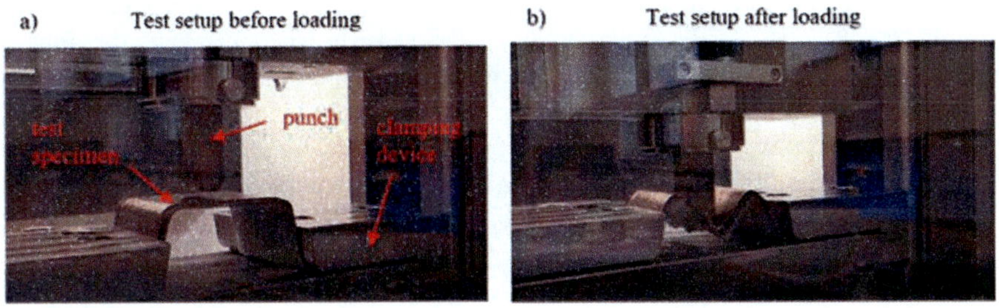

Fig. 4. Test setup to determine the mechanical properties of the test specimens.

3 Results

Figure 5 shows the maximum force and absorbed energy measured for test specimens manufactured using both joining approaches for a forming force of 1600 kN (Tests No 1 and 3) and 160 kN (Tests No 2 and 4). The diagram on the left shows the results obtained for a tool temperature of 120 °C, while the diagram on the right shows the results obtained for a tool temperature of 80 °C.

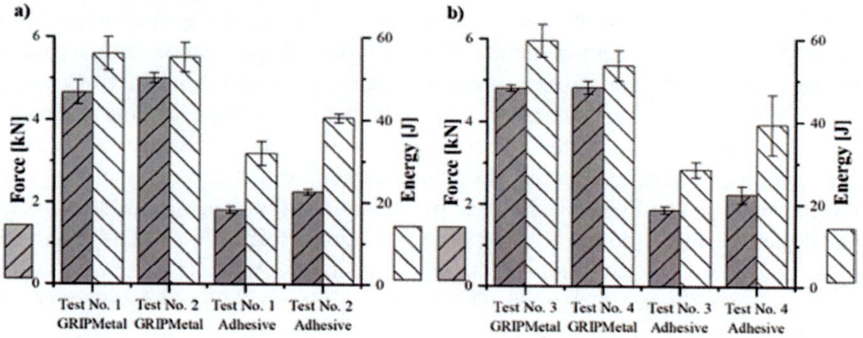

Fig. 5. Influence of the forming force on the mechanical performance obtained with GRIPMetal™ and an adhesive film for tool temperatures of a) 120 °C and b) 80 °C.

As shown in Fig. 5, specimens manufactured with GRIPMetal™ exhibit higher force at failure and absorb more energy than specimens manufactured with the adhesive film. Figure 6 shows that polymer matrix is pressed out more strongly for specimens prepared using the adhesive film. This leads to a thinning of the FRTP, which may decrease stiffness and cause premature failure of the specimens. Possible formation and propagation of delaminations in specimens manufactured with the adhesive film may also explain the difference in mechanical behavior obtained. Furthermore, at the bottom and the flank of the U-profile, the plastic flows out more intensively than at the flange. This is due to the temperature gradient in the semi-finished product before forming (see Fig. 3). At the outer areas of the semi-finished product, the material cools down faster than in the center, as it has direct contact to the die and the blank holder. As a result, the FRTP remains more viscous in the center of the semi-finished product during forming. The non-uniform pressing out of the polymer matrix causes a displacement of the fibres. This reduces the maximum bearable forces.

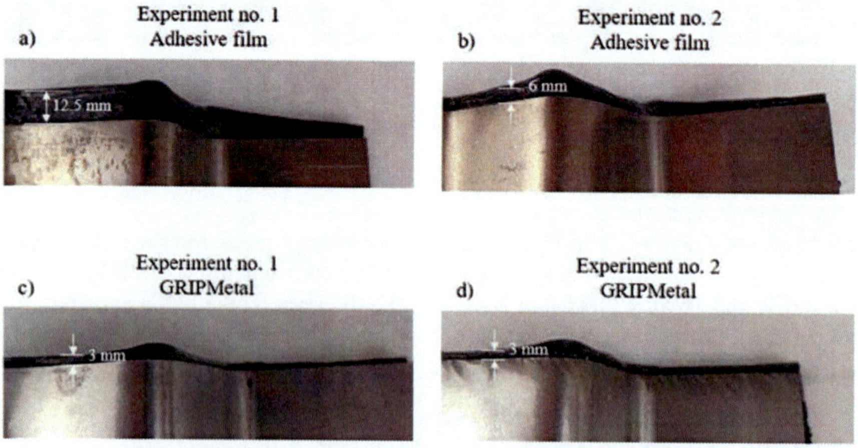

Fig. 6. The amount of plastic pressed out depending on the pressing force and the joining method.

With GRIPMetal™ the plastic is pressed out less and more uniformly. As a result, the fibres maintain the unidirectional orientation and the mechanical performance is higher. This is due to the hooks of GRIPMetal™, which prevent the plastic to flow out. Similar to Cold Metal Transfer pins (CMT-pins), the hooks have the further advantage that the loads can be transferred from the steel sheet into the fibres across all layers [17, 18]. In contrast, the joining approach with the adhesive film the loads are transferred from the steel sheet only into the first FRTP layer. As also shown in Fig. 5, variation of the forming force has no significant effect on the mechanical performance of the test specimens manufactured with GRIPMetal™. Also, the amount of plastic pressed out does not differ significantly between the test specimens with a pressing force of 160 kN and 1600 kN (Fig. 6). However, a reduction of the pressing force leads to a slight improvement of the mechanical performance of test specimens bonded with the adhesive film (Fig. 5). Due to lower pressing forces, the plastic is pressed out less, which leads to a reduction of the fibre displacement and thinning of the FRTP.

Figure 7 shows the maximum force and absorbed energy as a function of the tool temperatures (a and b) and the blank holder force (c). It can be seen that neither the tool temperature nor the blank holder force has significant influence on the mechanical performance for both joining approaches. In experiment No. 6, the FRTP separated from the steel sheet with the adhesive film (see Fig. 7d). For this reason, this series of experiment is not considered. For this condition, the melting temperature of the adhesive material corresponds to the set tool temperature of 120 °C. The combination of high blank holder forces with high tool temperatures possibly results in high frictional forces and high shear stresses in the adhesive material that exceed the maximum bearable stresses at this temperature.

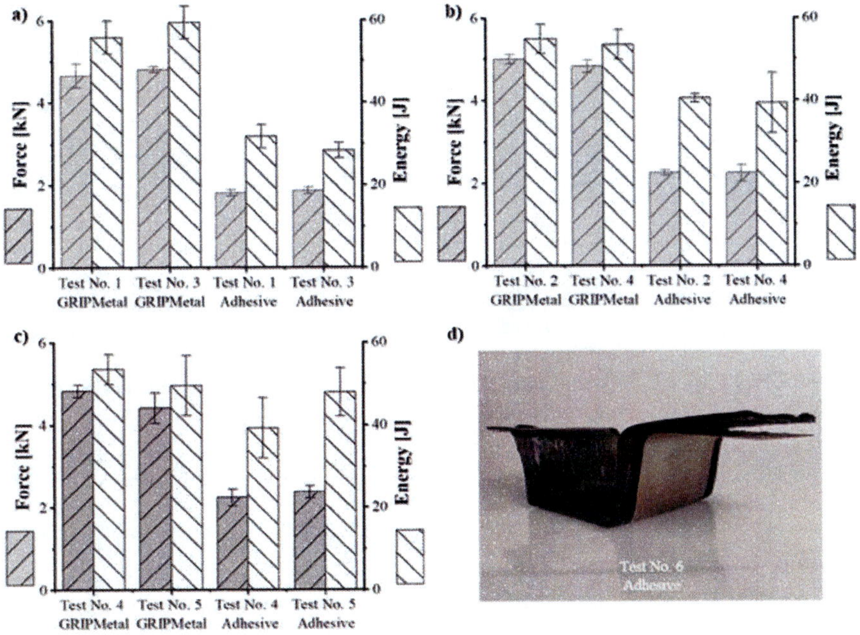

Fig. 7. Influence of a) and b) the tool temperature and c) the blank holder force on the mechanical performance of the joining methods with GRIPMetal™ and an adhesive film. (d) FRTP separated from the steel sheet and the adhesive film after forming.

4 Conclusion

In this paper, the mechanical performance of hybrid test specimens prepared using adhesive bonding or mechanical interlocking are investigated. For this purpose, hybrid semi-finished products were first formed into U-profiles using one-shot forming process. During forming trials, the forming force, blank holder force and tool temperature were varied in order to determine the effects of these parameters on mechanical performance. Profiles were then characterized mechanically to obtain the maximum force and absorbed energy. Results showed that the mechanical performance of the test specimens manufactured using mechanical interlocking are not sensitive to studied process parameters. On the other hand, for the test specimens manufactured with the adhesive film, the forming force was seen to have significant effect on the mechanical properties. For these specimens, an increase of the forming force resulted in a reduction of the force at failure and energy absorption. This was attributed to polymer matrix being pressed out transverse to the fibre direction, which resulted in thinning of the FRTP and to displacement of the fibres. For specimens manufactured with the adhesive, the objective should thus be to reduce the forming force to reduce the polymer matrix flow, while ensuring that enough pressure is applied to obtain a sufficient level of consolidation in the FRTP. This will be the objective of future activities. Finally, tool temperature and blank holder force showed no significant influence on the mechanical properties of the adhesive bonded specimens.

Acknowledgements. The authors gratefully acknowledged the financial support of the ZIM cooperation project ZF4096706, which is supported by the AiF within the program Zentrales Innovationsprogramm Mittelstand (ZIM) of the Federal Minitry of Economics and Technology. Furthermore, the authors would like to thank NRC, NUCAP, Compositence GmbH and the Institute of Machine Tools and Production Technology of the TU Braunschweig for the outstanding cooperation.

References

1. EUR-Lex: REGULATION (EU) 2019/631 OF THE EUROPEAN PARLIAMENT AND OF THE COUNCIL of 17 April 2019 setting CO_2 emission performance standards for new passenger cars and for new light commercial vehicles, and repealing Regulations (EC) No 443/2009 and (EU) No 510/2011, EUR-Lex, PE/6/2019/REV/1. https://eur-lex.europa.eu/ (2019). Accessed 17 Feb 2020
2. VuMa: Bevölkerung in Deutschland nach der Wichtigkeit von Umweltfreundlichkeit als Kriterium beim Autokauf, von 2015 bis 2018 (Personen in Millionen). Statista. https://de.statista.com (2018). Accessed 17 Feb 2020
3. Friedrich, H.E.: Leichtbau in der Fahrzeugtechnik. Springer Fachmedien, Wiesbaden (2013)
4. Trautwein, T., Henn, S., Rother K.: Gewichtsspirale. ATZ – Automobiltechnische Zeitschrift **113**(5), 390–395 (2011)
5. Uddin, N.: Developments in Fiber-Reinforced Polymer (FRP) Composites for Civil Engineering. Woodhead, Cambridge (2013)

6. Ickert, L.H.: VK-Metall-Hybridbauweise für die automobile Großserie. Dissertation, Institut für Kraftfahrzeug, RWTH Aachen University, Aachen (2014)
7. Lauter, C.: Entwicklung und Herstellung von Hybridbauteilen aus Metallen und Faserverbundkunststoffen für den Leichtbau im Automobil. Dissertation, Universität Paderborn, Paderborn (2014)
8. Fahrig M.: Composite insert as a structural reinforcement for A-pillars. lanxess.com (24 Sept 2019). https://lanxess.com/en/Media/Press-Releases/2019/09/Composite-insert-as-a-structural-reinforcement-for-A-pillars. Accessed 17 Feb 2020
9. Behrens, B.A., Hübner, S., Neumann, A.: Forming sheets of metal and fibre-reinforced plastics to hybrid parts in one deep drawing process. In: 11th International Conference on Technology of Plasticity, ICTP 2014, pp. 1608–1613 (2014)
10. Behrens, B.A., Hübner, S., Grbic, N., Micke-Camuz, M., Wehrhane T., Neumann, A.: Forming and joining of carbon-fiber-reinforced thermoplastics and sheet metal in one step. 17th International Conference on Sheet Metal, SHEMET17, pp. 227–232 (2017)
11. Gresham, J., Cantwell, W., Cardew-Hall, M.J., Compston, P., Kalyanasundaram, S.: Drawing behaviour of metal-composite sandwich structures. 13th International Conference on Composite Structures: ICCS/13, pp. 305–312 (2005)
12. Dröder, K., Brand, M., Meiners, D., Müller, S.: Mechanische Strukturierung für hochfeste Metall-Kunststoff-Hybride – HyTensile. Europäische Forschungsgesellschaft für Blechverarbeitung e. V. (2017)
13. Kühn M.: Prozessabhängige Eigenschaften strukturierter Warmumformstähle in hybriden Materialverbunden. Dissertation, Vulkan-Verlag, Essen (2019)
14. Demes, M., Kühn, M., Gebken T., Dröder, K.: Thermal behavior of polymer metal hybrids of hot stamped steel and fiber-reinforced thermoplastics. In: 4th Brazilian Conference on Composite Materials (2018)
15. AZoM: AISI 1010 Carbon Steel (UNS G10100). azom.com (21 Sept 2012). https://www.azom.com/article.aspx?ArticleID=6539. Accessed 17 Feb 2020
16. thyssenkrupp: Tiefziehstähle DD, DC und DX. thyssenkrupp-steel.com (July 2018). https://www.thyssenkrupp-steel.com/de/publikationen/. Accessed 17 Feb 2020
17. Stelzer, S., Ucsnik, S., Pinter, G. Composite-composite joining with through the thickness reinforcements for enhanced damage tolerance. In: 20th Symposium on Composites (2015)
18. Ucsnik, S., Scheerer, M., Zaremba, S., Pahr, D.H.: Experimental investigation of a noval hybrid metal-composite joining technology. Compos. Appl. Sci. Manuf. **41**(3), 369–374 (2010)

Thin Film Sensor Systems for Use in Smart Production

A. Schott(✉), S. Biehl, G. Bräuer, and C. Herrmann

Fraunhofer Institute for Surface Engineering and Thin Films IST, Bienroder Weg 54 E, 38108 Braunschweig, Germany
{anna.schott,saskia.biehl,guenter.braeuer, christoph.herrmann}@ist.fraunhofer.de

Abstract. For further success of the fourth industrial revolution, it is necessary to acquire detailed process data parallel to data processing. For measuring exact data of the production system, the sensor systems has to be integrated in the decisive area of the production process. This leads to an increasing demand for sensor systems that can directly be applied on tool or component surfaces, which are in high loaded contact with the workpiece or counterpart. For this reason, Fraunhofer IST develops thin film sensor systems by physical vapor deposition (PVD), plasma enhanced chemical vapor deposition (PECVD) and structuring technologies with high hardness, high wear resistance, low friction coefficient and a high load capacity that are very small and only few μm thick. These multisensory thin-film-systems can record force, temperature and their distributions directly in a production system or a tool surfaces with spatial resolution, so that both inline condition monitoring and predictive maintenance are possible. For such a multisensory thin film system the thermoresistive and piezoresistive properties of metal or hard coatings and diamond-like carbon layers (DLC) are used and has to be well adapted and arranged e.g. by structuring processes. Different thin film sensor types, their manufacturing as well as examples for application and the sensor performance in industrial production processes will be introduced. A thin film sensor on a tool segment for injection molding of natural fibre reinforced polymers will be shown. This thin film sensor system with high wear resistance allows measuring the temperature distribution on the mold surface and the monitoring of the melt front movement. A further interesting example are smart washers for static and dynamic control of the true clamping force of screw connections. Prospectively, these various thin film sensor systems could be a capable contribution to establish an intelligent, digitalized and smart production.

Keywords: Thin film sensor · Smart production · Thermoresistive sensor · Piezoresistive sensor · Injection molding

K. Dröder and T. Vietor (Eds.): *Technologies for economic and functional lightweight design*, Zukunftstechnologien für den multifunktionalen Leichtbau, pp. 220–231, 2021. https://doi.org/10.1007/978-3-662-62924-6_19

1 Introduction

In recent years the industry is moving toward Industry 4.0 with automated manufacturing, digitalized, intelligent and self-optimizing systems. Additionally, future forms of industrial production will be characterized by a high degree of product individualization, flexibility and resource efficiency. Therefore, it is essential to collect and analyze process data in real time and in-situ of production processes, proven by a hugely growth of the sensor market [1]. Moreover, sensors acquire vital information about the machine and its environment, either to monitor the operating status of the machine or to control its operation in real-time. Particularly, the demand for sensors, which can detect load conditions in the decisive operating area, has increased significantly. Therefore different materials for the production of piezoresistive thin-film load sensors as strain gauges are investigated. For high-temperature pressure sensors, strain-sensitive Pt-SiO2 nanocermet thin films can be used [2], as well as carbon-based resistive strain gauges, which are prepared on Ti using microdosing direct writing technology [3]. For example, metal-based thin films such as Pt/Cr and Pt/Cr2O3 systems on SiOx/Si substrates are used to manufacture temperature sensors [4]. For the next development step, such sensors have to be applied directly onto the component/tool surfaces in close contact with the counterpart respectively work piece. For this reason, Fraunhofer IST has developed a multifunctional thin film system to measure process data, such as force and temperature, locally with spatial resolution even during the process. Temperature and load are important and often even most critical parameters in many processes.

2 Thin Film Sensor and Diaforce®-Layer System

The multifunctional thin film sensor is a system (see Fig. 1) of different functional coatings based on the special piezoresistive and wear resistant amorphous hydrogenated carbon (a-C:H) layer DiaForce® with a hardness of about 24 GPa (Table 1), which is a further development of the well-known Diamond like Carbon (DLC) layer [5, 6].

Thin film sensor layer systems can be designed in huge variety due to the special requirements of application and parameter to be detected. In general, structured metal films e.g. chromium are applied to measure the temperature. The DiaForce® layer shows a drop of electrical resistivity under applied load, a so called piezoresistive behavior. Therefore, the DiaForce® layer is mainly applied to detect forces, loads or pressures. Additionally, the DiaForce® layer shows an exponential decrease of electrical resistance with increasing temperature. A comparable behavior similar to semiconductor materials has been detected, and therefore it could be used as a temperature sensor in some cases too.

Due to the different properties of different sensory thin films in combinations with the capability and flexibility of structuring techniques, complex thin film sensor systems can be designed aligned to the special measurements requirements in different applications. A typical sensor layer system for measurement load and temperature in local resolution is described as follows.

The sensor layer (5) is homogeneously and directly deposited onto steel substrates (6) in a plasma enhanced chemical vapor deposition process (PECVD) with a thickness of approx. 6 µm. On this layer a thin chromium layer (4) is deposited in a physical vapor deposition (PVD) process. Round structures are fabricated, in the areas where the load has to be measured, by photolithography and chemical wet etching process. Followed by an isolating intermediate layer of a silicon and oxygen modified carbon coating (3) (SiCON® developed at the Fraunhofer IST) with a hardness of about 8 GPa. The electrical circuit path from measurement area to contact area is structured on top. For temperature measurement, a chromium layer with meander structures (2) will be integrated and protected with a second SiCON® top layer (1) in the range of 3 µm, that shows although very good anti-adhesive behavior and a high wear resistance.

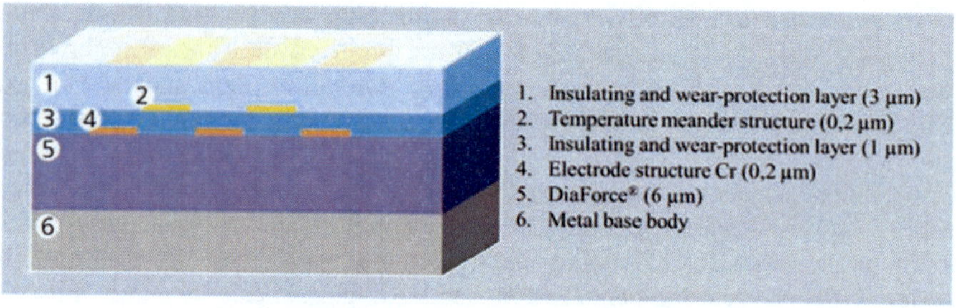

Fig. 1. Schematic of the multifunctional layer system.

Table 1. Specification of DiaForce® film.

Micorhardness Hu_{pl}	24 ± 4 GPa
Young's modulus	264 ± 39 GPa
Friction coefficient	≤0.17 against steel
Film thickness	5–6 µm
Specific resistance	$\approx 5 \cdot 10^5$ Ωm
$k^* = \Delta R/(\Delta F \cdot R_0)$	$\approx 10^{-4}$ N–1

3 Characteristic Resistance Dependencies of Thin Film Sensor Structures

3.1 Thermoresistive Temperature Sensors on Injection Mold Inserts

In plastic processing, like injection molding, the real temperature at the tool surface is a crucial parameter for quality survey and process optimization. Exemplified characteristics of temperature measurement will be described for this application (see Fig. 2).

The characteristic electrical resistance dependencies as a function of temperature is determined for each sensor structure. Typically, the meander shaped structures based on chromium have a positive temperature coefficient (PTC). For characterization of resistance dependencies, the metal bodies with the thin film sensor system are heated in an oven. As reference a temperature sensor Pt100 was used, which is applied to the sensor layer system with a heat-conducting paste. The sensors are measured in a four-wire circuit. A constant current of 1 mA is applied via two contacts and the voltage dropping across the sensor is measured via two further contacts, from which the sensor resistance can be calculated. Figure 3 shows typical characteristics of several sensors. The linear resistance dependence on temperature expected for metallic structures could be observed for the Cr-layer (blue line). For the DLC layer, with negative temperature coefficient (NTC sensor), a reference resistor was connected in series with the sensor structure and a constant voltage of five volts was applied to record the temperature resistance characteristics. The DLC coatings show a more exponential dependency of the resistance from temperature, similar to semiconductor behavior. Using this thermoresistive characteristic curve, the temperature can be determined by measuring the sensor resistance.

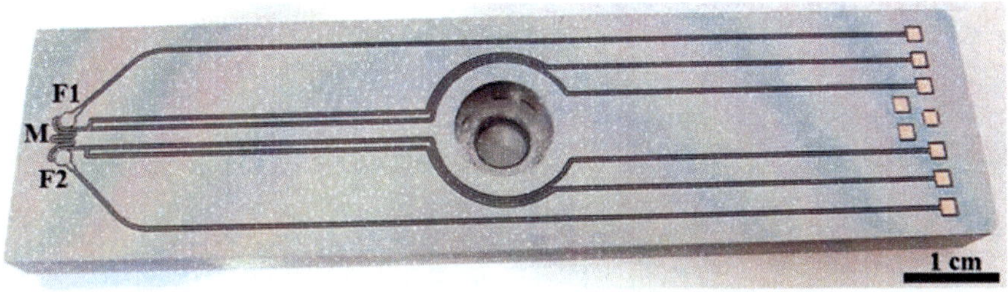

Fig. 2. Sensory insert for injection molding process with temperature sensor structures.

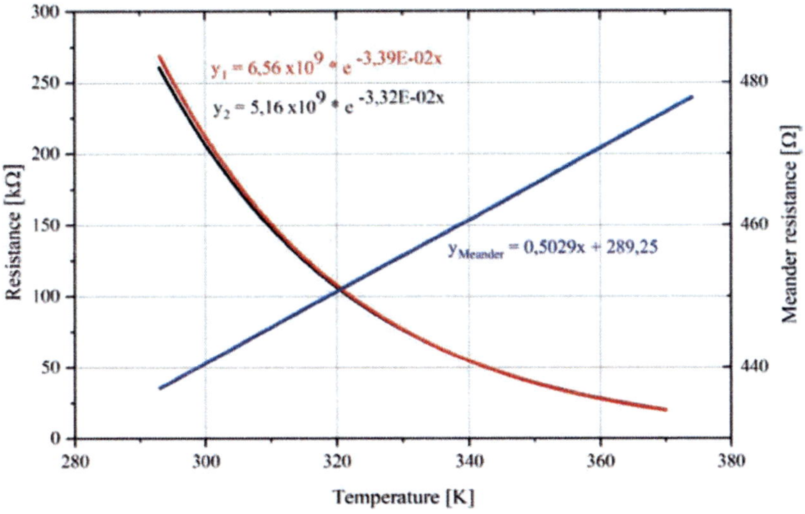

Fig. 3. Characteristic thermoresistance dependencies of sensor structures.

3.2 Piezoresistive Force Sensors for Forming Tools

The DiaForce® coating shows a drop of electrical resistivity under load, a so called piezoresistivity behavior. This behavior could be used for the introduction of a force sensor. Exemplified this thin film sensor is described as a sensor on tools for sheet metal forming.

To measure the piezoresistive behavior of the sensor structures on tools for sheet bending (see Fig. 4), loads between 0 and 1000 N were dynamically introduced in several loading and unloading cycles. Loading at the structures was carried out with a punch (load surface Ø = 1,4 mm). The sensor structures are connected in series with a reference resistor in the similar range as sensor resistance and a constant voltage of 5 V was applied. The piezoresistive resistivity shows typically k*-factors in the region of 10^{-3} N$-^{1}$ ($k^* = \Delta R/(\Delta F \cdot R_0)$). The larger the k* is, the greater the sensitivity of the sensor. In Fig. 5 the linear characteristic is shown in detail, with a resistance change of ca. 350 Ω per Newton. Due to the fact that DiaForce® shows a piezoresistive effect and, comparable to a semiconductor, as well a thermoresistive effect. The temperature effect has to be compensated to be able to detect loads in changing ambient temperatures with this thin film sensor system.

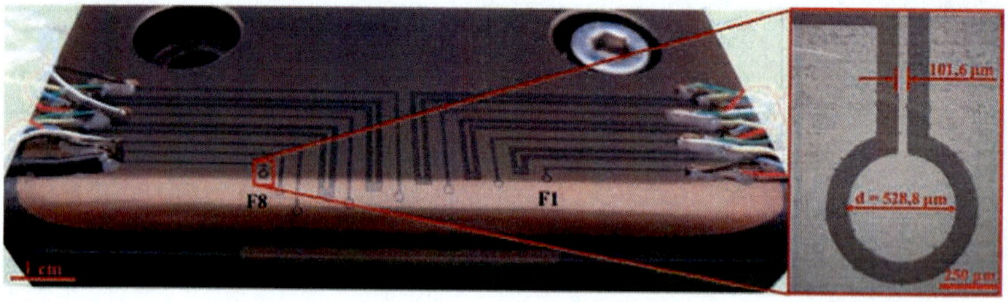

Fig. 4. Sensor tool with piezoresistive and thermoresistive sensor structure for sheet bending process.

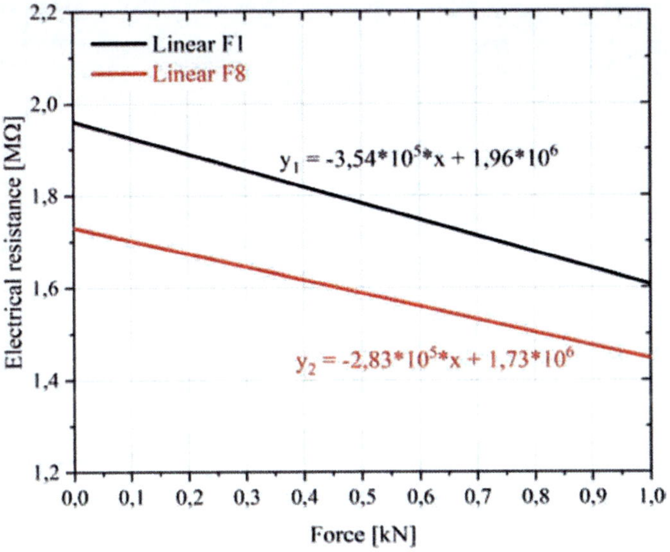

Fig. 5. Characteristic piezoresistance dependencies (linear fit) of sensor structures F1 and F8.

4 Application Examples for use in Smart Manufacturing

4.1 Sensory Tool for Sheet Bending Processes

During production of components from metal sheets by forming processes, defects could arise due to cracks, creases and necking. This leads to high rejecting rates. For better process control and process quality thin films sensor systems could be integrated that enables to minimize rejects. One example of gained data by a thin film sensor system is shown in Fig. 6 for a bending process. The circular force and temperature meander structures are arranged over the rounding of the bending tool. In the border area of the tool are the contact areas of the individual sensor structures. These developed sensor modules were integrated into a strip drawing test setup at the Fraunhofer IWU to evaluate the bending of an aluminium strip metal (AA6016). The sheet was preheated to a temperature of 200 °C and drew over the bending tool. In order to minimize friction between the thin film sensor on the tool surface and the aluminum sheet a special lubrication oil was used. When the heated sheet metal comes into contact with the thin film temperature sensors, a temperature increase is measured at first. The movement of the aluminum sheet during bending over the force sensors reduces their resistance. The rapid relief during the bending process leads to local resistance minima of the individual force sensors (see Fig. 6). Every step of the strip drawing test sequence could be detected separately.

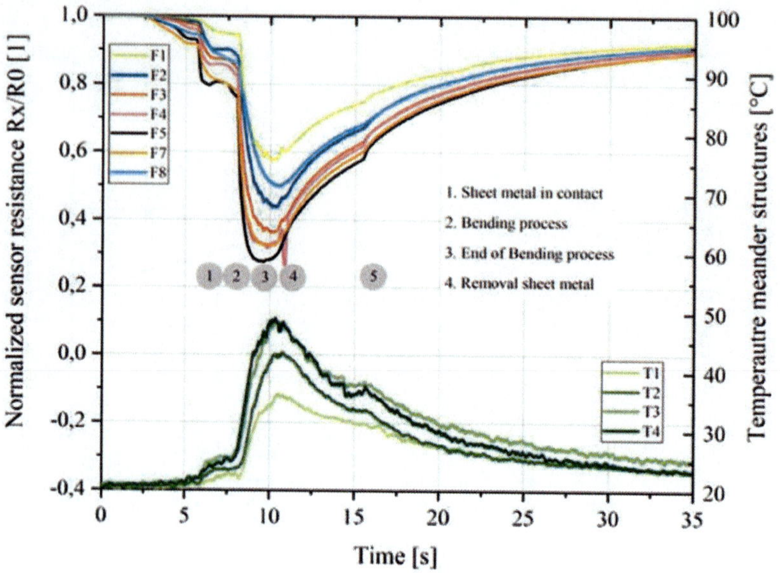

Fig. 6. Measurement results of individual sensor structures during forming process (sheet metal drawing force 10 kN, sheet metal temperature 200 °C) [7].

4.2 Sensory Module for Deep Drawing Processes

One of the forming techniques with the largest range of applications is deep drawing. In its simplest form a rigid tool, consisting of a drawing punch, a blank holder and a die is used for common processes. The key to produce high quality products is reducing rejects due to misshaped parts, cracks and wrinkles, which are often caused by process parameter instabilities. The sensory layer system and structuring of the sensor structures over the curved tool surfaces was particularly challenging with this deep-drawing tool (see Fig. 7A). The locally offset force sensor structures made it possible to record the feeding behavior of the sheet metal during the deep-drawing process successfully. After pre-characterizations, the sensory deep drawing tool has been installed into a deep drawing machine at the Fraunhofer IWU to form an aluminium sheet with a sheet thickness of 1 mm and a diameter of 100 mm. First level tests on a deep drawing process with the sensory deep drawing tools are shown in Fig. 7B. The tests were performed at room temperature as well as with pre-heated aluminium sheets. The movement of the aluminium sheet over individual sensor structure was measured during the deep-drawing process, which is indicated by a decrease in the electrical resistance of the sensor structures. A minimum is detected in the moment, when the deep drawing process is finished.

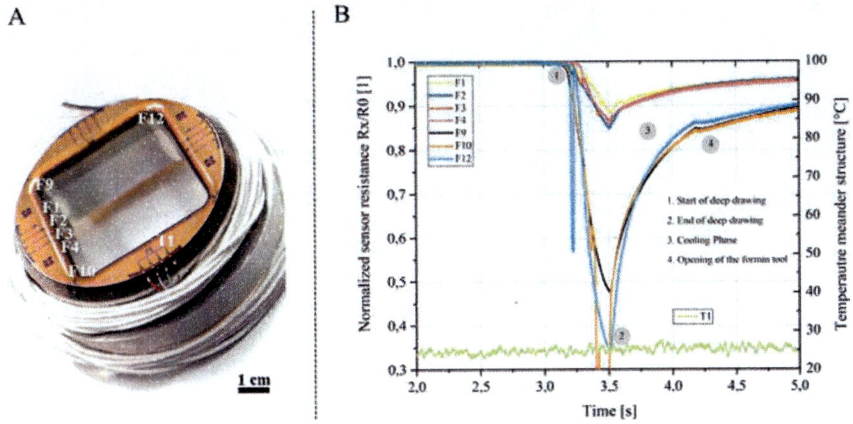

Fig. 7. **A** Deep drawing tool with thin film sensor system to measure force and temperature **B** Measurement results from a deep-drawing process [7].

4.3 Injection Molding Sensor Inserts

The integration of renewable raw materials as functional fillers in plastics is a highly focus objective in terms of ecological perspective and lightweight aspects. For plastic melts modified in this way, there are higher demands on the wear resistance of the molds and the process control must be re-optimized in order to minimize scrap rates caused by changed flow properties. Therefore, steel inserts (see Fig. 2) with thermoresistive and piezoresistive thin film sensors were developed, pre-characterized and integrated directly into the injection molding tool (see Fig. 8A). The measurement results of two different injection molding processes are shown in Fig. 8B. As compound in this test, a mixture out of polyvinyl chloride (PVC) as the basic material filled with 55% wood fiber was used. The first melt contact with the sensor inserts results in a steep temperature increase in every position for a well-formed part (continuous line). Compared with the second temperature distribution of a not completely filled form part (dotted line) it is evident that these sensor systems can be used for quality indication tool during manufacturing processes. After 300 injection processes the sensor systems were still without significant wear. Furthermore, slag powder- and talc powder-reinforced melts were tested with these systems.

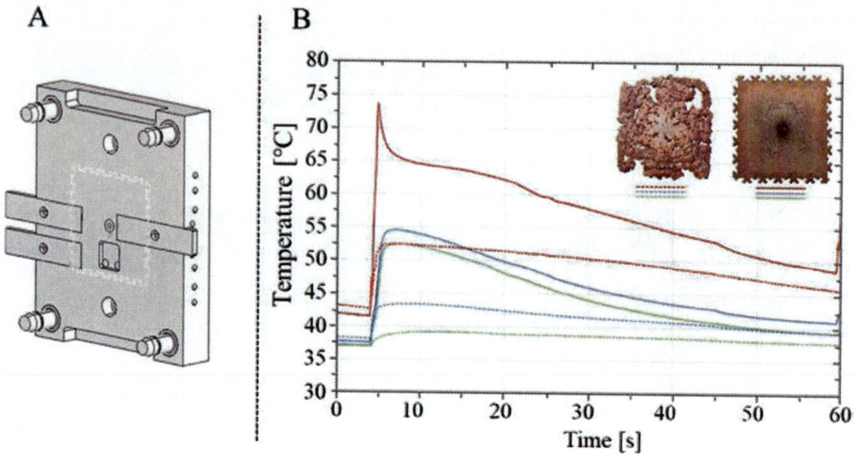

Fig. 8. **A** Injection molding tool with three sensor inserts **B** Temperature distribution as function of time for two processes. Continuous line is the perfectly fabricated part. [8]

Based on the very good results of the first design, the sensor design was extended so that 16 circular sensor structures are in line (see Fig. 9). This has made it possible to detect the flow front movement of the melt and its temperature distribution during the injection molding process in a time-dependent and dynamic way. Due to the direct contact of the melt with the sensor thin-film system, very good findings for process optimization can be obtained.

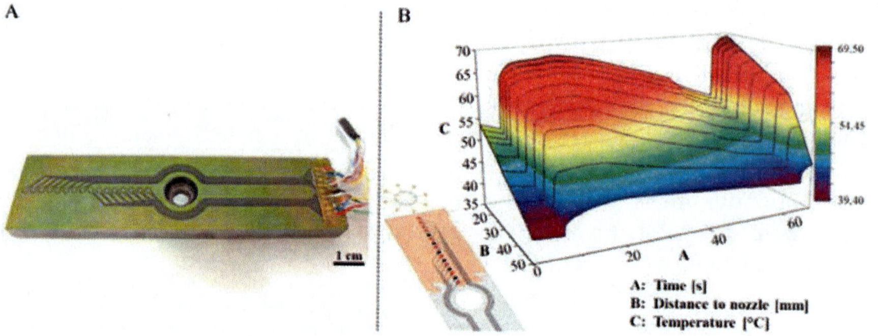

Fig. 9. **A** Sensor module with 16 thermoresistive sensors **B** Distribution of temperature in the mold as a function of nozzle distance and time. [9]

4.4 Smart Washer System

The smart washer is a universal measuring system with a wide range of applications. The possibility to measure static and dynamic forces continuously over a long period of time opens up the opportunity to use this system for monitoring screwed joints, but also especially for condition monitoring of process forces in production lines. A manual, subsequent adjustment of the components only becomes necessary when a drop in force is measured, so that e.g. maintenance and assembly cycles can be optimized. Figure 10A shows a sensory washer of size M12. Three circular force sensors are arranged evenly around the circumference so that tilting can also be detected. In addition, a temperature meander and temperature-compensating structures outside the load range can be seen. Temperature compensation, which is implemented here by a voltage divider structure, is necessary because the DiaForce® layer as a semiconductor has piezoresistive, but also thermoresistive properties. The force- and temperature-dependent resistance characteristics are recorded for each sensor structure by a preliminary characterization and are shown as an example in Fig. 10B. The thin-film sensors are wired to the measurement device and a voltage source applying a constant voltage. The temperature sensor needs additional a current source applying a few milliampere to the sensor. The resulting voltage drop across the sensors can be recorded and analyzed.

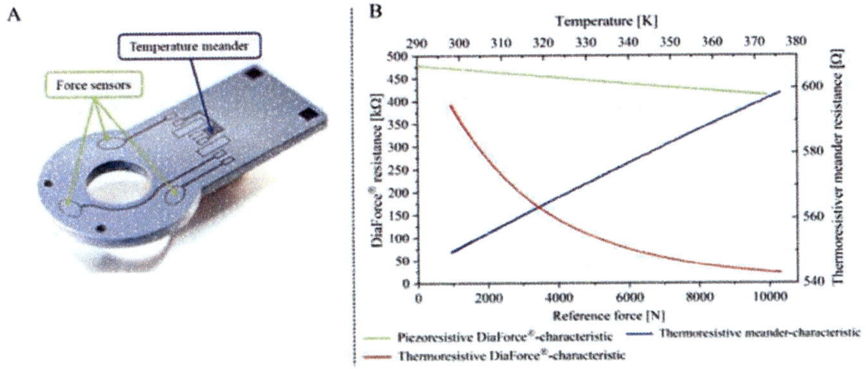

Fig. 10. **A** Smart sensor washer system with three force sensors and one temperature meander structure. **B** Load- and temperature-dependent characteristics of DiaForce® and meander structure.

5 Summary and Outlook

Thin film sensor systems with an outstanding property combination and a capable sensitivity for measurements in different industrial application were introduced. In order to gain process data such as force and temperature the thin-film sensor system is measuring directly in high loaded areas, which are hardly accessible with other sensor systems. The developed thin film sensor systems consist of several functional layers. The piezoresistive DiaForce® layer and the insulating intermediate and top layer of the layer system are based on the Diamond Like Carbon (DLC) layers developed by Fraunhofer IST. As part of a layer system with a total thickness of about 10 μm, the DiaForce® layer is applied together with the adhesion layers, metallic layers for electrical contacting and electrical insulation and wear protection layers in PECVD and PVD processes on hardened steel bodies or directly on tool surfaces. The sensors are structured by a combination of photolithography and wet chemical etching or by laser structuring. Due to high hardness and wear resistance of the coating system, such thin film sensors can be integrated directly in high loaded areas, e.g. on tools for sheet metal bending or deep drawing, and thus measure process data such as force, contact and temperature in spatial resolution. The possibility to design the thin film sensor systems onto special inserts instead onto complex shaped tools can reduce costs with constant utility value e.g. for polymer injection molding. Additionally, flexible structuring techniques, coating layer variations and different sensor designs allow aligning of thin film sensors to the regarded application and measurement task.

The topical project "Industry 4.0", a high-tech strategy of the German government, is intended to promote the digitalization of technical industrial processes [10]. The core element is the "smart factory", consisting of a large number of connected cyber-physical systems, in which individual processes of production, engineering, quality assurance and management are to be merged [10]. The basis for connected industrial processes lies in the sensory acquisition of basic data, such as the physical parameters force, pressure and temperature [11, 12]. Here, sensors represent the interfaces between the digital world of information technology and the engineering world of physical parameters. The presented thin-film sensor systems are ideally suited for inline monitoring of production processes, as data are acquired directly from the process zone and can then forwarded to higher-level system interfaces for monitoring, data processing and control. Moreover, the data are useful for design, verification of FEM analyses or for the creation of digital twins. In addition, the presented systems can be used to reduce prototyping time by about 30 percent, which enables a faster time-to-market. Flexible thin film sensor systems onto inserts or standard components, such as the smart washer, can be used as reliable and capable data acquisition tool for production process optimization and simulation. Prospective thin film sensor modules with wireless data transfer via Bluetooth, RFID or 5G technology will be used for further purposes like life cycle monitoring, predictive maintenance and many other sensory tasks for the topical frame of "Industry 4.0".

References

1. Frost & Sullivan: Top 50 Emerging Technologies & Growth Opportunities (2019)
2. Schmid-Engel, H.: Strain sensitive Pt-SiO_2 nano-cermet thin films for high temperature pressure and force sensors. Sens. Actuators A **206**, 17–21 (2014)
3. Wie, L.-J., Oxley, C.H.: Carbon based resistive strain gauge sensor fabricated on titanium using micro-dispensing direct write technology. Sens. Actuators A **242**, 389–392 (2016)

4. Garraud, A., Combette, P., Giani, A.: Thermal stability of Pt/Cr and Pt/Cr_2O_3 thin-film layers on a SiNx/Si substrate for thermal sensor applications. Thin Solid Films **540**, 256–260 (2013)
5. Bewilogua, K.: History of diamond-like carbon films – from first experiments to worldwide applications. Surf. Coat. Technol. **242**, 214–225 (2014)
6. ISO 20523:2017: Carbon based films – classification and designations (2017)
7. Cornet-Projekt SensoFut 'Sensorized Future – Sensing of temperature and pressure in harsh enviroments', Förderkennzeichen 88 EBG (2015)
8. Biehl, S., Meyer-Kornblum, E.: Novel sensor modules for efficient manufacturing of natural fiber reinforced plastics. Eurosensors 2018 Conference, Graz (2018)
9. SmartNFR: Entwicklung intelligenter Schichtsysteme zur Prozesssteuerung und Steigerung der Verschleißfestigkeit von Maschinen- und Werkzeugkomponenten bei der Verarbeitung von naturfaserverstärkten Kunststoff. Förderkennzeichen 163 EBG (2019)
10. Botthof, A., Hartmann, E.A.: Industrie 4.0 als Chance für die Wettbewerbsfähigkeit von Arbeit. Springer Vieweg, Heidelberg (2015)
11. Gevatter, H.-J.: Grünhaupt, U: Handbuch der Mess- und Automatisierungstechnik in der Produktion, 2nd edn. Springer, Berlin (2006)
12. Hesse S., Schnell, G.: Sensoren für die Prozess- und Fabrikautomation, 7nd edn. Springer Vieweg (2018)

Check for updates

Radomes – Process Influences on the Integration of Radar Sensors

Teresa Bonfig[1(✉)], Joachim Sterz[2], Jan P. Beuscher[2], and Klaus Dröder[2]

[1] Volkswagen AG, Brieffach 011/17774, 38436, Wolfsburg, Germany
teresa.bonfig@volkswagen.de
[2] Technische Universität Braunschweig, Institute of Machine Tools and Production Technology, Langer Kamp 19b, 38106, Braunschweig, Germany
{j.sterz,j.beuscher,k.droeder}@tu-braunschweig.de

Abstract. Intensive research is currently being focused on autonomous driving. Nowadays, comprehensive assistance systems already support the driver and ensure increased safety in road traffic. Core of these systems are well developed sensors that detect the surrounding area and thus ensure the safety of all road users. An important technology in this context is the radar sensor. When integrating it into the vehicle the performance of this sensor is highly dependent on its environment. In this regard the design of the cover of the radar sensor a so-called radome plays an important role. Radomes are supposed to ensure the reliability of the radar sensor by their protective function under consideration of the vehicle design. However, the material of radomes influences the high frequency electromagnetic waves.

Therefore, a functional design of radomes requires fundamental knowledge of the interaction of materials and radar waves and thus of the influence of the manufacturing process, the process parameters and the part design. In this paper the structural and process-related influences of injection moulding on damping behaviour of engineering plastics are investigated. A strong relation between thickness and signal damping is observed. Furthermore, a relation between the thermodynamic process parameters of the injection moulding process and the damping behaviour is demonstrated.

Keywords: Function integration · High frequency · Polymers · Radar sensors · Radome

1 Introduction

The progress in autonomous driving and on-going development of driver assistance systems will require more electronic components to be realized in future generations of vehicles. For the implementation of these functions, automotive radar sensors with the frequency fixed in the area of 75 to 81 GHz are particularly important [1]. The

K. Dröder and T. Vietor (Eds.): *Technologies for economic and functional lightweight design*, Zukunftstechnologien für den multifunktionalen Leichtbau, pp. 232–238, 2021. https://doi.org/10.1007/978-3-662-62924-6_20

wavelength of the electromagnetic wave resulting from this frequency highly influences the interaction with vehicle parts. Therefore, the performance of the integrated radar sensor is strongly dependent on its environment.

Ideally, the wave is not affected by the exterior boundaries of the vehicle. For this purpose the structure, geometry and material of the radome plays an important role. Radomes are supposed to ensure the reliability of the radar sensor with its electromagnetic constraints under consideration of the vehicle design. These two often-competing criteria usually require a compromise. There are different strategies for the installation of radar sensors in automotive engineering. Depending on the function, the installation must be carried out in areas that influence the design of the vehicle. Sensors that scan the long range are placed in the front and rear of the vehicle at bumper level.

For outer skin parts made of plastic, the polymer type, additives, thickness and distance affect the interplay of reflection, transmission and absorption by high frequency electromagnetic waves passing through. In addition, the manufacturing parameter and microstructure of the polymer can influence the radomes electromagnetic properties, which is examined in this paper for the injection molding process. Under defined specifications it is possible to use polymers for radomes without strongly disturbing the electromagnetic wave. Therefore, knowledge of the influencing parameters on the transmission is necessary to ensure the function of the radar sensor. The reflections of a component have a decisive influence on the radar transparency. They lead to a reduction of the range by the lost energy of the wave as well as to a disturbance of the radar sensor. Therefore, it is of special significance to reduce reflections. A common approach is to change the thickness of the part to a certain material dependent value, which is described in [2, 3]. The adjustment of the wall thickness is investigated more detailed in the following. Compensation methods for coated polymer radomes with high reflecting paints are described in [2]. Further possibilities of keeping a component reflection-free by using metamaterials or multi-layer structures are studied in [4–7].

2 Materials and Methods

In the automotive industry, thermoplastics are used due to their thermal process ability and acceptable process times. Thermoplastics can be purchased at low cost depending on the application requirements and allow easy handling. Bosch uses a mixture of polycarbonate and polyethylene terephthalate to produce radomes [2] while Poly Fluoro recommends polytetrafluoroethylene due to its excellent properties. [8] The thermoplastics polyamide 6 (PA 6) and polypropylene (PP) have already prevailed over other plastic types in many applications in the automotive industry, so that both plastic types are used in this study to proof there suitability for radomes [9, 10]. The test specimens are manufactured using an ENGEL VICTORY SPEX 120 injection moulding machine. High productivity and less post-processing of the parts characterize injection moulding systems. In the course of many research projects focusing on mechanical strength, it has been shown that thermodynamic parameters such as mold temperature and pressure of the process have a great influence. This knowledge is incorporated into the current investigations. Table 1 shows an extract of the data sheets of the examined plastics, PA 6 and PP.

Table 1. Datasheet of PP and PA 6 [11, 12]

Plastic	MFR Temperature (dry) [°C]	Tensile Modulus (dry) [MPa]	Yield Stress (dry) [MPa]	Yield Strain (dry) [%]	Density (dry) [kg/m^3]
PP	230	1450	33	9	900
PA 6	275	3500	90	4	1130

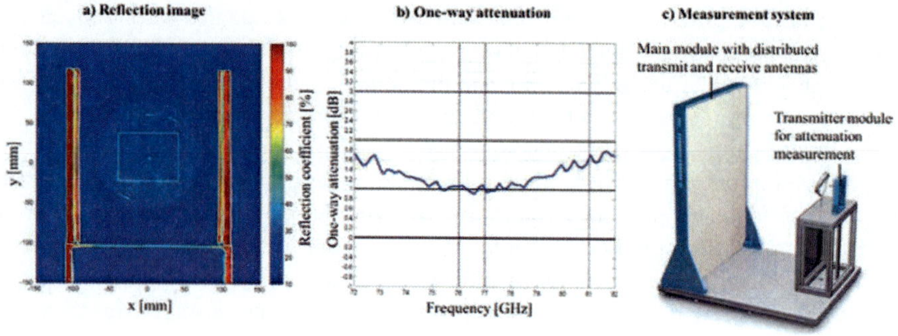

Fig. 1. QAR Measurement system c) with reflection image a) and on-way attenuation b) [13]

The used measuring equipment is the Quality Automotive Radom (QAR) Tester from Rohde & Schwarz (Fig. 1c). With this device, the one-way attenuation and reflection image of a component can be displayed. The high-frequency one-way attenuation is spectrally measured in the frequency range of 72 to 82 GHz, which includes the important interval of 75 to 81 GHz. The wide frequency band allows a more precise detection of the reflection minimum. The transmission is determined through the entire structure of the component and resonances and multiple reflections can be detected as shown in Fig. 1a). The adapted thickness of a radome can be determined via frequency shift. A further measurement result of the tester is the spatially resolved image of the high-frequency reflection. With the image obtained, wave matches, stronger reflecting and thus blinded regions can be found and are shown in Fig. 1b). This also allows conclusions to be drawn about thickness or material issues in the production process. In addition, an average value of the reflection over a selected area of the component and the frequency range can be examined. Furthermore, the homogeneity of the reflection can be observed [13].

3 Results

When measuring the average value of the reflection over the area of the PA 6 and PP samples with different thicknesses, the expected sinusoidal shape of the curve can be detected (Fig. 2, 3). Furthermore, the reflection factor can be simulated based on the Maxwell equations using MATLAB with the permittivity value and the thickness of the sample. The reproduction of the curves from the experiments by simulation shows that the relative permittivity values of PP is around 2.1 and the value of PA 6 is around 2.7. This dimension of the values in the GHz range are close to the values in literature [3]. For PA 6 the datasheet value at 100 Hz is 4.1 and at 1 MHz 3.3 [10]. Therefore, the permittivity of it decreases with increasing frequency. Permittivity values for PP are not illustrated in the datasheet [11]. Beyond the sinusoidal waveform the simulation shows that as the thickness of the sample increases, the absolute value of the maximum and minimum reflection also rises. However, the amplitude remains the same. The measurement points do not show this trend clearly. The deviation is explained by the fact that the reflection is not measured spatially. The use of the mean value of the reflection leads to an imprecise reflection value averaged over a larger frequency band and the area of the sample.

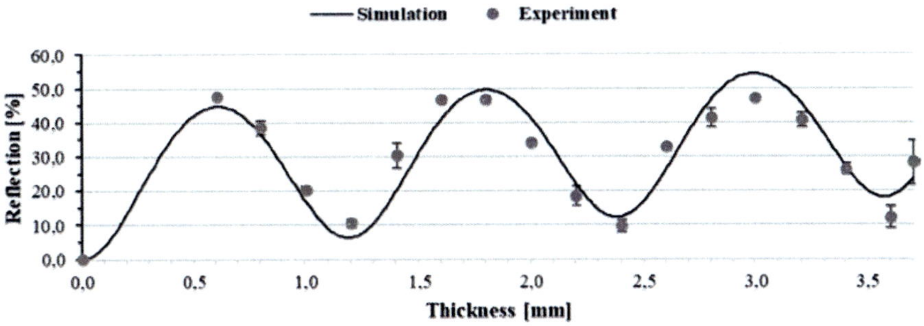

Fig. 2. Experiment and Simulated reflection of PA6 over the thickness

The two plastics show their minima and maxima at different thicknesses (1,2 mm at PA 6 and 1,4 mm at PP) and they have various distances between each other. A close match between the experimental and calculated position of these minima und maxima is found. Moreover it can be seen that the samples wall thickness has a very high influence on the interplay of reflection and transmission.

The position of the minima and maxima of reflection at different frequencies and the varying distance of those for the two polymers results from their material parameter permittivity. The experimental and simulated results are in close match.

The thickness of a polymer part has a very high influence on the reflection and transmission in the GHz range. For an adapted thickness of a component, the reflections at the two interfaces must be matched to each other. The reflection at the first boundary surface must not interfere constructively with the reflection at the second. If this constructive interference occurs, the waves overlap in such a way that they strongly blind the sensor. In the case of destructive interference, the waves overlap in such a way that they cancel each other out. The reflections back onto the sensor are therefore low. Furthermore, there are no multiple reflections in the material. The component is hence adapted to the wave in this material thickness. This can be seen in the sinusoidal course of the reflection over the thickness; see also [2].

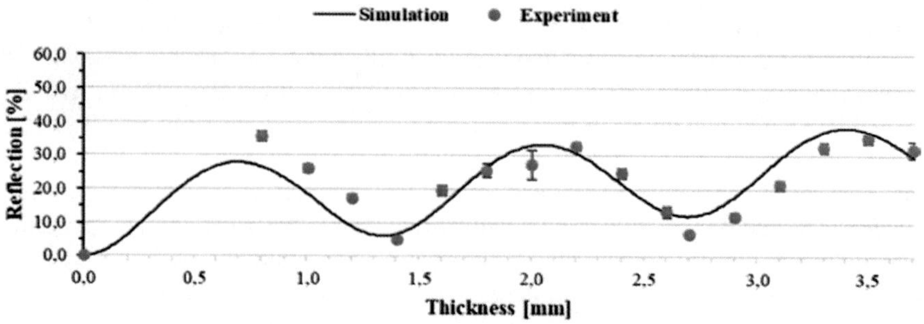

Fig. 3. Experiment and Simulated reflection of PP over the thickness

This knowledge about the thickness influence allows components to be designed for improved reflection. The reflection is adjusted by shifting the minimum of the reflection factor to the desired application frequency of the radar sensor. This is, therefore, possible by changing the thickness of the part. In most cases the center frequency of the radar sensor is at 76.5 GHz.

In a further step, the damping behaviour is investigated as a function of chosen injection moulding parameters. The focus is set on thermodynamic influencing factors such as mould temperature, tooling temperature and pressure. Figure 4 shows the results for the plastic PA 6.

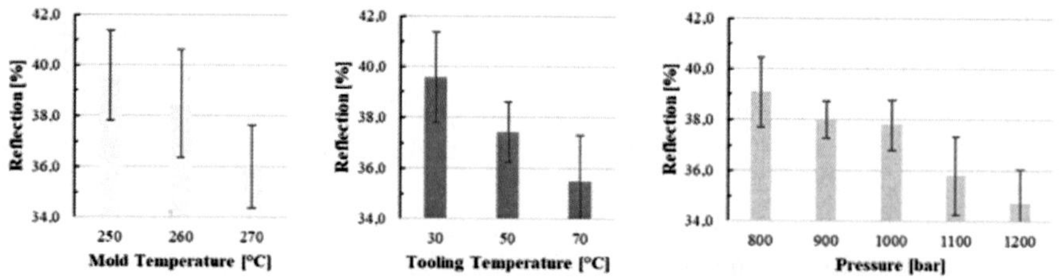

Fig. 4. Results of the Damping Behavior of PA 6 depending on the Process Parameters Mold Temperature, Tooling Temperature and Pressure

An increase in the temperature parameters (mould temperature and tooling temperature) leads to a reduction in reflection. The same progression is shown by an increasing of the pressure, which also leads to a reduction of the reflection. This behaviour can also be observed with the plastic PP. A Comparison with other injection molding parameters shows that they have less influence on the reflection and transmission. Neither the cooling time nor the holding pressure time seem to have a measurable influence for both plastics. One explanation for this behavior is the presence of air bubbles. If the injection molding parameters are selected in such a way that air bubbles are present in the sample, they create further boundary layers. However, if these air bubbles reach a particular size, they are detected by the radar and thus increase the reflection. Figure 5 shows this behavior. On the left side of the figure, a high reflection is shown including many air bubbles, while on the right side of the figure, there are no air bubbles and the reflection is therefore low.

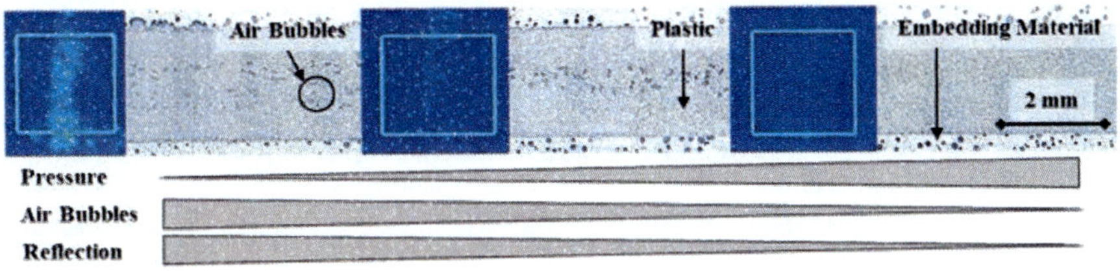

Fig. 5. Reflection and Air Bubbles depending on increasing pressure.

However, further results show a contrary behaviour. An increased temperature and pressure leads to a reduction of air bubbles but also to an increase of reflection. Therefore, this behaviour cannot be explained by micrographs. It is assumed that one explanation lies in a possible change in the crystal structure of the plastic. This Hypothesis has to be verified in further investigations.

4 Summary and Outlook

In this paper, experimental and numerical results on the reflection and damping behaviour in the GHz range of injection moulded samples as a factor of thickness and thermal process parameters are presented. Using PA 6 and PP as an example material the thickness of the sample and the manufacturing process parameters show influences of the damping behaviour. The influence of the thickness is significant and varies in a very high interval (5% minimum to 38% maximum at PP and 9% minimum to 48% maximum at PA 6). The influence of manufacturing parameters shows a lower impact. Thermodynamic manufacturing parameters such as temperature and pressure also have an effect on damping while other manufacturing parameters such as cooling time do not seem to have an effect on the damping. However, the interval is only 2–5% at the plastics PP and PA 6. One explanation is the presence of air bubbles, which can lead to further detectable boundary layers. However, since this behaves contrarily at higher temperatures and pressures, research is being carried out to find a further explanation.

As the crystallinity of thermoplastics has a high influence on mechanical strength, this is also taken into account in further tests on damping behaviour. Furthermore, the existing simulation model, which currently only takes into account the permittivity of the plastics, will be extended by implementing the influences of the injection moulding process.

Acknowledgements. This research and results published are based on the support of many ambitious students and assistant scientists, in detail: Alper Balkan, Nader Kasmi, Artur Kimmel, Qizhao Li, Rayan Mikhael, Maryam Rahimi-Karimabad and Lucas Sobiech. All results presented in this paper were developed at the BMBF Forschungscampus Open Hybrid LabFactory.

References

1. Ramasubramanian, K.: Moving from Legacy 24 GHz to State-of-the-Art 77-GHz Radar. Texas Instruments (2017)
2. Pfeiffer, F.: Analyse und Optimierung von Radomen für automobile Radarsensoren. Cuvillier, Elektrotechnik 31 (2010)
3. Bonfig, T., Körner, E., Kroll, L.: Influences of molecular structure and additives of polymers on relative permittivity for radar sensor integration. In: Technologies for Lightweight Structures, 4th International MERGE Technologies Conference – IMTC 2019 Lightweight Structures, 18th–19th September 2019, pp. 81–82. Verlag Wissenschaftliche Skripten, Chemnitz (2019)
4. Biber, S., Richter, J., Martius, S., Schmidt, L.-P.: Design of artificial dielectrics for anti-reflection-coatings. In: 33rd European Microwave Conference, 4–6 October 2003, pp. 1115–1118. London: Horizon House, Munich (2003)
5. Fitzek, F., Rasshofer, R.H.: Automotive radome design – reflection reduction of stratified media. Antennas Wirel. Propag. Lett. **8**, 1076–1079 (2009). https://doi.org/10.1109/LAWP.2009.2032571
6. Öziş, E., Osipov, A.V., Eibert, T.F.: Metamaterials for Microwave Radomes and the Concept of a Metaradome. Review of the Literature. Int. J. Antennas Propag. **2017**(2), 1–13 (2017). https://doi.org/10.1155/2017/1356108
7. Fitzek, F., Abou-Chahine, Z., Raßhofer, R.H., Biebl, E.M.: Automotive radome design – Fishnet structure for 79 GHz. In: IEEE German Microwave Conference, 15–17 March 2010, pp. 146–149, Berlin (2010)
8. Poly Fluoro Ltd. Homepage: https://polyfluoroltd.com/blog/polymer-radomes-radar-enclosures/. Accessed 8 July 2020
9. Schröder, B.: Kunststoffe für Ingenieure – Ein Überblick. Springer Vieweg, Wiesbaden (2014)
10. Abts, G.: Kunststoff-Wissen für Einsteiger. Hanser, München (2016)
11. Albis Homepage: https://www.albis.com/en/products/download/doc/en/SI/basf/UltramidB3S.pdf. Accessed 8 July 2020
12. Albis Homepage: https://www.albis.com/en/products/download/doc/en/SI/LyondellBasell/MoplenHP501H.pdf. Accessed 8 July 2020
13. Rohde & Schwarz GmbH & Co. KG. Homepage: https://scdn.rohde-schwarz.com/ur/pws/dl_downloads/dl_common_library/dl_brochures_and_datasheets/pdf_1/QAR_bro_en_5215-5388_12_v0100.pdf. Accessed 20. Feb 2020

Reports from the Research Clusters

Fiber Orientation Evaluation of Intrinsically Manufactured Metal-CFRP Hybrid Structures by Data Fusion of Pulsed Phase Thermography and Laser Light Section

Lucas Bretz(✉), Adrian Gärtner, Benjamin Häfner, and Gisela Lanza

Karlsruhe Institute of Technology (KIT), Wbk Institute of Production Science, Kaiserstraße 12, 76131 Karlsruhe, Germany
lucas.bretz@kit.edu

Abstract. The intrinsic production of metal-CFRP (carbon fiber-reinforced polymers) hybrid structures allows a load-path-optimal design of connecting components. However, the intrinsic production of complex hybrid 3D parts is prone to defects such as delaminations or fiber misalignments. Fiber misalignments of the woven fabric especially occur in the region of the metal insert. Pulsed phase thermography (PPT) is widely used for non-destructive testing of fiber-reinforced polymers, especially for delaminations. Laser light section (LLS) is suitable for generating a 3D cloud of points (CoP) of the component to be measured. The usage of two laser light section sensors allows in-line measurements of complex 3D geometries by preventing shades. An in-line quality assurance of the finished hybrid component is performed by data fusion of LLS and PPT. The resulting 3D thermographic data can be used to locate and quantify defects. In this contribution, a method for detecting individual fiber bundles in woven fabrics and quantifying the 3D fiber alignment based on the fused LLS and PPT data is presented. This could allow for the simultaneous detection of delaminations and fiber misalignments using the same measurement technology.

Keywords: Thermography · Data fusion · In-line quality assurance · Intrinsic production · Hybrid metal-CFRP structure

1 Introduction

The trend in lightweight design is moving towards multi material design to improve components regarding various requirements [1, 2]. Thus, the demand of hybrid lightweight material components increases, mainly driven by the automotive industry. The connection of metals and fiber reinforced polymers (FRPs) is still a challenge. Conventional connection concepts in downstream processes, such as bolt connections, are predominantly used [3]. However, drilled holes disrupt the flow of forces because

K. Dröder and T. Vietor (Eds.): *Technologies for economic and functional lightweight design*, Zukunftstechnologien für den multifunktionalen Leichtbau, pp. 241–252, 2021. https://doi.org/10.1007/978-3-662-62924-6_21

of damaged continuous fibers. Thus, friction and form fit, which can be realized in intrinsic production processes, are preferred for the connection of FRPs. An intrinsic hybridization of materials requires no additional process steps compared to downstream processes while improving the mechanical properties and reducing costs [2]. However, the intrinsic production of metal-CFRP hybrids is prone to internal (delaminations, folds, resin accumulations, fiber orientations) and external (geometric deviations, fiber orientation) manufacturing defects [4]. In-line quality control of hybrid components is challenging because no measurement technology exists which can determine all relevant features.

2 State of the Art

2.1 Laser Light Section

Laser light section sensors are widely used in industrial and scientific applications for length measurements. The contactless measurement and the flexibility regarding the geometry of the part qualify the sensor for in-line quality assurance. High geometrical deviations of the specified preform geometry shall be identified before the cost-intensive infiltration process [5]. Thus, the whole preform is to be detected as a 3D cloud of points (CoP). Shading effects, occurring because of complex 3D geometries, can be reduced through the usage of multiple lasers and detectors. The CoPs of the individual pairs of laser and detector need to be merged to acquire one dense CoP which contains the whole scanned geometry. System settings and parameters need to be optimized dependent on the specimen geometry for good measurement results [4].

2.2 Thermography

Thermography can be subdivided into active and passive thermography. The part under inspection is externally stimulated for the investigation of the thermal response. Active thermography proved to be useful in non-destructive testing (NDT) of composites and hybrid materials [6, 7]. Pulsed phase thermography (PPT) acquires phase information within a shorter time window (time efficiency) compared to modulated thermography. An image series is recorded and represents the temperature evolution of the specimen due to the thermal stimulus. Phase and amplitude information are obtained through the application of the discrete Fourier transform (DFT). Phase information is more robust against variations in surface emissivity and reflections from the environment. It has deeper probing capabilities than amplitude information and is influenced more sensitively by thermal material properties than geometrical features. [8] Thus, thermography is especially suited for the detection of internal errors, such as delaminations or pleats [9–11]. Schwarz et al. investigated defects in different CFRP layers and between interfaces of aluminum-CFRP hybrid structures using active thermography [12]. Thermography proved to be useful for the detection of critical delaminations in the insert region of hybrid components [13]. The structural alignment of carbon fibers is visible in thermographic images [12, 14].

2.3 Data Fusion of Laser Light Section and Thermography

Schäferling developed a data fusion procedure for thermographic images and CoP data [15]. The following section describes the procedure:

A thermographic camera and a laser light section system (LLSS) are mounted on a 3-axis-portal, giving high precision to the measurement. Initially, the raw data of the LLSS and the PPT are recorded sequentially. Acquired thermograms need to be corrected for image distortion errors through camera calibration. Cylinders serve as registration bodies. Thus, the external camera parameters are determined. The tool center point coordinates of the portal are recorded. As a result, the external camera parameters of further images can be derived from the translation of the portal and the determined position of the registration bodies. Ultimately, the data fusion procedure is performed through the mapping of points from the CoP in world coordinates to the image plane, using the determined camera model equation. The color information of the thermographic image is allocated to the respective 3D point. A flowchart of the procedure is given in Fig. 1.

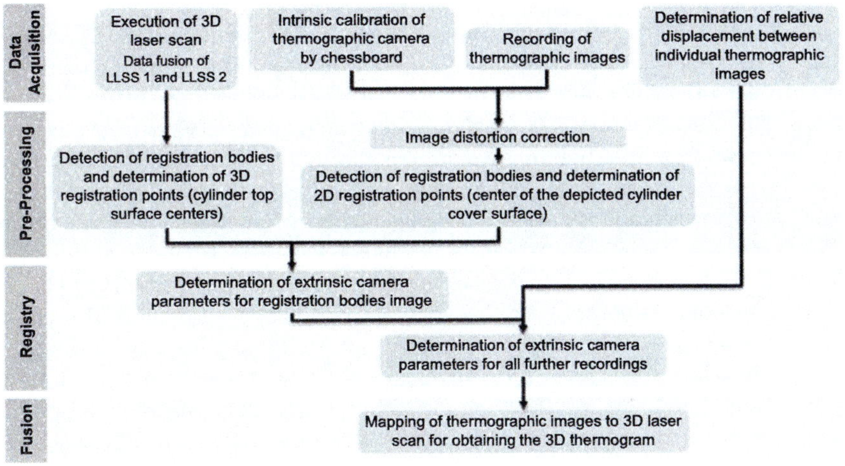

Fig. 1. Data fusion process of laser light section and active thermography [12]

The data fusion approach enables a 3D defect localization based on frequency-dependent depth information and CoPs. Thus, potential geometric deviations of the specimen under investigation are taken into account for the assessment of internal defects. The multi-sensor system, attached to a portal with a motion kinematics in x-, y- and z-direction, is displayed in Fig. 2.

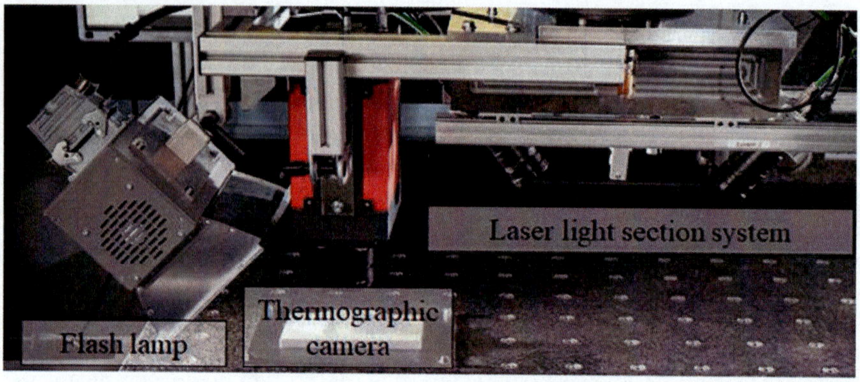

Fig. 2. Multi-sensor system including active thermography (left) and laser light section (right) on a 3-axis-portal (Sandra Göttisheimer, Karlsruhe Institute of Technology (KIT), 2017)

2.4 Fiber Orientation Analysis

Simulations of the filling behavior in the resin transfer molding (RTM) process, including load transmission elements, during the design phase allow predictions of the local fiber orientation and content [16]. Various methods exist for post-process measurements. Fiber orientations on planar preform geometries were measured using a 2D camera and optimal lightning conditions through a dome light [4]. They were extracted from the images through the usage of existing image processing algorithms [17, 18]. The investigations focused on planar geometries. Local fiber orientations and the insert placement with respect to the fibers was extracted for a planar hybrid specimen [4]. Large 3D geometries were investigated in the aircraft industry [17, 19]. Fujita and Nagano proposed a method for fiber orientation evaluation of sheet-like specimen based on in-plane thermal diffusivity and applied it to unidirectional and short fiber reinforced specimen [20]. Fernandes and Maldague used the anisotropic properties of carbon fibers by observing the local elliptical heat propagation by applying pulsed thermal ellipsometry (PTE), fused the thermal data with a CoP, and determined the fiber orientation [21]. Amjad et al. presented a method based on digital image correlation to track fibers in continuous fiber reinforced on microscale cross sections [22]. Fiber orientations were determined within a timespan of a few hours. Schöttl et al. quantified local fiber orientations based on volumetric images from computed tomography (CT) scans [23].

Overall, the quality assurance of hybrid parts is challenging and a single measurement technology cannot detect potential internal and external defects within the production takt time. The fiber orientation of continuous fibers in hybrid specimen can be impeded by inserts. Fiber orientation measurements of 3D specimen is still time consuming and challenging, especially for larger areas. The hybrid parts pose a challenge for CT measurements because of the hybrid material design itself, leading to different optical path length, and the high aspect ratio (thickness compared to length or width). A quick measurement technology is desirable for in-line measurements.

3 Hybrid Specimen Design

3.1 Demonstrator

The interfaces between the intrinsically connected components determine the performance of the hybrid composite. Thus, particular attention needs to be paid to the interfaces and defects need to be avoided. A shape-optimized metal insert was initially developed for a 2D geometry. The metal insert was overmolded with a thermoplastic, customizing the decisive boundary layer to the load case. Significant 3D stresses occur in 3D specimens under uniaxial tensile load. The specimen design, given in Fig. 3a), was chosen to achieve an even stress distribution [24]. The CFRP layers are illustrated in black, the thermoplastic in red and the metal insert in gray. An individual CFRP layer is approximately 0.5 mm thick, the metal insert 4 mm and the overmolded thermoplastic layer 2 mm on each side. Figure 3b) shows a manufactured component. The layer structure, consisting of four layers of CFRP and the insert laying in-between, is depicted in Fig. 3c).

Fig. 3. a) 3D design of the intrinsically manufactured metal-CFRP hybrid structure; b) manufactured specimen; c) numbered layer structure

3.2 Integration of Representative Defects

Hybridization presents a challenge, especially in the contact areas between the metal insert and the CFRP. Defects, such as delaminations, folds and, fiber misorientations, vary in shape, size and position and impede the performance and reliability of the produced part [4, 11, 25]. For the experimental investigation of defective specimens with the multi-sensor system, different specimens were infiltrated with purposely inserted defects representing common flaws which may occur during infiltration or curing. A flexible PTFE (Teflon, 10 mm wide and 0.127 mm thick) stripe was inserted into the laminate between layers no. (2) and (3). Pleats of the same width were created in layer no. (3). A global fiber misorientations of 10° was introduced to layer no. (1). All preform layers were laser cut to precisely achieve the desired global fiber orientation. The defects were integrated into separate preforms (cf. Fig. 4a) to c)), combinations of defects in one specimen were not produced. An exemplary detection of an integrated delamination in a 3D phasegram is presented in Fig. 4d). The scope of this paper is the investigation of a method for the extraction of fiber orientation information within the same measurement process as for delamination or pleat detection.

Fig. 4. Preform with integrated representative defects: a) delamination; b) 10° global fiber misorientation; c) pleat; d) Exemplarily detected delamination using a 3D phasegram [26]

4 Evaluation of 3d Fiber Orientation

4.1 Parameter Settings of the Multi-sensor System

An equidistant movement of the laser light section system over the specimen under investigation was necessary because of a limited depth of field in the millimeter range [27]. Because of the complex 3D shape of the specimen and a non-zero gradient of paths along the y-direction in the insert region, it was not possible to record a CoP of the specimen with an optimal distance between laser light section system and specimen. Figure 5 illustrates the height profile of the specimen with different potential paths (highlighted in cyan) along the specimen. The green path was chosen as a compromise in the insert region in order to minimize height differences along the path (blue) with respect to a plane parallel to the x–y-plane (displayed with red lines).

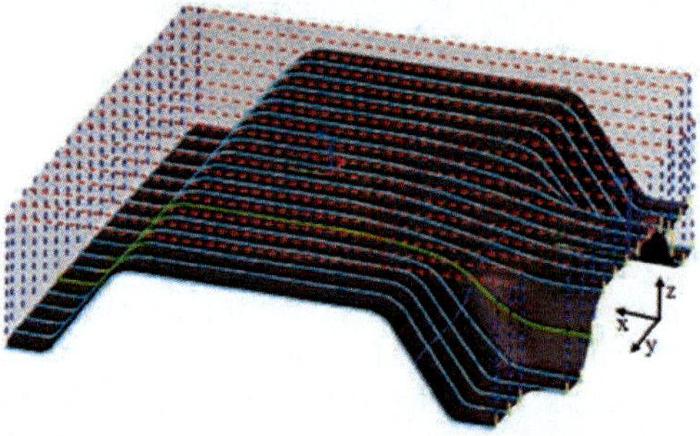

Fig. 5. Chosen height tracking of the portal according to the z-values of the green path

Optimal system settings needed to be identified. The triangulation angle of 30° was fixed for the multi-sensor system. The parameters laser angle, sidewall angle, camera threshold, laser intensity, exposure time were tested and evaluated regarding missing data points in the CoP as well as the root mean square error (RMSE) with respect to a reference CoP. The reference CoP was generated from the CAD file of the specimen. Optimal settings were identified to generate reliable COPs for finished hybrid metal-CFRP components. Table 1 gives the chosen parameter settings.

Additionally, a filter for eliminating outliers in the CoP compared to the CAD file was integrated to increase the robustness of the 3D mapping process of the thermographic image.

Table 1. Optimal parameters for the inspection of the finished hybrid specimen with the laser light section system

Parameter	Value
Laser angle	4°
Sidewall angle	10°
Camera threshold	150 W
Laser intensity	4 V
Exposure time	1200 μs

Different mounting configurations for the active thermography were tested. A flash mounting position of 160 mm below the portal attachment and a flash angle of 40° with respect to the horizontal axis led to the most homogeneous excitation of the specimen. This configuration resulted in a distance of 165 mm between flash lamp and specimen. The object distance to the camera was kept constant at 350 mm. Recording time was 60 s after excitation for the DFT, resulting in a minimum available frequency of 0.017 Hz in ampligrams and phasegrams.

4.2 Algorithm for Fiber Orientation Analysis

The intersections of the carbon fiber bundles in the plain woven fabric appeared darker than the surrounding material in thermographic images. These gray value differences built the foundation for the fiber tracking, using the commercial software MATLAB. The fiber tracking algorithm was applied on a 2D image, because of easier image processing. The specimen does not contain any undercuts, which allowed for an invertible mapping from the 3D CoP to a 2D gray value image. This 2D image was essentially a stitched image of the individually recorded thermographic images. The image contrast was improved using the function *imadjust()*. A local thresholding filter according to Bradley [28] was applied to obtain a 2D binary image. Here, the gray value of a pixel was set to zero if it is five percent darker than its eight neighboring pixels. Afterwards, the image was inverted because the darker fiber intersections in the plain weave represent the image regions to be detected. Singular pixels were deleted to increase robustness. The function *bwconncomp()* was applied to identify pixels with connected corners and edges. The center of every identified connected pixels was determined using the function *regionprops()*. The average distance between the identified centers was calculated, hereinafter called grid distance g. The starting points for the fiber bundle tracking needed to be determined. Dependent on the intended tracking direction (0° or 90° direction), the starting points were identified through their minimal u- or v-coordinate in the image coordinate system. The calculated grid distance g needed to be maintained between different starting points. The points aligned to the fiber bundle which should be tracked, were determined through a search region S_i (determined by the grid distance) in a distance $d=g$ along a search vector s_i to the previous point (initially the starting point). The search region S_i was a square with a side length of $g/4$. The search vector s_i was constantly updated and determined by the normed vector between the previous two center points along the identified fiber bundle. Initially, $s_{i=0}$ was set to the expected fiber orientation and constantly updated afterwards. If no center point was detected in S_i, the search was continued in S_{i+1} in the distance d along s_i from S_i. A flow chart is given in Fig. 6.

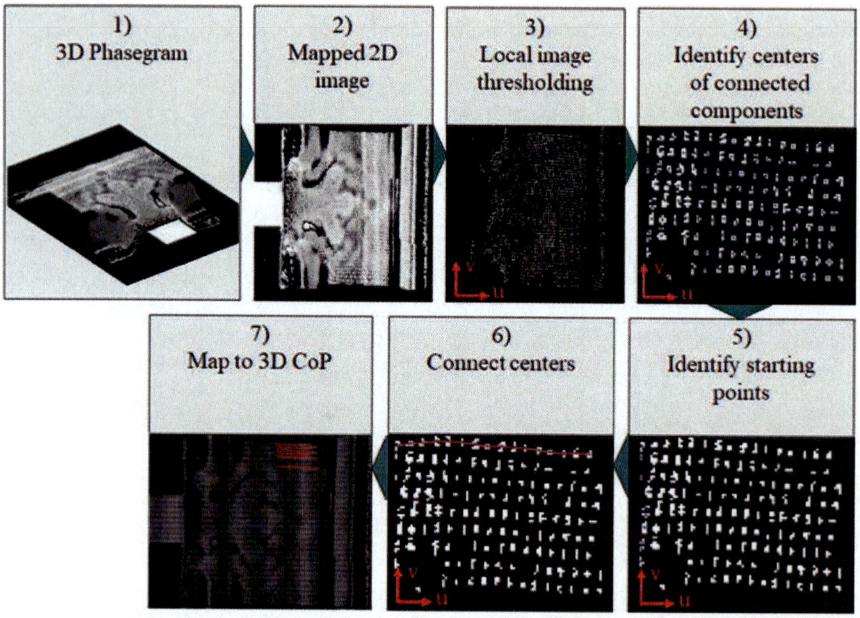

Fig. 6. Flow chart of fiber bundle tracking algorithm: The 3D phasegram (1) is mapped to a 2D representation (2) and binarized (3). The center of connected, binarized pixels is identified (4) and those with the lowest u-coordinate serve as starting points (5) for the subsequent connecting along the fiber direction (6). Results are finally mapped to 3D again (7).

4.3 Results

The fiber orientations were exemplarily evaluated in the first layer for two different specimens, with (specimen no. (1)) and without (specimen no. (2)) an intentionally inserted global fiber misorientation. However, fiber orientations can be identified in deeper layers with means of active thermography by changing recording parameters [29]. Two different regions of interest (ROI) were chosen for testing of the algorithm. The first one was located on the top horizontal surface, where no influences of the specimen curvature or the insert was expected. The second ROI, representing a more challenging region, was set to the region with the strongest part curvature, close to the metal insert. The ROIs are highlighted in Fig. 7a). Extracted fiber bundle orientations for both specimens in the first ROI are given in Fig. 7b) and c), with an average in-plane (x–y-plane) orientation of +6° and −2° with respect to the horizontal axis (x-axis). The 3D phasegram further allows the determination of orientations in curved areas. The changing orientation of the fiber bundle illustrated in Fig. 7d) is exemplarily highlighted at three points using a green arrow and the respective 3D vectors are given in Table 2.

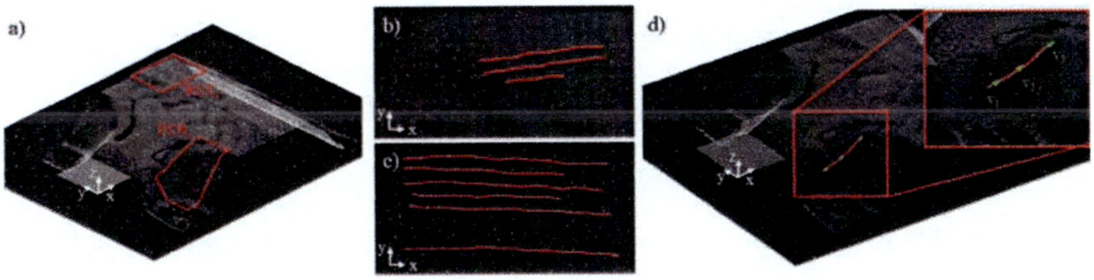

Fig. 7. a) 3D phasegram including ROIs; b) Tracked fibers for specimen no. (1) in ROI_2, +6° misorientation; c) Tracked fibers for specimen no. (2) in ROI_2, −2° misorientation; d) Tracked fibers in ROI_1 with strong curvature

Table 2. Evaluated fiber bundle orientation given for illustrated normed vectors in Fig. 7d)

Name	Euclidean Vector (x,y,z)	Resulting projected orientation (°)
v_1	(0.95, 0, 0.32)	0
v_2	(0.90, 0.13, 0.42)	8.5
v_3	(0.83, 0.12, 0.54)	8.5

Reasonable tracking results, as presented in Fig. 7, were so far not obtained in every case. Good results were achieved as long as the next fiber bundle intersection follows the direction of the current search vector s_i. Results including mistracked fibers are given in Fig. 8a). The search vector s_i failed to follow the local waviness in fiber bundle (6), highlighted within the green box in Fig. 8b). Furthermore, the 3D specimen geometry posed a significant challenge for the thermographic image acquisition, leading to an image which was not as strong in image contrast as of a 2D specimen is due to different energy input and object distances. The local image thresholding of images with poor contrast led to smaller and missed intersection points, exemplarily depicted in the binary image in Fig. 8b).

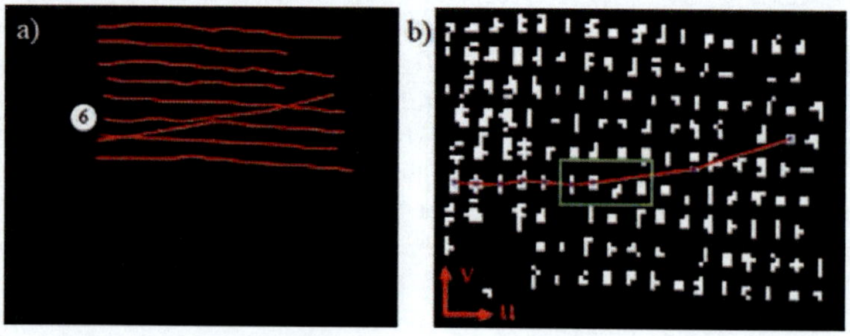

Fig. 8. a) Mistracked fibers in specimen no. (2); b) Starting points and identified neighbors during fiber tracking procedure of fiber no. 6

5 Conclusion and Outlook

An existing data fusion approach for laser light section and thermographic data could be successfully applied to a 3D metal-CFRP hybrid specimen. Parameter settings were identified to obtain reasonable results from the respective sensors. An algorithm was proposed to track fiber bundles in the 3D structure of the specimen. The algorithm allows to evaluate local fiber bundle orientations in 3D structures. Thus, thermography, as an established method for detecting delaminations, was applied to quantify fiber misorientations. A comparison between the intended fiber orientation in order to optimize the flow of force during the design process [30] and the actual produced fiber orientation in the part could be performed. However, the algorithm was prone to misdetections when the thermographic image had poor image contrast. Thermographic image data is to be improved in following research. It lacked of contrast in the contour regions of the insert. Thus, it could not be applied close to the edges of the metal insert. Using modulated thermography could be one possibility to improve image quality, but it is more time consuming. Additionally, the results presented need to be evaluated regarding their measurement uncertainty according to the Guide to the expression of uncertainty in measurement (GUM). Computed tomography (CT) can serve as a reference measurement.

Acknowledgements. The authors thank the German Research Foundation (DFG) for funding the collaborative research program "Priority program 1712".

References

1. Bader, B., Türck, E., Vietor, T.: Multi material design. A current overview of the used potential in automotive industry. In: Technologies for economical and functional lightweight design, pp. 3–13 (2019)
2. Fleischer, J., Nieschlag, J.: Introduction to CFRP-metal hybrids for lightweight structures. Prod. Eng. **12**(2), 109–111 (2018)
3. Schürmann, H.: Konstruieren mit Faser-Kunststoff-Verbunden. Springer, Berlin (2007)

4. Berger, D., Brabandt, D., Bakir, C., Hornung, T., Lanza, G., Summa, J., Schwarz, M., Herrmann, H.-G., Pohl, M., Stommel, M.: Effects of defects in series production of hybrid CFRP lightweight components – detection and evaluation of quality critical characteristics. Measurement **95**, 389–394 (2017)
5. Lanza, G., Brabandt, D.: Sustainable automated production of fiber reinforced plastics (FRP) through inline quality assurance. In: 10th Global Conference on Sustainable Manufacturing, pp. 123–127. Istanbul (2012)
6. Zaiß, M., Jank, M.-H., Netzelmann, U., Waschkies, T., Rabe, U., Herrmann, H.-G., Thompson, M., Lanza, G.: Use of thermography and ultrasound for the quality control of smc lightweight material reinforced by carbon fiber tapes. Procedia CIRP **62**, 33–38 (2017)
7. Vaara, P., Leinonen, J.: Technology survey on NDT of carbon-fiber composites. In: Publications of Kemi-Tornio University of Applied Sciences, Serie B, Reports 8 (2012)
8. Ibarra-Castanedo, C., Maldague, X.: Pulsed phase thermography reviewed. Quant. InfraRed Thermography J. **1**(1), 47–70 (2004)
9. Herrmann, H.-G., Schwarz, M., Summa, J., Grossmann, F.: Non destructive testing for evaluation of defects and interfaces in metal carbon fiber reinforced polymer hybrids. Int. J. Mater. Metall. Eng. **11**(8), 554–560 (2017)
10. Pohl, M., Stommel, M., Baumann, F., Berger, D., Lanza, G., Summa, J., Schwarz, M., Herrmann, H.-G.: Entwicklungsstrukturen Intrinsischer Hybride. Wt-Online **7**(8), 546–550 (2017)
11. Berger, D., Zaiß, M., Lanza, G., Summa, J., Schwarz, M., Herrmann, H.-G., Pohl, M., Günther, F., Stommel, M.: Predictive quality control of hybrid metal-CFRP components using information fusion. Prod. Eng. **12**, 161–172 (2018)
12. Schwarz, M., Schwarz, M., Herter, S., Herrmann, H.-G.: Nondestructive testing of a complex aluminium-CFRP hybrid structure with EMAT and thermography. J. Nondestr. Eval. **38**(1), 35 (2019)
13. Summa, J., Becker, M., Grossmann, F., Pohl, M., Stommel, M., Herrmann, H.-G.: Fracture analysis of a metal to CFRP hybrid with thermoplastic interlayers for interfacial stress relaxation using in situ thermography. Compos. Struct. **193**, 19–28 (2018)
14. Oswald-Tranta, B., Tuschl, C., Schledjewski, R.: Flash and inductive thermography for CFRP inspection. In: SPIE Proceedings 11004, pp. 21–30. Baltimore, Maryland, United States (2019)
15. Schäferling, M: Development of a data fusion-based multi-sensor system for hybrid sheet molding compound. Dissertation, Karlsruhe Institute of Technology (KIT), Karlsruhe (2019)
16. Magagnato, D., Seuffert, J., Bernath, A., Kärger, L., Henning, F.: Experimental and numerical study of the influence of integrated load transmission elements on filling behavior in resin transfer molding. Compos. Struct. **198**, 135–143 (2018)
17. Orth, A.: Entwicklung eines Bildverarbeitungssystems zur automatisierten Herstellung faserverstärkter Kunststoffstrukturen. Dissertation, RWTH Aachen University, Aachen (2008)
18. Mersmann, C.: Industrialisierende Machine-Vision-Integration im Faserverbundleichtbau. Dissertation, RWTH Aachen University, Aachen (2013)
19. Göttinger, M., Weimer, C., Miene, A.: Inline-Preformprozesskontrolle in der CFK-Fertigung. In: Deutscher Luft- und Raumfahrtkongress (2009)
20. Fujita, R., Nagano, H.: Novel fiber orientation evaluation method for CFRP/CFRTP based on measurement of anisotropic in-plane thermal diffusivity distribution. Compos. Sci. Technol. **140**, 116–122 (2017)

21. Fernandes, H.-C., Maldague, X.: Fiber orientation assessment in complex shaped parts reinforced with carbon fiber using infrared thermography. Quant. InfraRed Thermography J. **12**(1), 64–79 (2015)
22. Amjad, K., Christian, W.J.R., Dvurecenska, K., Chapman, M.G., Uchic, M.D., Przybyla, C.P., Patterson, E.A.: Computationally efficient method of tracking fibres in composite materials using digital image correlation. Compos. Part A Appl. Sci. Manuf. **129**, 105683 (2020)
23. Schöttl, L., Dörr, D., Pinter, P., Weidenmann, K.A., Elsner, P., Kärger, L.: A novel approach for segmenting and mapping of local fiber orientation of continuous fiber-reinforced composite laminates based on volumetric images. NDT & E Int. **110**, 102194 (2020)
24. Günther, F., Bretz, L., Jost, H., Schwarz, M., Stommel, M., Lanza, G., Herrmann, H.-G.: Herausforderungen im Entwicklungszyklus von einfachen (2D) zu komplexen (3D-) CFK-Hybridbauteilen mit thermoplastumspritzen Metalleinleger. In: 23rd Internationales Dresdner Leichtbausymposium, Dresden (2019)
25. Summa, J., Schwarz, M., Herrmann, H.-G.: Evaluating the severity of defects in a metal to CFRP hybrid-joint with in situ passive thermography damage monitoring. In: Proceedings 5th International Conference on Integrity, Reliability & Failure, pp. 117–126. Porto (2016)
26. Bretz, L., Günther, F., Jost, H., Schwarz, M., Kretzschmar, V., Pohl, M., Weiser, L., Haefner, B., Summa, J., Herrmann, H.-G., Stommel, M., Lanza, G.: Design and quality assurance of intrinsic hybrid metal-CFRP lightweight structures. In: Proceeding of 4th International Conference on Hybrid Materials and Structures, pp. 144–156. Karlsruhe (2020)
27. Brabandt, D., Lanza, G.: Data processing for an inline measurement of preforms in the CFRP-production. Procedia CIRP **33**, 269–274 (2015)
28. Bradley, D., Roth, G.: Adaptive thresholding using the integral image. J. Graph. Tools **12**(2), 13–21 (2007)
29. Schwarz, M.: Multimodale zerstörungsfreie Charakterisierung der Grenzflächen von Metall-CFK-Hybridstrukturen. Dissertation, Saarland University, Saarbrücken (2019)
30. Kretzschmar, V., Günther, F., Stommel, M., Scheuermann, G.: Tensor spines – a hyper-streamlines variant suitable for indefinite symmetric second-order tensors. In: IEEE Pacific Visualization Symposium (PacificVis), pp. 106–110 (2020)

Combined External and Internal Hydroforming Process for Aluminium Load Introduction Elements in Intrinsic Hybrid CFRP Contour Joints

Raik Grützner[1](✉), Veit Würfel[2], Roland Müller[1], and Maik Gude[2]

[1] Fraunhofer-Institute for Machine Tools and Forming Technology IWU, Reichenhainer Straße 88, 09126 Chemnitz, Germany
raik.gruetzner@iwu.fraunhofer.de

[2] Institute of Lightweight Engineering and Polymer Technology, Technische Universität Dresden, Holbeinstr. 3, 01307 Dresden, Germany

Abstract. This paper presents the intrinsic processing technology of a hybrid carbon fibre reinforced thermoplastic/aluminium hollow structure with the focus on a novel two-step hydroforming process of the aluminium load introduction element. An overview in the design of the hydroforming tooling with a combined external and internal pressure application is given. Concepts for sealing the multi-part forming tool for a pressure up to 350 MPa were investigated and validated. Using structurally relevant combinations of meso structures and macro contours, numerical sensitivity analyses were used to determine the relationships between geometric parameters of the shape elements, the material and the process parameters. The most important influences on the hydroforming operation and the geometric limits that can be achieved are presented.

Keywords: Hydroforming · Intrinsic hybrid · Contour joint · Multi-scale structuring

1 Introduction

In order to meet the requirements of future lightweight applications material specific benefits need to be combined and utilized by distinct multi-material-design approaches. Carbon fibre reinforced polymers (CFRP) enable to improve the performance and efficiency of lightweight constructions, due to their high specific mechanical material properties. Structures with highly orientated loading directions like rods, tubes or profiles benefit the most from the use of the anisotropic material behaviour of CFRP. Nevertheless, these structures need to be joined in widely used monolithic or hybrid differential designs. In classical joining techniques for fibre reinforced materials like bonding, bolting, riveting, welding or screwing, the joining operation is performed after the manufacture of each single component in a subsequent process step.

K. Dröder and T. Vietor (Eds.): *Technologies for economic and functional lightweight design*, Zukunftstechnologien für den multifunktionalen Leichtbau, pp. 253–266, 2021. https://doi.org/10.1007/978-3-662-62924-6_22

Many times this is linked to labour intensive operations like surface preparation and machining, which requires additional effort during the manufacturing process and may cause process-induced damage to the laminate.

In this work, the efficient intrinsic manufacture of a hybrid carbon fibre reinforced thermoplastic hollow structure with a contoured aluminium load introduction (LI) element is studied. Due to the possibility of an intrinsic manufacturing process of both materials post-processing operations like trimming or joining can be eliminated. An industrial scale production is pursued by the use of thermoplastic unidirectional (UD) tape material preformed in an automated braiding process and hydroformed aluminium LI elements with a multi-scale structured contour design. In order to design high performing and low cost LI elements a deeper understanding in the hydroforming process and the limits of achievable contouring are necessary. Numerical sensitivity analyses are used to investigate the influence of geometric parameters and process parameters on shaping. The aim is to identify potential and process limits for reproducible production.

Contour joints prove to be a good alternative to classical joining technologies for composite/metal hybrid hollow structures like tension–compression struts or drive-shafts. They are able to transfer high mechanical loads with high material utilization rates [1]. Hufenbach et al. studied the capabilities of a contour joint on the basis of an interlocking joining strategy in CFRP hydraulic cylinders in which the axial loads are transmitted from a metallic cylinder flange into the fibre reinforced tube structure [2]. Similar joining strategies by an interlocking principle are shown for composite drive shafts and torsional load applications in [3].

An integral bladder assisted moulding (IBM) process is presented in [1] in which forming and consolidation of a braided hybrid commingled yarn preform into a metallic functional element is done simultaneously in one single step. Subject of investigation was interlocking on the macro level, no additional strengthening by micro or meso structuring was studied.

In [4, 5] non rotational symmetric aluminium CFRP hybrid hollow structures were intrinsically manufactured and joined by rotational moulding. An inner elastic core served as a pressure intensifier to improve the impregnation of the fibres with the thermoset resin. Micro structuring via laser subtraction processes was studied in [6, 7] to either increase the interface area or create interlocking elements resulting in an increase in bonding strength. The advantages of intrinsically joined multi-scale structured aluminium LI elements and carbon fibre reinforced thermoplastic hollow structures were studied in [8, 9]. The intrinsic manufacturing and joining process consists of two separate process paths where two semi-finished parts are manufactured and then intrinsically joined in the IBM process. The multi-layer unidirectional thermoplastic tape-preform was produced by thermoplastic tape braiding [10]. Shaping of the metallic LI element was achieved by knurling or press forming on the meso scale and hydroforming was used to generate the desired macro contour for a form-fit connection. A combination of meso structuring and macro contouring improved the quasi-static pull-out strength of the joint compared to sole meso and macro contouring.

Most studies on metallic LI elements focused on the interface between the material partners or the manufacturing processes to intrinsically join the two material partners. Functionalization on the micro and meso scale was done in upstream processes which required additional effort. We have developed an efficient two-step hydroforming process to shape multi-scale metallic LI elements for tubular CFRP structures. Therefore, the benefits of multi-scale structuring can be combined with an efficient manufacturing method which is scalable for the use in industry. Focus in the generic investigation is a CFRP tension–compression strut with an aluminium Al 6060 LI element and braided unidirectional carbon fibre reinforced Polyamide (PA) 6 tapes – Celanese CF/PA6 Celstran® CFR-TP PA6 CF60-03.

2 Design of Multi-Scale Structured Load Introduction Elements

The design of a reliable CFRP/metal hybrid contour joint requires the information about several mechanical and physical aspects of the materials and manufacturing process used. The shape and structure of the LI element dominates the load bearing capability as well as the failure behaviour for the chosen material combination of aluminium and carbon fibre reinforced PA 6.

Tailoring of the functional areas in a fibre appropriate design is necessary to transfer the targeted high loads homogeneously from the metallic LI element into the hollow CFRP part. In this work surface structures on different scale levels from meso to macro are analysed. The discretization of the structuring levels are defined as smaller than 1000 µm for the meso level and greater than 1000 µm for the macro level. The multi-scale load transfer strategy consisting of macro-form closure with a graded design and meso structuring adapted to the braiding architecture is shown in Fig. 1.

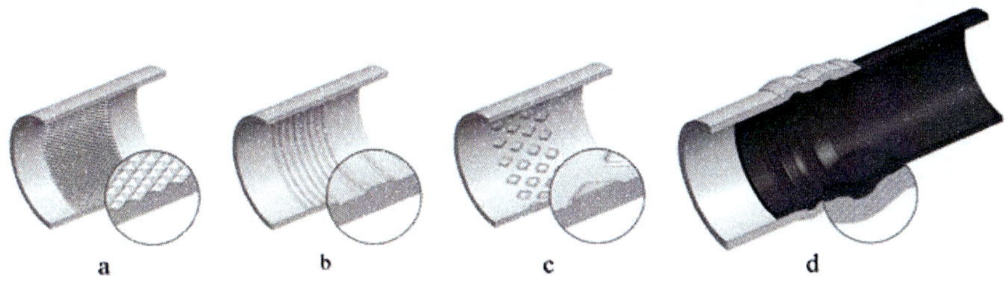

Fig. 1. Meso structure designs: Pyramid structure (a), groove structure (b), rhombus structure (c); Hybrid contour joint with groove meso structure and macro contour (d)

Three suitable designs of meso structures regarding the efficient manufacturability and mechanical behaviour with shapes either interlocking with single fibres or complete tape yarns were identified. One is a structure of a pyramidal design, which interacts with the single fibres and is produced by knurling in an upstream process step [10]. The rhombic shape is designed to interact with the whole tape. Both meso structure designs are adapted to the thermoplastic tape width of 3 mm and braiding angles of the $\pm 30^\circ{}_2/0^\circ{}_7/\pm 30^\circ{}_2$ laminate of the CFRP component. The design with circumferential grooves induces undulations on the meso scale in the laminate in which the undercuts and therefore the interlocking of the laminate is oriented in load direction.

The shape of the macro contour was optimized by the use of a parameterized Finite Element (FE) model of the hybrid tension–compression strut in a design of experiment (DoE) approach developed by Barfuß et al. The quasi-static pull-out force of the CFRP part is calculated in multiple FE simulations with linear elastic material behaviour and the CUNTZE [11] failure mode criterion is used to find the best design with the highest load bearing capability in the pareto optimization. The interface behaviour from meso to macro structuring is either modelled geometrically or with a cohesive zone description. The load bearing capability is increased with a higher amount of undercuts as well as by grading of the contour to achieve a more homogeneous load introduction. A design guideline taking the interface and manufacturing restrictions into account has been developed for future applications [12].

3 Two-Stage Hydroforming Process of Load Introduction Element

In order to manufacture metallic LI elements with a multi-scale structuring scalable for industrial application a novel two-phase hydroforming process has been developed. It consists of a combination of external and internal hydroforming to shape the desired meso and macro structures on the LI element. During the process, a tubular aluminium profile is inserted into a macro contoured outer tool together with the meso structured inner tool (Fig. 2). The high pressure fluid is orange in the figure.

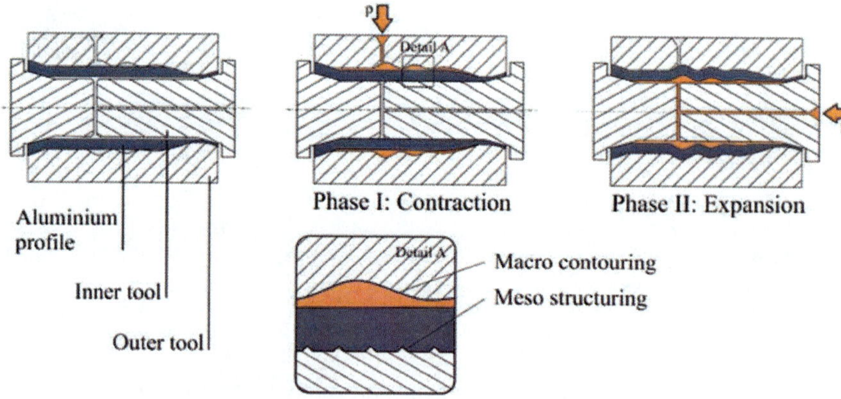

Fig. 2. Principle representation of the two-stage hydroforming process

First, external pressure is applied to the aluminium profile resulting in a diameter reduction by plastic deformation. Subsequently the shape of the meso structure on the inner tool is formed into the inner surface of the aluminium profile. In the following step the metal tube is expanded by internal pressure application in which the macro contour of the outer tool is formed. In this process sequence the meso structure stays intact and perpendicular to the macro contour. The combination of meso structuring and macro contouring in one process reduces manufacturing effort, shortens handling and cycle times and enables the LI element to be manufactured without the need for additional tools.

3.1 Process Simulation of Meso Structuring

To derive design guidelines for hydroformed metallic LI elements, numerical sensitivity analyses were carried out for the structuring designs on the meso level. The influence of geometrical parameters of the form elements and process parameters on the forming of the structuring was investigated and stable process windows were identified.

The **structuring with circumferential grooves** can be simplified as a 2D rotationally symmetric FE model. For the simulation of the hydroforming process the software DEFORM (SFTC) was used. The tube was modelled with elastic–plastic material behaviour and the inner tool as a rigid body. The required material properties of the aluminium material Al 6060 were determined in state T4 by tensile tests on tube samples at room temperature. The friction between tube and tool is described according to the shear friction model. Here the frictional shear stress is linked to the proportionality factor m in proportion to the shear yield point. Based on the experience with the simulation of solid forming processes, the friction factor was defined as $m = 0.3$ in all numerical analyses.

As shown in Fig. 3 on the left, the groove is described by the width W_{Gr}, the radius R_{Gr} and the depth D_{Gr}, whereby the depth was constant in the investigations with $D_{Gr} = 1.0$ mm. Other constant parameters are the tube radius $R_{iTube} = 22.5$ mm and the length of tube $L_{Tube} = 20.0$ mm. The width W_{Gr} and the radius R_{Gr} as geometry parameters, the wall thickness of the tube t_{Tube} and the yield stress k_f as material parameters, the radius of the tool R_{iTool} and the pressure p as process parameters were varied. Table 1 lists the variable parameters with the selected range of variation.

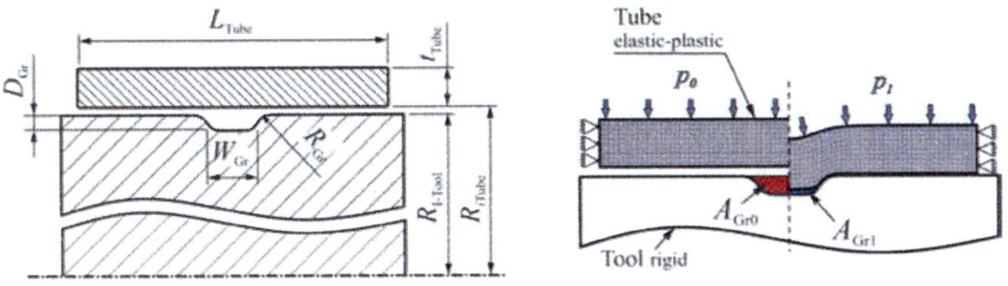

Fig. 3. Geometry parameters of groove (left), FE model with target value before and after hydroforming process (right)

Table 1. Design parameters and their levels for the DoE investigation.

Variable Parameters	*Unit*	*Steps*	*Minimum*	*Middle*	*Maximum*	*Interval*
Width of groove, W_{Gr}	mm	6	2.0	3.5	5.0	0.5
Radii of groove, R_{Gr}	mm	3	0.5	1.0	1.5	0.5
Tube wall thickness, t_{Tube}	mm	3	2.0	2.5	3.0	0.5
Radius inner tool, $R_{I\text{-}Tool}$	mm	3	21.5	22.0	22.5	0.5
Pressure, p	MPa	3	100	200	300	100
Flow stress factor, k_f	-	3	0.8	1.0	1.2	0.2

In order to efficiently perform sensitivity analysis for the large number of influencing parameters, a design-of-experiment (DoE) is required. Based on Latin Hypercube Sampling, 100 parameter sets with the lowest possible correlation of the parameters were defined. The target value is the form of groove F_{Gr}, which is calculated from the ratio of the maximum area of A_{Gr0} and the formed area A_{Gr1}.

$$F_{Gr} = (A_{Gr0} - A_{Gr1})/A_{Gr0} \tag{1}$$

A correlation analysis between the influencing parameters and the result variables is used to create the best possible correlation model (metamodel). In the used software OptiSLang it is listed as “Metamodel of optimal prognosis” (MoP). In the procedure to generate the MoP, an analysis of different linear and polynomial regressions is performed. The evaluation of the approximation quality of a metamodel is performed via the coefficient of prognosis (CoP), which estimates the variance that can be explained by the metamodel on the basis of cross-validations [13]. The CoP indicates the percentage of the determined result variable variation that can be described by the metamodel. The CoP(x) value of the individual parameters represents the percentage of the parameter variance in the result variable. In addition, a CoP(x) diagram is being developed to provide a summary representation of the influence of selected parameters on all result variables. It shows the relevance of a parameter to several result variables and the direction of action development of the result variables by the parameter change. A positive CoP(x) value means an increasing result variable value with increasing parameter value.

As a result of the sensitivity analysis for structuring with circumferential grooves, Fig. 4 on the left lists the individual influencing parameters in the CoP diagram. The strongest influence on the forming of the circumferential groove is shown by the parameters pressure and groove width. In contrast, the variation of the yield stress and the tube wall thickness causes only a slight change in the formation. The parameters radius of the inner tool and radii of the groove have no significant influence.

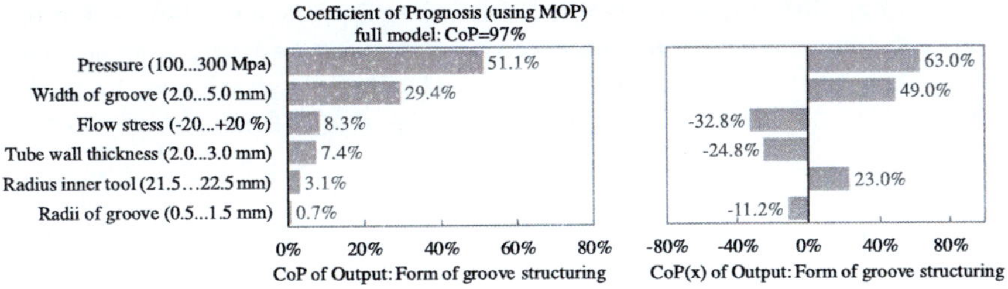

Fig. 4. Influence of parameter variations (left) with respective directions of action (right) on the forming of the circumferential groove

To evaluate the effect of the parameters, Fig. 4 on the right shows the CoP(x) values with the respective directions of action. An increase in pressure (CoP(x) of 63%) and an increase in groove width (CoP(x) of 49%) in the area under consideration has a positive effect on the forming of the circumferential groove. An increase in the yield stress (CoP(x) of −32.8%) and the tube wall thickness (CoP(x) of −24.8%) leads to a significant reduction in forming.

Figure 5 shows the form of the circumferential groove as a function of the main influencing parameters pressure and width of groove for the tube wall thicknesses investigated. All other parameters are assigned mean values.

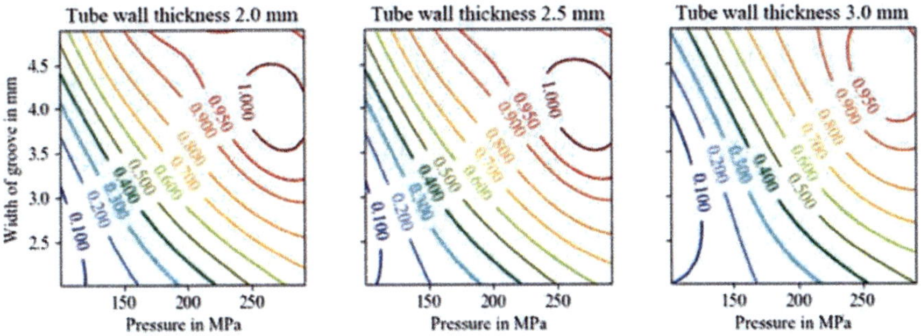

Fig. 5. Contour plot of the metamodels on the forming of the circumferential groove

For the tube wall thicknesses 2.0 mm and 2.5 mm only slight differences in the forming of the circumferential groove can be detected. Complete forming is achieved from a pressure of 250 MPa and a groove width of 3.5 mm. From a groove width of 4.5 mm the forming is slightly reduced again. Depending on the pressure, the aluminium material first comes into contact with the groove base. This support prevents the free forming and thus the radial material flow. A significantly higher pressure is required for further forming. In the investigated parameter space, a complete groove filling cannot be achieved with a tube wall thickness of 3.0 mm.

The **rhomb structuring** requires 3D modelling for the simulation of the forming process. In the numerical investigations a 10° segment with two rhombic quarters and symmetry boundary conditions was used. The shape of the rhomb is adapted to the 30° braided fibre angle of the preform. The definition of material and friction is identical to the modeling of structuring with circumferential grooves. In Fig. 6 on the left the FE model after external hydroforming (1) and after subsequent internal hydroforming (2) is displayed. With the aim of determining the design scope for hydroformed rhombic form elements, a sensitivity analysis of the influencing parameters on the target specific form of rhomb was carried out on the basis of a Latin Hypercube Sampling DoE. Besides the rhomb forming due to the external pressure, especially the back forming due to the internal hydroforming process was investigated. A metamodel was created for both process steps. The parameters were varied: Length of rhomb L_{Rhomb}, the external pressure p_E, the tube wall thickness t_{Tube}, the radii of rhomb R_{Rhomb} as well as the internal hydroforming tool diameter D_{IP} and the internal pressure p_I. In all variants the depth of the rhombic shape in the inner tool with $D_{Rhomb} = 1.0$ mm was constant.

As a result of the sensitivity analysis, the external pressure with a CoP of >70% on the overall model was determined as the most important parameter on the forming of the rhomb. This is followed by the rhomb length, the tube wall thickness and radii. Depending on the edge radius, the work hardening in the radii after forming in the external pressure hydroforming process counteracts a back forming of the rhombus in the internal pressure hydroforming process. The internal pressure and the tool diameter show a very small influence on the final shape of the rhombus within the investigated limits. Figure 6 on the right shows an example of the correlation between the length of the rhomb and its forming in the external pressure hydroforming process at different pressures for a tube wall thickness of 2.5 mm.

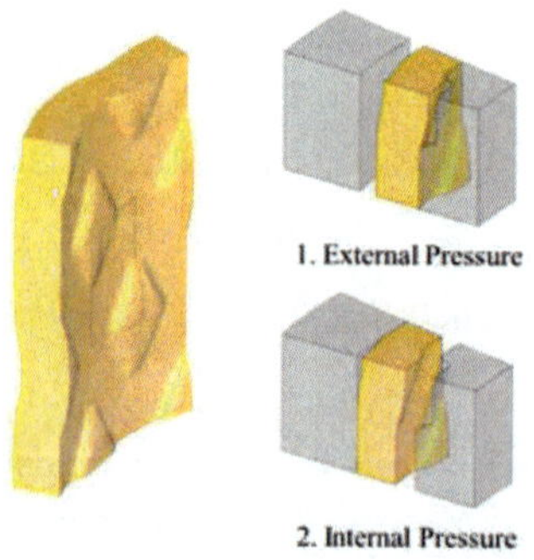

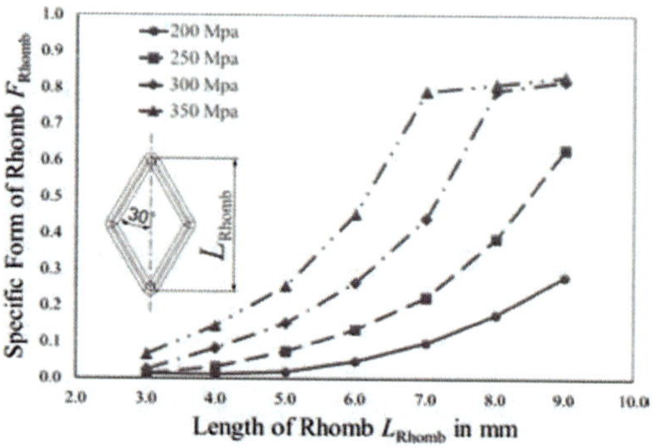

Fig. 6. FE quarter model of rhombic forming in the combined hydroforming process (left), Form of rhomb F_{Rhomb} in correlation to the length of rhomb L_{Rhomb} and external pressure p_E for a tube wall thickness $t_{Tube} = 2.5$ mm; $R_{Rhomb} = 0.5$ mm (right)

The forming quality increases exponentially with increasing length. The amount of the pressure determines the gradient of the curve. Even at a maximum pressure of 350 MPa, a complete filling of the tool cavity ($F_{Rhomb} = 1.0$) cannot be achieved. This is due to the already described contact of the aluminium material with the base of the rhombus and the material flow hindered by this support to form the radii of the tool wall. For the examined lengths no adequate shaping of the rhombs can be achieved up to a maximum pressure of 250 MPa. Only from a length of 8.0 mm and a pressure of 300 MPa is the rhombic cavity in the inner tool filled to 80%. At a pressure of 350 MPa, this filling level is already achieved at 7.0 mm length. By increasing the length of the rhomb and further increasing of the pressure, it is only possible to slightly increase the shaping of the rhombus.

3.2 Tool and Process Development

With the aim of an efficient production of the metallic load introduction elements with multi-scale structuring, an implementation of external and internal hydroforming in one tool was aimed at. Tool concepts known from literature for external hydroforming of tubes and profiles use a closed external tool [14]. With this concept, the removal of the formed workpiece takes place in axial direction and therefore only requires an axial sealing of the tool ends. However, for the realization of the macro contour in the hydroforming process the outer tool must be designed in two parts with a parting line in the closing direction of the tool for part removal. A major challenge is the sealing of the tool halves. To form the meso structures in the external hydroforming process, the parting plane must be sealed for forming pressures of up to 350 MPa. For this purpose, suitable sealing concepts based on metallic flat and round seals were developed, evaluated with regard to production and tested. Axial sealing is achieved, as is usual in hydroforming processes, by means of a sealing cone on the end caps of the axial cylinders. The tube ends are plastically formed to the inversely conical contour of the tool inserts. In the developed tool concept, the tightness in the external hydroforming process is guaranteed by surface pressure between the seal and the tool halves as well as between the seal and the tube shell surface in the axial transition area. For reasons of simple production of the sealing groove in the tool as well as the long service life of the seal, a round seal made of Cu-ETP R240 with a diameter of 3.5 mm was used. In Fig. 7 on the left, the lower half of the tool is positioned with a round seal and sealing caps. DIN ISO 3601-2 for determining the installation spaces for O-rings was used as the basis for designing the sealing groove. Due to the high tool stress caused by the superimposition of pressing force, axial force and forming pressure, the tools were designed numerically. As a result of the simulations, the depth of sealing groove and the radii in the transition area were adapted in such a way that local stress peaks are reduced and thus tool breakage can be safely avoided.

The combined hydroforming process requires two independently operating pressure intensifiers which are linked by the press control system. The supply of the medium during the external hydroforming process takes place via a high-pressure connection in the upper tool. In the internal hydroforming process the supply and pressurization is realized via a bore in the axial cylinder of the filling side. A hydraulic 3-column hydroforming press from company Dunkes GmbH with a maximum press force of 15,000 kN is used for the test.

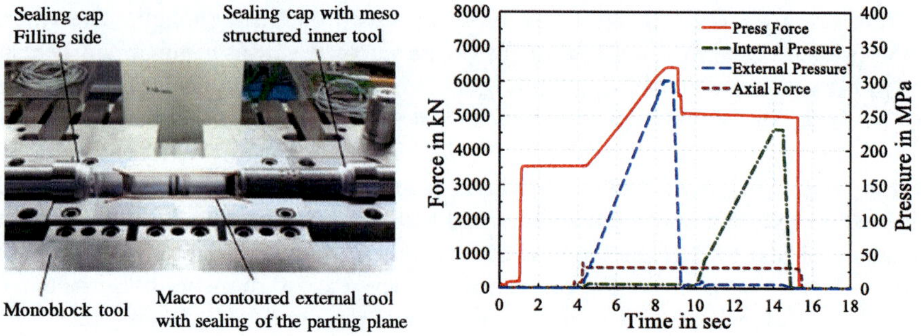

Fig. 7. Hydroforming tool (left), process diagram of the combined hydroforming process (right)

As an example, Fig. 7 on the right shows a process diagram of the combined external-internal pressure hydroforming process. In the first step the tool is closed and the basic press force of 3500 kN is applied. Then the axial cylinders are moved into position and seal the tube ends with a constant force of 700 kN. While the pressure intensifier 1 builds up the external pressure, the press force increases proportionally. After reaching the specified maximum pressure of 300 MPa, the external hydroforming process is terminated, the pressure is released and the press force is reduced to 5000 kN. Afterwards the pressure intensifier 2 for the hydroforming process applies an internal pressure of 230 MPa to shape the macro contour of the part. After the internal pressure is released, the axial cylinders are simultaneously retracted and the press is opened in the final step. As a result of the process development, cycle times of <16 s were achieved for the combined external-internal pressure hydroforming process in one tool.

3.3 Experimental Validation

On the basis of the process limitations for forming the meso structures determined by means of sensitivity analyses, the load introduction element was designed in relation to the macro contour [12]. For the transmission of high loads a demonstrator geometry with two macro undercuts was selected, taking into account the strength and ductility of aluminium. The combined hydroforming process is implemented experimentally on aluminium tube specimens with an outer diameter of 50 mm and a wall thickness of 2.5 mm.

To increase the undercut, an axial and circumferential distribution of the meso form elements on the flanks of the macro contour was carried out. From the given width of a macro undercut of 13.8 mm, the circumferential groove with symmetrical distribution in the axial direction results in a width $W_{Gr} = 3.45$ mm at an average distance of $D_m = 6.9$ mm. This results in four circumferential undercuts over the entire length of the joining area. The grooves are provided with an edge radius of $R_{Gr} = 1.0$ mm.

For a design of the rhombus structuring, a total of 72 rhombuses are geometrically arranged in the joining area. In the axial direction four rows with 18 rhombuses each, offset by 20°, are distributed around the circumference. The length of the individual rhombs and the average row spacing is 6.9 mm. The edge radii of the rhomb are $R_{\mathrm{Rhomb}} = 0.5$ mm.

FE simulations were carried out for the tool and process design for both structuring variants of meso and macro level. The forming production of the multi-scale structuring of the load introduction element in the combined hydroforming process was investigated with regard to occurring high local strains. The experimental validation of the simulation models is done by comparing measured and calculated formations of the structures on meso and meso macro level. Despite the high local plastic strain, no material failure could be observed in the hydroforming processes. Thereby structural heights of the meso structure of 0.5 mm were realized and combined with the macro contour height of 2.0 mm. In Fig. 8 the cross sections of the experiment and the simulation are displayed for comparison.

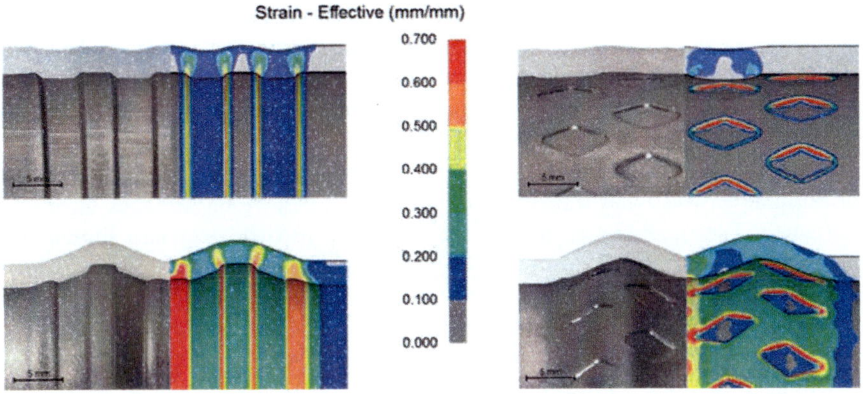

Fig. 8. Comparison of experiment and FE model with plastic strain for groove and rhombus meso structure (top) and with superimposed macro contour (bottom)

The high accuracy of the simulation models is evident. Remaining small differences result, among other things, from neglecting the tool elasticity, material anisotropy and from the simplified modeling of friction conditions. Thus it could be proven that the chosen modeling strategy allows a reliable prediction of the expected structural shape for the different geometry-, material- and process parameters.

4 Conclusion

The design and manufacturing approach for multi-scale structured metal LI elements for tubular composite structures is presented. In the two-step hydroforming process aluminium LI elements can be manufactured and functionalized with an inner meso structure and macro contour less than 20 s without the need of an additional upstream process like knurling, press forming or laser subtraction.

In numerical sensitivity analyses the influences of the manufacturing parameters have been studied to identify the process limitations in a DoE approach. Pressures of up to 350 MPa are required to form the desired circumferential groove and rhombic meso structures. The smaller the dimensions of the meso structures are the more difficult the forming is and the higher is the required external pressure in the hydroforming process. For structures of smaller scale like the pyramidal design upstream processes like knurling still remain necessary.

The very high pressures pose a significant challenge for the sealing of the parting plane in the required hydroforming tooling. Numerically designed grooves in the tooling in combination with round copper seals proved to be a process stable solution. Good agreements of simulation and experiment in the forming of the meso structures and macro contour were found after measuring the formed LI elements. Therefore, accurate predictions of the forming capabilities by hydroforming can be made and the development process of hydroformed and multi-scale structured parts can be accelerated.

Further investigations in the hardening on the aluminium structure during the hydroforming process and the resulting influence on the joint strength have to be investigated. The moulding of the CFRP part in the IBM process into similar meso structured LI elements, manufactured by press forming, has been studied [10]. The forming into the hydroformed meso structured LI elements as well as the resulting joint strength is subject of future investigation.

The developed two-step hydroforming process for the manufacture of LI elements for hybrid carbon fibre reinforced thermoplastic hollow structures which can be intrinsically joined enables an industry scale production due to the highly automated and fast processes. Thereby new fields of applications are accessible. In Fig. 9 a generic node structure of hybrid CFRP hollow profiles and an aluminium cast node is shown demonstrating the possibilities of hybrid components as well as their joining operations in multi-material lightweight constructions.

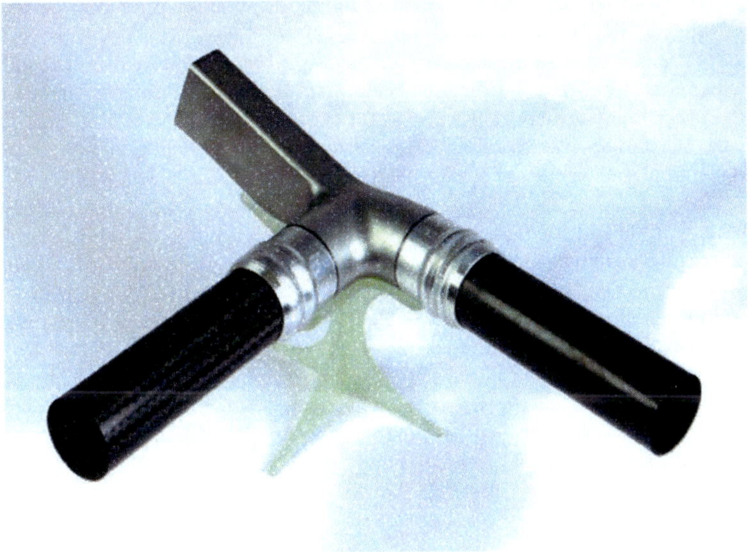

Fig. 9. Application example: Generic lightweight node structure with hybrid components

Acknowledgements. The authors gratefully acknowledge the funding of this work by the German Research Foundation, DFG, within the priority programme 1712: Intrinsic hybrid composites for lightweight structures – basics of manufacturing, characterization and design.

References

1. Hufenbach, W., Gude, M., Helms, O. et al.: Novel lightweight aircraft components in multi-material design with adapted load-introduction systems. In: European Conference on Materials and Structures in Aerospace (EUCOMAS), Berlin, 26–27 May 2008
2. Hufenbach, W., Kroll, L., Gude, M., et al.: Integrated load introduction system for composite lightweight structures in aerospace applications. In: 27th SAMPE Europe international conference, Paris, 27–29 March 2006
3. Gude, M., Lenz, F., Gruhl, A., et al.: Design and automated manufacturing of profiled composite driveshafts. Sci. Eng. Compos. Mater. **22**, 187–197 (2014)
4. Fleischer, J., Koch, S., Coutandin, S.: Manufacturing of polygon fiber reinforced plastic profiles by rotational molding and intrinsic hybridization. Prod. Eng. **9**, 317–328 (2015)
5. Nieschlag, J., Ruhland, P., Daubner, S., et al.: Finite element optimisation for rotational moulding with a core to manufacture intrinsic hybrid FRP metal pipes. Prod. Eng. **12**, 239–247 (2018)
6. Wang, Z., Bobbert, M., Dammann, C., et al.: Influences of interface and surface pre-treatment on the mechanical properties of metal-CFRP hybrid structures manufactured by resin transfer moulding. Int. J. Automot. Compos. **2**(3/4), 272–298 (2016)
7. Baburaji, E.G., et al.: Enhancement of adhesive joint strength by laser surface modification. Int. J. Adhes. Adhes. **27**, 268–276 (2007)

8. Barfuss, D., Grützner, R., Hirsch, F., et al.: Multi-scale structuring for thermoplastic-metal contour joints of hollow profiles. Prod. Eng. **12**, 229–238 (2018)
9. Würfel, V., Grützner, R., Hirsch, F., et al.: Hybrid fibre reinforced thermoplastic hollow structures with a multi-scale structured metal load introduction element. In: Proceedings Hybrid Materials and Structures, Karlsruhe, 28–29 April 2020
10. Garthaus, C., Barfuss, D., Witschel, B., et al.: Tape braiding: high-performance fibre-reinforced thermoplastic profile structures. JEC Compos. Mag. **96**, 62–64 (2015)
11. Cuntze, R.: The failure mode concept – a new comprehensive 3D-strenght analysis concept for any brittle and ductile behaving material. In: Proceedings of European Conferences of Spacecraft Structures. Materials and Mechanical Testing, pp. 269–287, Braunschweig (1998)
12. Barfuss, D., Garthaus, C., Gude, M., et al.: Design of multi-scale structured Al-CF/PA6 contour joints. Int. J. Automot. Compos. **2**(3/4) (2016)
13. Most, T., Will, J.: Sensitivity analysis using the metamodel of optimal prognosis. In: Weimarer Optimization and Stochastic Days, Weimar, 24–25 November 2011
14. Siegert, K., Guel-López, P.: Hydroforming of tubes with external high pressure. In: International Conference Hydroforming, Fellbach, 12–13 October 1999

Mesoscale Surface Structures in CFRP-Metal-Hybrid Joints – Aspects of Design and Manufacturing

Fabian Günther[1(✉)] and Markus Stommel[2]

[1] Chair of Plastics Technology, TU Dortmund University, Leonhard-Euler-Str. 5, 44227 Dortmund, Germany
Fabian.guenther@tu-dortmund.de
[2] Leibniz Institute of Polymer Research, Hohe Str. 6, 01069 Dresden, Germany
stommel@ipfdd.de

Abstract. In light weight construction, the concept of multi-material design allows to benefit from the advantageous properties of different materials. Thereby the joining technique for dissimilar materials must always be taken into account. The joining of carbon fibre reinforced polymers and metals is an example where innovative approaches complement or replace the usual joining technology. A special approach is a metal insert overmoulded with a multifunctional polymer, whereby the surface structures improve the intrinsic properties of the compound. These inserts are produced in an injection moulding process.

In this contribution, the forced demouldability of undercuts from mesoscale surface structures is investigated for reinforced and non-reinforced Polyphthalamid (PPA). Simulations predict very high deformations rates for the forced demoulding of the undercuts. In experimental tests, the demouldability of four parts with different angles of inclination (0°, 15°, 30° and 45°) and pins with height of 0.5 mm, 1 mm and 2 mm are examined. Thereby the influence of injection moulding parameters such as melt and mould temperatures is studied by a full factorial Design of Experiments (DoE) analysis. The results show that even undercuts from a 45° inclination angle can be demoulded under certain conditions. The statically evaluation points out that the material reinforcement as well as mould and polymer temperature are the main factors for demouldablitiy of pin undercuts. Finally, an overview of the demoulability of pins is given as a function of the material, pin heights and part inclination angle.

Keywords: Injection moulding · Pins · Undercuts · Force demouldability · DoE

K. Dröder and T. Vietor (Eds.): *Technologies for economic and functional lightweight design*, Zukunftstechnologien für den multifunktionalen Leichtbau, pp. 267–278, 2021. https://doi.org/10.1007/978-3-662-62924-6_23

1 Introduction

Lightweight design requires an application and function-oriented material combination. Combining different materials involves the issue of bonding dissimilar materials. The connection approaches differ according to the application cases, component geometries and load cases. Combinations of different joining concepts of force fit, form fit and adhesion are widely used. For carbon fibre reinforced polymers (CFRP)-metal parts, the issue is already known and different approaches besides common bolt joints or adhesion are invented. For example, metallic inserts that join intrinsically to the CFRP during the Resin Transfer Moulding consolidation step of the composite [1–4]. Intrinsic means that the connection is gained within the curing process of the composite. This brings two advantages: on the one hand, there is no need of further process steps for joining. On the other hand, the connections do not damage any fibres as it is the case for afterwards assembled bolt joints. A special hybrid connection is shown in Fig. 1 and introduced in [2–4]. This hybrid connection consists of a thermoplastic overmoulded metal inserts. The thermoplastic jacket assumes different functions, as easy and cost efficient manufacturing for large-scale production, protection against corrosion, damping and reduction of the jump in stiffness. The connection combines different joining mechanisms as adhesion and form-fit. From the mechanical perspective, the form-fit can be used to locally improve the bonding strength at different scale levels. On the macro scale, arm-like rows are designed in the way that load is optimal transmitted to the part [5]. The form-fit can be achieved not only on macroscale through structural undercuts (cm) and on microscale through surface roughness (μm), but also through a structured surface in mesoscale form (mm). Mesoscale structures are introduced in [6, 7] as Integral Micro Mechanical Interlocking (IMMIs) or metallic pins and further structures. Mostly, the production of such structures is costly, demanding and difficult because of individual metal welding processes. On the contrary, the thermoplastic overmoulding of metallic inserts allows a high production volume combined with a great freedom in shape and surface design to gain fibre-friendly and load optimized inserts shapes. However, the freedom of design in the injection moulding process is limited. The positioning of pins on the differently inclined, draping-friendly insert surfaces can lead to very distinct undercuts in the injection mould. Design guidelines for the design of the component and mould design have been elaborated many times [8–11]. Side action or moving cores are used to gain necessary undercuts in injection moulding part if force demoulding is not possible. However, forced demoulding is the only reasonable mean of production for the pins due to the complexity and disproportionate tooling costs.

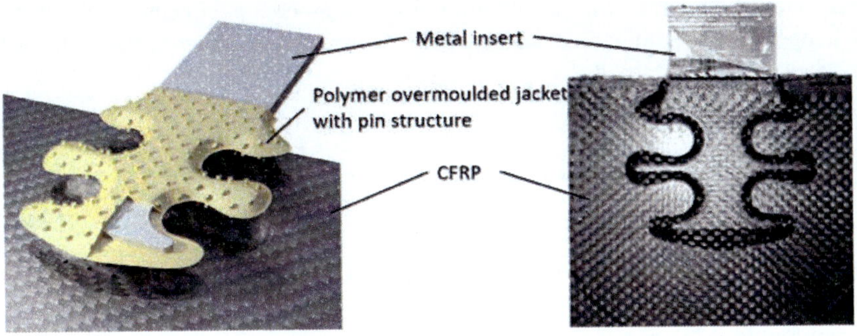

Fig. 1. Metal-CFRP-Joining technique with thermoplastic overmoulded jacket and pin structure

2 State of the Art

The insert shape of the hybrid connections depends on the complexity of the composite part. Hence, the inclinations of the insert surface to the parting line of the mould can be at least up to 45°. The positioning of pins at these surfaces cause distinct undercuts in the mould which can be challenging for the demoulding. Recent research in demoulding processes focuses mainly on micro and nano structures from injection moulding and hot embossing processes [12–14]. Here, demoulding forces introduced by friction and stress because of shrinkage or different manufacturing technologies are examined in different studies. Milling effect influences are studied for similar straight holes (pillars in the mould) by Masato et al. [15].

Forced demoulding of mesoscale undercuts is used for example to eject parts with threads as shown in [8] whereby a ratio of wall thickness to diameter is mentioned. However, the strip off of threads is hard to compare to pin structures. Moreover, the material properties have a significant effect on the deformation and demoulding properties. In [16] maximum undercuts for different materials are listed as short-term load for clip-on or snap-on connection. The limit for Polyamide (PA) is mentioned as smaller than 2%. Further in [8] for PA66 (Zytel® nylon), for example, limits of 6–10% undercuts and for reinforced polymers 1–2% undercuts depending on the geometry and mould temperatures are specified at which forced demoulding is possible. Otherwise forced demoulding cause damages as mentioned in [9] due to the disturbed demoulding movement by undercuts. In addition, shrinkage and high mould temperature can be a reason for the increase of demould forces. Another possible damage effect because of pin structure in the injection moulding process can be the diesel effect [9]. This causes burning marks on the polymer. The diesel effect emerges by trapped and compressed air in blind holes of the mould. The high process pressures heat the air to a degree at which combustible gases leaving the polymer melt self-ignite.

In this work, finite element analysis (FEA) is used to examine the deformation rates during the ejection of the undercuts under certain inclinations to get a theoretical overview of the expected deformations due to the pin height. With experimental tests the effects through pin structures on the polymer surface in the injection moulding process are further studied in detail with respect to process parameter as polymer and tool temperature, cooling time and packaging pressure. Moreover, the influence of reinforced and unreinforced material is examined.

3 Simulation of Forced Demoulding Deformations Through Undercut

FEA is used to gain an overview of geometric based deformation during the ejection phase of parts with surface structure undercuts in the mesoscale size. The issue is simplified to a 2D model where the maximum cross section of the mesoscale pins are represented. Different inclination of 15°, 30° and 45° with pins of 0.5 mm, 1 mm and 2 mm length and 1.5 mm in diameter each are implemented as shown in Fig. 2. Between the pins and the mould, a frictionless hard contact is assumed since only deformations are considered and the resulting stresses are not subject to the analysis. In the region of interest, a mesh refining is used with shell elements CPS4/CPS3.

Table 1 shows the implemented material data of the polymer pin and the metal mould counterpart which are implemented as linear elastic. The simulation has also been run with the mould as a rigid body with similar results. As boundary conditions, the rear side of the mould is fixed and the movement of the polymer part is restricted to the ejection direction (no lateral movement to x-direction see Fig. 2). The load is considered as a displacement of 3 mm into the ejection direction in negative y-direction.

Table 1. FEA linear elastic material parameter: polymer and metal

Material	E-modulus [MPa]	Poission's ratio
Polymer	3.400	0.4
Metal	210.000	0.28

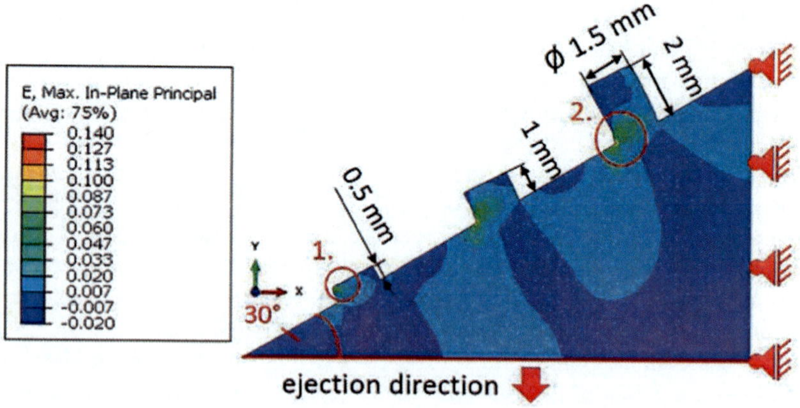

Fig. 2. Maximum principal strain in a demoulding part with 30° inclination angle and three different pin heights at 0.5 mm ejection displacement

Figure 2 shows exemplarily the resulting strain (E, Max.) in the polymer part. There are two main areas which get the highest loads (marked with red circles in Fig. 2). The upper edge has large load magnitudes when it comes to the final stage of ejection. Here, the load can lead to plastic deformations and visible damage. The second area at the pin bottom is the most loaded part of the mesoscale pin undercuts. This area risks that the pin will be plastically deformed and stretched or even broken off during ejection. Table 2 lists the occurring strain rates of the different examined structures. For the 15° angle, the deformations at the pin bottom seems to be moderate at least for small pins. The 2 mm pin shows already maximum strain values of more than 14% which is above a certain theoretical material limits for force demoulding as mentioned in [8]. For the 30° angle the strain is already much higher even for the small pins with more than 16%. The simulations with 1 mm and 2 mm pins aborted after 0,85 mm ejection due to numerical issues. The abortions are caused by the imposed boundary conditions of displacement and the node intersection at the contact faces of the mould and part. But even after only 0,85 mm ejection, the decrease in strain is already high which announces challenging circumstances for the experimental test. For an angle of 45°, large deformation are expected which only very soft and yielding materials can achieve. In the experiments, the viscoelastic material behaviour of the polymer at higher temperatures and the combination with the process parameters are interesting to examine during the demoulding process.

Table 2. Strain results for different inclination angles and pin heights from FE simulation

		Pin heights		
		0,5 mm	1 mm	2 mm
Inclination angle	15°	2%	7,5%	>14%
	30°	>16%	>25%*	>25%*
	45°	>23% **	>34%***	>34%****

simulation aborted at * 0,85 mm, ** 0,33 mm , *** 0,44 mm , **** 0,46 mm ejection

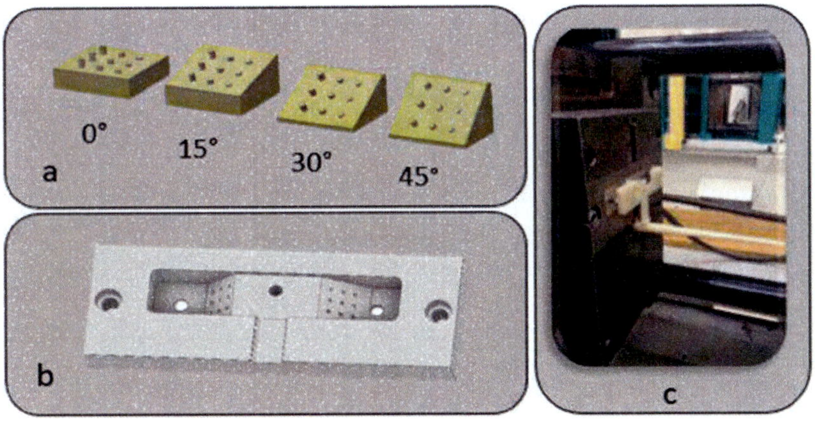

Fig. 3. a) Parts with different inclination angles, b) mould insert for 30°- and 45°-samples and c) moulded part in the injection moulding machine

4 Experimental Ejection Tests

For the experimental demoulding tests, four parts with different inclinations (0°, 15°, 30° and 45°) towards the ejection direction are designed. Each part has three times three pins of a diameter of 1.5 mm and with the length of 0.5 mm, 1 mm, and 2 mm on the surfaces (cf. Fig. 3).

When designing the moulds, there are various difficulties to consider. If each part is manufactured individually, tool costs and production time are high. If all parts are manufactured in one mould, the entire tool may be blocked if parts are difficult to remove from the mould. If several cavities are used, independent demoulding must be possible so that the components do not influence each other during demoulding. Thin runners that break off in case of overload can guarantee the unabstracted demoulding of other parts. At least, sufficient ejector pins must be provided to ensure uniform demoulding as required.

Two moulds with two cavities each were produced for the tests (cf. Fig. 3). The 0°- and 15°-parts were combined, which have a lower potential for ejection difficulties. Another mould is made for the 30°- and 45°-parts as shown in Fig. 3. The test is run on a commercial Arburg S270 with an external tempering of the mould. The tests are run with a Polyphthalamid (PPA) with and without short fibre-reinforcement. The chosen material, Vestamid® M1000 and M1033 with 30% glass fibres (PPAGF30), is the same as used in the project [3, 17]. While testing reinforced and non-reinforced materials, the influence of the stiffening of the material on the production of the pins is investigated. In Addition, four main injection moulding process parameters are varied to examine if there exist an effect of injection moulding on the pin demoulding. These factors are polymer melt and mould temperature, cooling time and packaging pressure (cf. Table 3). Considering the material variation and the different mould cavities a full factorial DoE is run with 128 different parameter settings. The influence on the undercut demoulding of the parameters A–F from Table 3 is analyzed. For the injection moulding tests, the parameters of a stable process were determined first, see Table 3. From this, appropriate upper and lower limits for the parameter variation are determined, within which the production of the pins or the influence on them can be verified. Different damaged stages have been classified for the DoE analysis as the quality criteria κ shown in Eq. (1) with the factor φ (cf. Table 4) and index i for the pins one to nine.

$$\kappa = \sum\nolimits_{i=1}^{9} \varphi \cdot Pin_i \tag{1}$$

In order to evaluate the statistical test procedure, the state of each of the nine pins of each sample is determined by visual inspection and multiplied by the factor given in Table 4. Thus, a valuation factor between 0 and 45 is output after a visual inspection. Examples of damaged samples can be seen in Fig. 4.

Table 3. Injection moulding process parameter

	Process parameter	Low	Stable process	High
A	Material	PPAGF30	–	PPA
B	Inclination angle [°]	0 15	30	45
C	Polymer temp. [°C]	340	345	350
D	Mould temp. [°C]	180	190	200
E	Cooling time [s]	50	55	60
F	Packing pressure [bar]	600	800	1000

Table 4. Damage status of pins after demoulding

Factor φ	Damage status
0	No damage
1	Slight damage to the upper pin edge
2	Slight bending of the pin
3	Bending and damage to the upper edge
4	Bending and slight damage at the bottom of the pin
5	Broken pin

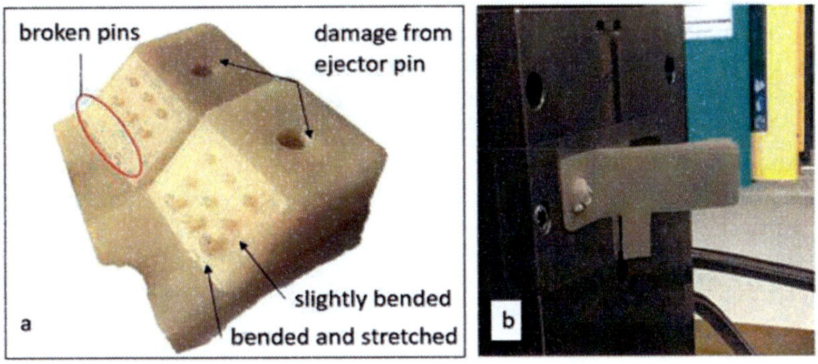

Fig. 4. a) Exemplary damages on 45°-samples, b) undemoulded, damaged sample in the mould

5 Results and Evaluation

The evaluation of the injection moulding tests for the 0°- and 15°-parts by means of visual inspection shows a demoulding without difficulties for all parameter variations. For each parameter set, up to three samples are moulded. In the most cases, the results are very similar. If not, the worst damage state per set is taken into account. Both the demoulding of the 2 mm pins and the undercut of the pins on the 15°-part are uncritical. This applies to the PPA as well as to the PPAGF30, so that here no influence of the injection moulding parameters can be recorded, neither negative nor positive.

The results of the tests with 30° inclination for PPA show that pins can be produced with almost all settings and this does not lead to ejection failure of the components. However, small deformations to the long pins are evident in some cases. This is the case at high temperatures for the polymer melt and mould. Then the polymer is very easily plastically deformable at the time of ejection. This leads to a stretching and bending of the pins. Furthermore, the upper edge of middle and long pins are deformed in some cases. For the reinforced samples, the demoulding comes along with more difficulties. While the short pins are still demouldable, middle and long pins show deformation and further damages for certain process parameters. At least a few samples have been undemouldable. Again, high melt and mould temperature have been the reason for these problems.

Contrary to what is expected for the 45°-parts due to the large deformation of the pins, which occurs in the simulations, these are also manufacturable. This applies to the tests with unreinforced PPA at low temperatures of the polymer melt. At high melt temperatures and in combination with high mould temperatures, demoulding is critical for the 45°-parts. While for some parameter configurations only slight deformations of the 1 and 2 mm pins are involved, the part is stuck in the mould at high mould temperatures and short cooling times. Figure 4b) shows a part in the mould penetrate by the ejector pins. This shows how high the demoulding forces are due to the undercut. The high temperatures do not cause the pins to fail, but the plastic part itself is deformed by the high force that the ejector pin exerts on the component. The ejector pin pierced through the component instead of ejecting it. This shows that a sufficiently long cooling time, depending on the mass and mould temperature, is necessary to apply the required ejector force for forced demoulding of the pins without damaging the part. For the tests with a 45° angle and PPAGF30, no parameter configuration could be determined within the given scope that would allow non-destructive demoulding. Since demoulding was always destructively, no differentiation can be made between the pin lengths in the 45° tests. It is therefore possible that 0.5 mm pins could be demoulded under certain conditions.

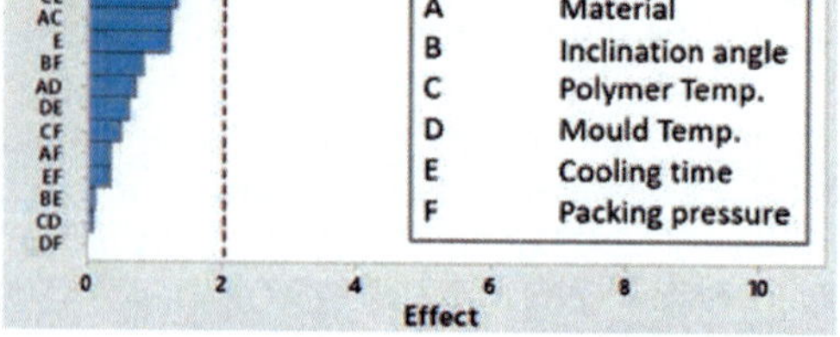

Fig. 5. Pareto chart of the effects

Figure 5 shows the Pareto chart of effects for the six varied parameters of Table 3 as well as their interactions. Only the 64 tests with 30° and 45° inclination angle are considered here. It can be seen that the choice of material or reinforcement (A) and the inclination angle of the part (B) have the main effect on the forced demoulding behavior. Furthermore, the polymer temperature (C), mould temperature (D) and the interactions of BC, AB and BD have a significant effect to the tests. In contrast, the variation of the injection moulding parameters the packing pressure (F) and cooling time (E) with in their chosen parameter levels show a minor influence on the demouldability of the parts.

Figure 6 shows the main effect chart of the quality criteria. The smaller the criteria value the better the demouldability and the less damage occurs to the pin undercuts. The first chart shows again the distinct influence of the material reinforcement. In the second and third diagram, a higher temperature of the polymer and the mould tends to make demoulding more difficult, as already described. With cooling time and back-pressure, the effect is very small, while the tendency is for better demoulding at the lower value.

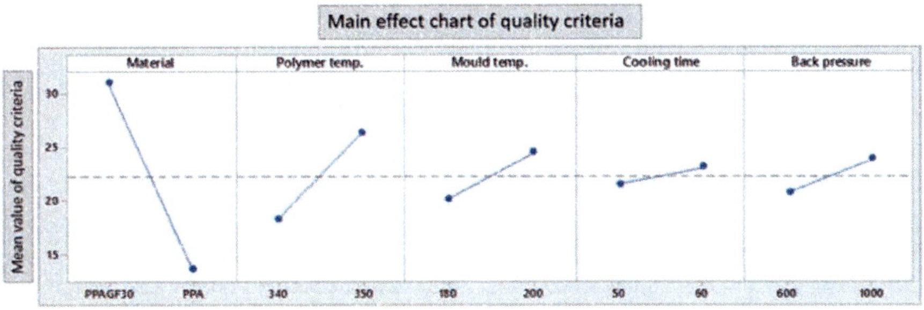

Fig. 6. Main effect chart of quality criteria

In order to sum up the results, Fig. 7 illustrates the demouldability of different samples classified into five statuses. Green (++) shows the demouldable pin undercuts while red (−−) depicts the undemouldable cases. In between there are gradations which represent different degrees of damage from demoulding or which under certain conditions are not or difficult to demould.

Pin	PPA			PPAGF30		
	Heights [mm]					
Angle	0.5	1	2	0.5	1	2
0°	++					
15°						
30°	++	++	+	+	0	-
45°	+	+	-	-	--	--

++	easy to demould
+	demouldable (small deformation)
0	limited demouldablity (deformation, little damage)
-	conditionally demoulded (deformation, damage)
--	undemouldable

Fig. 7. Evaluation of ejection results for different inclinations, materials and pin heights

6 Conclusion

In this work, the forced demouldability of mesoscale surface structures with undercuts is investigated for reinforced and non-reinforced PPA in an injection moulding process. Therefore four parts with different angles of inclination (0°, 15°, 30° and 45°) and pins with height of 0.5 mm, 1 mm and 2 mm are moulded and ejected. Frist, in simulations the necessary maximum deformations are determined. The results show strain values up to more than 30% which are much higher than the deformability of PPA and especially reinforced PPA at room temperature. These calculations show that demouldability can be critical under certain conditions. However, since polymer viscoelastic behavior is temperature dependent, the process parameters are important for demouldability. The experimental tests confirm this assumption. Up to a 30° inclination angle, forced deformation is uncomplicated with unreinforced PPA and only a few process configurations are critical with reinforced PPAGF30. Going on step back to the simulation results, it can be state that the simplified two-dimensional simulation with linear elastic material behavior overestimated the expected strain and possible risk of damage. But while the quantitative prediction through simulation is not exact, the qualitative prediction of the damaged area through marks at the upper edge or bending at the lower pin shaft is given.

The ejection tests with 45° angle show for PPA a general demouldability for certain process parameter even with large undercuts. This can be attributed to the interaction of viscoelastic material properties and the process parameters. For reinforced PPAGF30, no parameters for demoulding the samples could be found, as high demoulding forces are required. While high temperatures are necessary for sufficient (viscoelastic) deformation of the undercuts, component strength is too low and major damage occurs when the ejector pin penetrates the material in the mould. The statistical evaluation of the tests shows that material (reinforcement) and the mould angle are the main factors for the demouldability. The temperature of polymer and mould are also significant while packing pressure and cooling time had minor effects on these tests. In this analysis, it is clear to say that longer pins are more difficult to demould and they suffer more damage such as bending deformation. Since not all the parts could be demoulded and has to be removed destructively from the mould, the influence of the pin lengths in the 45° tests are not clear to state. It is possible that 0.5 mm pins are demouldable under certain conditions for PPAGF30. Therefore, it is depict as "conditionally demoulded" in Fig. 7. However, it can be deduced from the results that injection moulding with short fibre reinforcement has an effect on the strength of the mesoscale pins structure. Finally, the visual inspection shows no indication of a diesel effect in any of the tests. Therefore, the diesel effect is not an issue for the different depths of the pin cavities with PPA.

For future work, the influence of the pin length can be better worked out if only one pin length is injection molded at a time. In addition, the process parameters can be varied to a greater extent depending on the material. Furthermore, an investigation of the fibre orientation in the pins as well as the reinforcing effect of the short fibres in the pins is interesting.

Acknowledgements. The authors would like to thank the LKT technical employee, Mr. Dührkoop, for his help in making the moulds and setting up the injection moulding experiments. In addition, the authors acknowledge the funding of the collaborative research program "Priority Program 1712" by the German Research Foundation (DFG).

References

1. Fleischer, J., Nieschlag, J.: Introduction to CFRP-metal hybrids for lightweight structures. Prod. Eng. Res. Devel. **12**(2), 109–111 (2018). https://doi.org/10.1007/s11740-018-0825-0
2. Pohl, M., Stommel, M.: Intrinsic CFRP-metal-hybrids with rubber interface for the improvement of the damping behaviour. Prod. Eng. Res. Devel. **12**(2), 153–159 (2018). https://doi.org/10.1007/s11740-018-0792-5
3. Pohl, M., et al.: Development structures for intrinsic hybrids. WT Werkstattstechnik **107**(7–8), 546–550 (2017)
4. Berger, D., et al.: Predictive quality control of hybrid metal-CFRP components using information fusion. Prod. Eng. Res. Devel. **12**(2), 161–172 (2018). https://doi.org/10.1007/s11740-018-0816-1
5. Bretz, L., Günther, F., Jost, H., Schwarz, M., Kretzschmar, V., Pohl, M., Weiser, W., Häfner, B., Summa, J., Herrmann, H-G., Stommel, M., Lanza G.: Design and quality assurance of intrinsic hybrid metal-CFRP lightweight structures. In: Hybrid – Materials And Structures (2020)
6. Messler, R.W.: Joining of Materials and Structures: From Pragmatic Process to Enabling Technology, pp. 1–790 (2004)
7. Nguyen, A.T.T., Brandt, M., Orifici, A.C., Feih, S.: Hierarchical surface features for improved bonding and fracture toughness of metal-metal and metal-composite bonded joints. Int. J. Adhes. Adhes. **66**, 81–92 (2016). https://doi.org/10.1016/j.ijadhadh.2015.12.005
8. General Design Principles for DuPont Engineering Polymers. E.I. du Pont de Nemours and Company, USA (2000)
9. Goodship, V.: Practical Guide to Injection Moulding. Rapra Technology Limited, Shrawbury (2004)
10. Feistauer, E.E., dos Santos, J.F., Amancio-Filho, S.T.: A review on direct assembly of through-the-thickness reinforced metal–polymer composite hybrid structures. Polym. Eng. Sci. **59**(4), 661–674 (2019). https://doi.org/10.1002/pen.25022
11. Dröder, K., Brand, M., Kühn, M.: Numerical and experimental analyses on the influence of array patterns in hybrid metal-FRP materials interlocked by mechanical undercuts. Procedia CIRP **62**, 51–55 (2017). https://doi.org/10.1016/j.procir.2016.06.121
12. Guo, Y., Liu, G., Xiong, Y., Tian, Y.: Study of the demolding process – implications for thermal stress, adhesion and friction control. J. Micromech. Microeng. **17**(1), 9–19 (2007). https://doi.org/10.1088/0960-1317/17/1/002
13. Lucyshyn, T., Struklec, T., Burgsteiner, M., Graninger, G., Holzer, C.: Research in manufacturing of micro-structured injection molded polymer parts. AIP Conference Proceedings, 2015
14. Saha, B., Toh, W.Q., Liu, E., Tor, S.B., Hardt, D.E., Lee, J.: A review on the importance of surface coating of micro/nano-mold in micro/nano-molding processes. J. Micromech. Microeng. **26**(1) (2015). https://doi.org/10.1088/0960-1317/26/1/013002

15. Masato, D., Sorgato, M., Parenti, P., Annoni, M., Lucchetta, G.: Impact of deep cores surface topography generated by micro milling on the demolding force in micro injection molding. J. Mater. Process. Technol. **246**, 211–223 (2017). https://doi.org/10.1016/j.jmatprotec.2017.03.028
16. Menges, G., Michaeli, W., Mohren, P.: Spritzgießwerkzeuge: Auslegung, Bau, Anwendung, 6th edn. Hanser, München (2007)
17. Günther, F., Pohl, F., Kretzschmar, V., Scheuermann, G., Stommel M.: Optimizing Mechanical Interlocking Interface of CFRP-(Termoplastic)/Metal)-hybrids. In: Hybrid–Materials and Structures 2018 – Proceeding, pp. 254–260. https://hybrid2018.dgm.de/ (2018). Accessed 26 Feb 2020

Nondestructive and Destructive Testing on Intrinsic Metal-CFRP Hybrids

Hendrik Jost[1(✉)], Michael Schwarz[1,2], Felix Grossmann[1], Jonas Sauer[1], Alexander Hell[1], and Hans-Georg Herrmann[1,2]

[1] Chair for Lightweight Systems, Saarland University, Saarbruecken, Germany
hendrik.jost@izfp-extern.fraunhofer.de
[2] Fraunhofer Institute for Nondestructive Testing (IZFP), Campus E3.1, 66123 Saarbruecken, Germany

Abstract. Motivated by legislative and environmental restrictions, cost efficient lightweight design for automotive applications is becoming more and more important. While designing the automotive body completely out of CFRP is too expensive for series production, multi-material hybrid structures represent a good compromise between cost and weight reduction.

The presented approach is a CFRP specimen containing an intrinsic aluminium inlay that is overmoulded with a thermoplastic polymer reducing the stiffness difference between the used different materials.

Defects like delamination and fibre cracking often are not clearly visible, but have a significant influence on the mechanical properties of fibre-reinforced composites in general and more particularly of such an intrinsic hybrid as presented in this paper. In order to identify and evaluate these defects nondestructive testing can be a powerful tool. Furthermore, a combination of nondestructive and destructive testing methods can be used to develop a better understanding of the damage mechanisms of such a hybrid specimen under various environmental conditions.

In this approach, the presented specimens are tested under quasistatic and cyclic tension load using in situ passive thermography and digital image correlation. In this way, damage progression like delamination growth can be monitored.

Besides characterising hybrid samples during mechanical load, the samples are also tested before and after mechanical testing. Active thermography characterises defects in the CFRP component as well as defects at the interface of CFRP and thermoplastic whereas EMAT (electromagnetic acoustic transducers) characterise the interface between metal and thermoplastic.

By combining EMAT and active thermography before and after mechanical testing the whole hybrid structure with its multiple interfaces can be examined for defects. The results are validated with computer tomography (CT).

Keywords: Hybrid materials · CFRP · Thermography · Ultrasound · EMAT · Mechanical testing · Nondestructive testing

K. Dröder and T. Vietor (Eds.): *Technologies for economic and functional lightweight design*, Zukunftstechnologien für den multifunktionalen Leichtbau, pp. 279–289, 2021. https://doi.org/10.1007/978-3-662-62924-6_24

1 Introduction

The hybridisation of materials of different classes is a strong tool in lightweight design in general. In particular, the approach of a CFRP metal hybrid presented here shows good mechanical performance combined with a significant lightweight potential and could be used for series production as well. Obstacles for the use in industry are missing appropriate techniques for damage inspection and the lack of a deeper understanding of the damage mechanisms.

To face these obstacles the applicability of the nondestructive testing methods thermography, ultrasonic inspection and x-ray-tomography on the developed hybrid samples are regarded in this paper [1–5].

In a first step, 2D specimens are tested for an investigation of general usability, then the successful approaches are examined on new specimens with a 3D shape to determine their performance under more realistic conditions.

The general workflow in testing is, first, a nondestructive investigation followed by tensile testing combined with in situ nondestructive testing. After that, the mechanically treated specimens are again investigated nondestructively. Besides these defect-free specimens, there are specimens with included artificial defects like pleats or misalignment. With this workflow, it is possible to find out which defects can be detected, to learn about the effects of those defects on mechanical behaviour and to get more knowledge about the damage mechanisms. First, it is necessary to briefly explain the specimen design as well as the used methods for a discussion of the results.

2 Intrinsic Metal CFRP Hybrids

The 2D hybrid sample and its individual parts are shown in Fig. 1. It has an inlay of a 4 mm thick aluminium alloy (EN AW-6082) [6] with 6 arms. The characteristic form is manufactured in a stamping process and generates a tight fit. In the next step, this part gets overmoulded in an injection moulding process. The used thermoplastic is a polyphthalamide with 30% glass fibre content (Vestamid® HT plus M1033) and is named below by its acronym PPA GF30 [7]. In the last step, the overmoulded inlay is draped in the middle of four layers of C-fibre canvas fabric and is cured in a RTM process for 20 min at a temperature of 75 °C and a pressure of 8 bar up to a thickness of 1 mm. The used fibres are Torayca T300/FT300 [8], they are arranged symmetrically as a [0/90°, ±45°]s laminate and a 2 K hot curing epoxy system from SIKA (Biresin CR170/CH 150-3) is used [9]. The resulting hybrid has its joint strength mainly because of the mechanical interlocking. The thermoplastic layer improves the connection due to a reduction of the difference in stiffness.

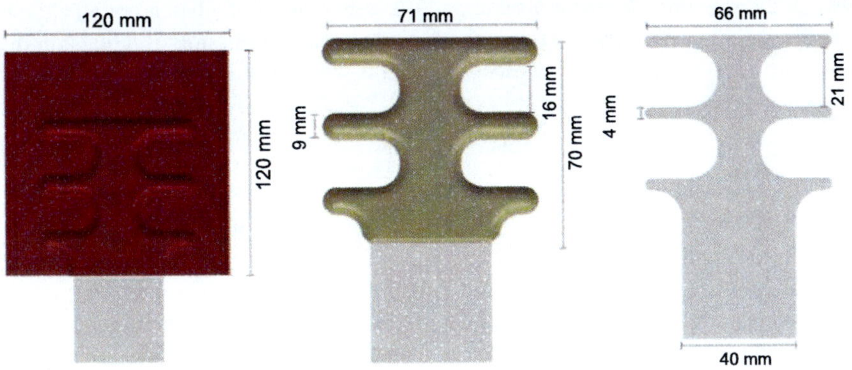

Fig. 1. Schematic drawing of the 2D hybrid sample (left), the overmoulded inlay (middle) and the aluminium inlay (right).

The 3D hybrid sample and its individual parts are shown in Fig. 2. In every part of the hybrid, there are three-dimensional structures. As in the 2D sample, the 3 mm thick aluminium inlay of the same aluminium alloy is manufactured in a stamping process and bent afterwards. All other steps are done in an equivalent manner to the 2D samples.

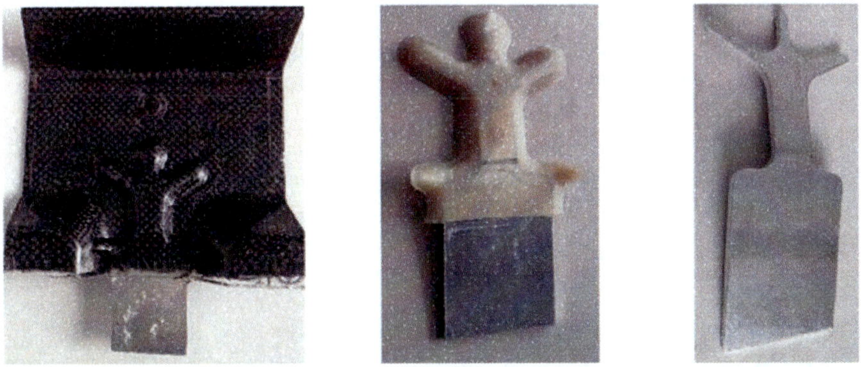

Fig. 2. Photos of the 3D hybrid sample (left), the overmoulded inlay (middle) and the aluminium inlay (right).

3 Infrared Thermography

One of the examined nondestructive testing methods is infrared thermography. Infrared thermography in general is a non-contact technique to record the temperature distribution of a surface [10]. The physical background is the characteristic of a solid above the temperature of 0 K to emit electromagnetic radiation. The correlation between the specific spectral emission and the temperature of a black body is given by Planck's radiation law where $M_{\lambda,S}$ is the specific spectral emission of a black body, C_1 and C_2 are constants, λ is the wavelength and T is the absolute temperature [11]:

$$M_{\lambda,S} = \frac{C_1}{\lambda^5 \left[e^{\frac{C_2}{\lambda \cdot T}} - 1 \right]}$$

In this paper, different infrared thermography techniques are used for different purposes. Before and after mechanical testing, this is active flashlight pulse thermography, while during mechanical testing it is in situ passive thermography. In quasistatic experiments simple difference pictures are regarded, in dynamic experiments, Lock-In thermography is used. While explaining those methods in detail would go beyond the scope of this paper, the extinction techniques and contrast mechanisms are explained briefly here and further information can be found in [5, 10–13].

For active flashlight thermography, the extinction occurs through a flashlight that heats up the specimen surface. The heat diffuses through the sample and inhomogeneities act as heat barriers and lead to a heat contrast on the surface which can be observed with the infrared camera. It follows that inhomogeneities in different depths can be seen in different time frames.

For passive thermography the extinction mechanism is the mechanical energy that is brought in through tension testing. The contrast occurs on the one hand through damage events with a sudden release of heat energy that can be measured on the specimen's surface with an infrared camera. During dynamic mechanical testing, on the other hand, a thermal wave is induced through the sinusoidal load that can be measured on the sample surface. Local differences in amplitude and phase of the thermal wave can show inhomogeneities like delamination in situ.

4 Electromagnetic Acoustic Transducer (EMAT)

The next examined nondestructive testing technique is a special ultrasonic technique called electromagnetic acoustic transducer (EMAT). EMAT is a contactless ultrasonic testing method that works without coupling agents. The ultrasonic waves are generated directly in the atomic lattice of the specimen due to the Lorentz Force in conductive materials or magnetostriction in ferromagnetic materials [14]. Thus, it is possible to generate guided shear horizontal waves (SH-waves) with a stress and strain orientation parallel to the surface [15].

Polymer coatings or adhesive layers that are bonded with the (echoed through) metal damp this SH-wave. This damping makes it possible to detect a peeling of the coating or the adhesive layer from the metal qualitatively due to higher signal peaks. The used measurement arrangement is sketched in Fig. 3.

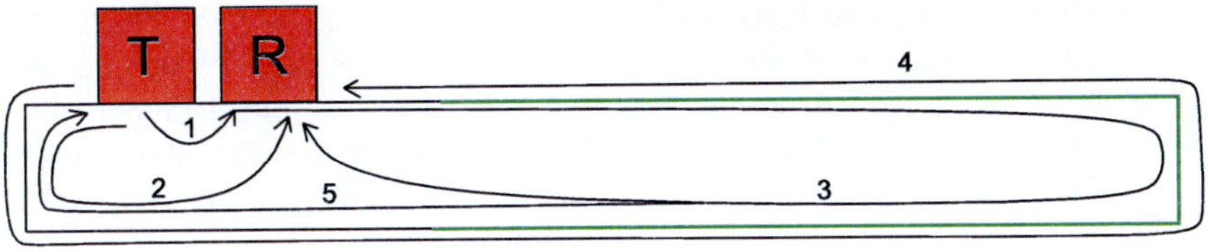

Fig. 3. Basic measurement arrangement in EMAT experiments; arrows represent the different paths for the guided SH-waves from transducer (T) to receiver (R) in a specimen with a polymer coating (green).

The velocity of the SH-waves differs depending on whether they are symmetric or asymmetric. For aluminium, the symmetric velocity is 3150 m/s and the asymmetric one is 2740 m/s. Different paths from transducer to receiver are possible and with the knowledge of the velocity of SH-waves and the run time in aluminium it is possible to relate the peaks with the wave's path through the specimen. For the measurement of the contact between the metal and the polymer coating, the information of the peaks caused by the paths 3, 4 and 5 through the specimen are relevant. Further information about EMAT can be found in [4, 14–16].

5 Results

Active thermography:
For the examination of the 2D samples, active flashlight thermography is used. As an example for an artificial defect, a specimen with a 10° fibre misalignment in the first layer is shown in Fig. 4a. The fibre structure can be seen clearly and thus, the misalignment can be recognized easily. Figure 4a and b show a specimen with a pleat in the first layer after fatigue testing. The pleat can clearly be seen as a brighter stripe due to a heat build-up 14 ms after the flash. Waiting for 1.6 s, a white halo can be seen all around the inlay. That can be assigned to a mechanical induced delamination between the two middle layers to each other and the inlay.

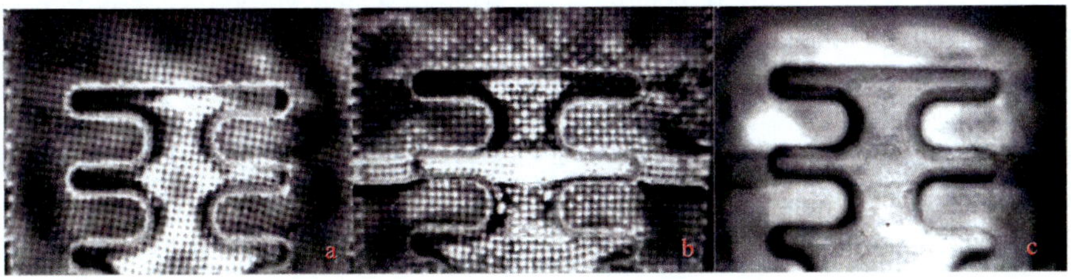

Fig. 4. Flashlight thermography: **a)** 2D specimen with 10° fiber misalignment before mechanical testing 80 ms after the flash; **b)** 2D specimen with artificial pleat after fatigue testing 14 ms after the flash **c)** 2D specimen with artificial pleat after fatigue testing 1.6 s after the flash.

Active thermography appears to be suitable for this specimen to detect defects in the CFRP layers as well as in the interface of CFRP and the thermoplastic. Furthermore, it allows a first inside view to the damage mechanism as you can see that the damage primarily occurs due to the massive delamination around the inlay.

The results of the flashlight thermography of the 3D-specimens are shown in Fig. 5. The misalignment in Fig. 5a can be detected in the plane surface as well as in the curved part in the right and left bottom of the figure. The black areas on the edges come into existence because of the diffuse scattering of the flashlight on these edges. In Fig. 5a, b specimen with artificial pleat in the first layer after quasistatic tension testing is pictured. The pleat can be observed weakly on the top of the figure. The white area above the inlay is a delamination between inlay and the middle CFRP layer.

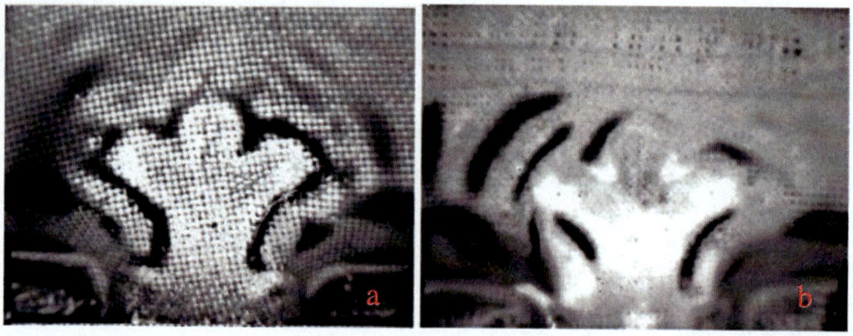

Fig. 5. Flashlight thermography: **a)** 3D specimen with 10° fiber misalignment before mechanical testing; **b)** 3D specimen with artificial pleat after quasistatic tension testing.

The results found in 2D-specimens thus are transferable to the those in 3D-specimens, but the results for the damage mechanisms are limited, because you only can observe them after mechanical testing. To get a deeper look at that, it makes sense to use a method that observes the damage while occurring.

Mechanical testing and passive thermography:
The quasistatic tension testing is carried out with a traverse speed of 2 mm/min on a servo-hydraulic testing machine. Due to the complex specimen geometry a presentation as force–displacement-diagram is useful as it is shown in Fig. 6.

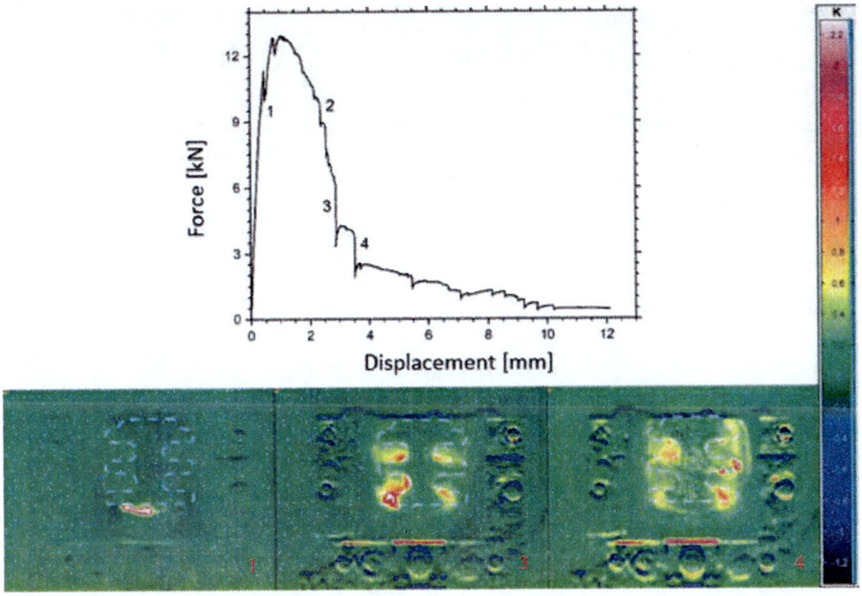

Fig. 6. Force–displacement-diagram of a 2D-specimen and thermography difference pictures of the specimen while tension testing to the times of the marked damage events 1, 3 and 4.

As can be seen in the force–displacement-diagram, the damage occurs stepwise with precise observable damage events. These damage events can be approximately localized because of the sudden release of energy that leads to a heating of the surrounding area on the surface. The first damage takes place below the inlay (1). Further damage happens between the arms (3) of the inlay until it comes to an overall failure. The development of damage in quasistatic tension testing thus can be observed with comparatively simple methods.

To observe damage mechanisms in dynamic fatigue testing, more sophisticated methods have to be used because the forces are smaller leading to a less obvious damage progression. In Fig. 7, the Lock-In-thermography phase images of a 2D-specimen are shown.

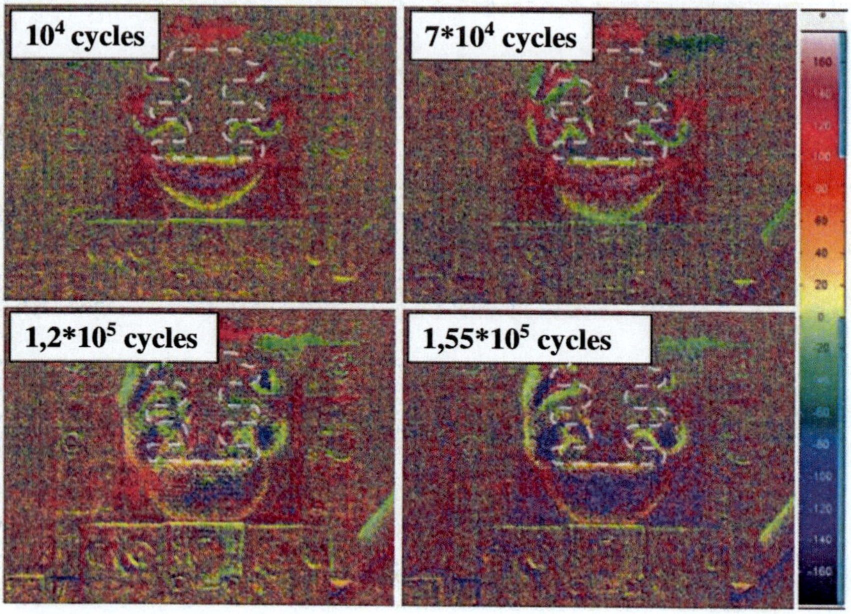

Fig. 7. Lock-In-thermography: Phase images of a 2D-specimen after the given cycles: 75% of the maximal force of quasistatic testing; R = 0.1.

In this figure, the damage progression can be retraced clearly. The green areas show where damage occurs to the given cycles, while the dark purple areas show the already delaminated areas. In an early state, delamination starts below the inlay. This delamination progress proceeds, while additional delamination growth between and around the arms of the inlay starts. The end of the life cycle is reached when the delaminated areas grow together and the failure occurs.

Thus, thermography, passive as well as active, are powerful tools to gather information about defects and damage mechanisms in the CFRP laminate itself and the interface between CFRP and the thermoplastic. However, defects on the interface metal-thermoplastic cannot be seen. In this regard EMAT can show its strengths.

EMAT:
Comparing a 2D-specimen before and after mechanical testing (see Fig. 8), a big difference can be observed in the peak intensities. While the defect-free specimen damps the ultrasonic waves almost completely, the specimen after mechanical testing shows less damping of those waves. This suggests that the thermoplastic detaches from the metal during mechanical testing as a part of the damage mechanisms.

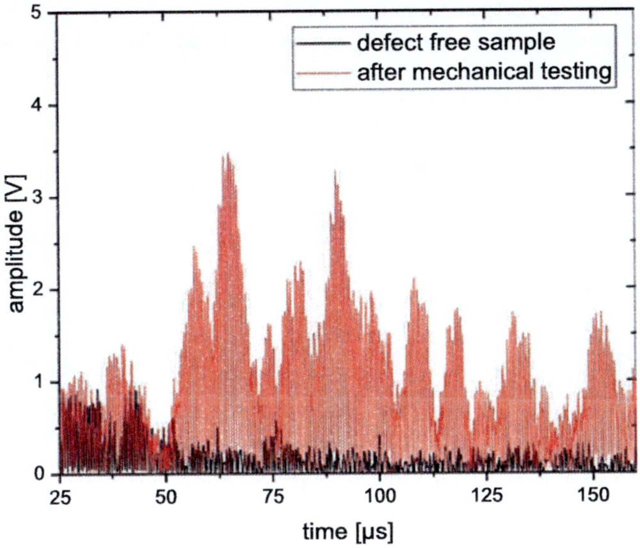

Fig. 8. EMAT on a 2D sample: before (black) and after mechanical testing (red).

The only part of the hybrid where we have no knowledge so far concerning the damage mechanisms is the metal inlay. That information can be provided through x-ray tomography.

X-ray tomography:
In Fig. 9, the x-ray projections of three 2D-specimens after a quasistatic tension testing are shown. All specimens have a similar behaviour. The middle arms are bent up, but not broken and the metal inlay is pulled out, while visible damage in the other two components has occured.

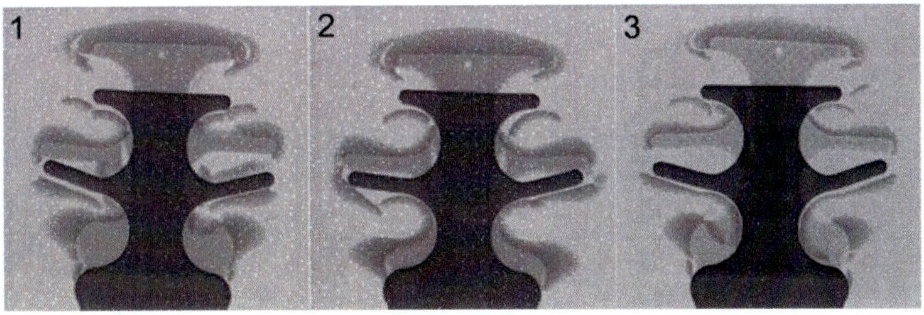

Fig. 9. X-ray projections of a 2D specimen after quasistatic testing.

A different picture presents itself after fatigue testing (Fig. 10). The arms are just bent a little bit, but cracks in the metal can be observed. The damage mechanisms in quasistatic and fatigue testing are clearly different.

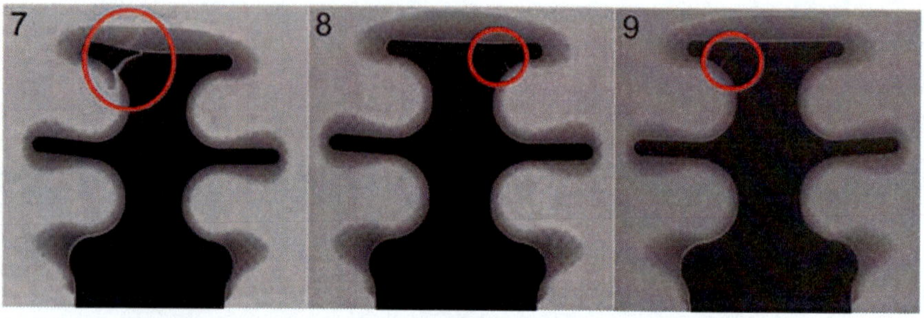

Fig. 10. X-ray projections of a 2D specimen after fatigue testing.

6 Conclusion

Nondestructive testing on a hybrid specimen consisting of more than one material is a challenging task that cannot be accomplished by one testing method alone. But the combination of different methods makes it possible to characterise defects in the material as well as to get a deeper knowledge about damage mechanisms. The used combination of infrared thermography, EMAT and x-ray tomography provides all the information needed in the used materials itself as well as at the interfaces between them.

Acknowledgements. We kindly thank the “Deutsche Forschungsgesellschaft” (DFG) for the financial support within the framework of SPP1712 and our project partners.

References

1. Summa, J., Becker, M., Grossmann, F., Pohl, M., Stommel, M., Herrmann, H.G.: Fracture analysis of a metal to CFRP hybrid with thermoplastic interlayers for interfacial stress relaxation using in situ thermography. Compos. Struct. **193**, 19–28 (2018)
2. Berger, D., Brabandt, D., Hornung, T., Bakir, C., Lanza, G., Summa, J., Schwarz, M., Herrmann, H.G., Pohl, M., Stommel, M.: Effects of defects in series production of hybrid CFRP lightweight components – detection and evaluation of quality critical characteristics. Measurement **95**, 389–394 (2017)
3. Schwarz, M., Summa, J., Quirin, S., Herrmann, H.G.: New approaches in nondestructive characterisation of defects in metal – CFRP hybrids. Mater. Sci. Forum, 976–982 (2015)

4. Quirin, S., Neuhaus, S., Herrmann, H.G.: Testing ultrasonic SH waves to estimate the quality of adhesive bonds in small hybrid structures. In: 2nd International Conference Euro Hybrid – Materials and Structures, ed. by the Institute of Materials, Kaiserslautern, 216–223 (2016)
5. Malheiros, F.C., Ignacio, L.H., Figueiredo, A.A.A., Herrmann, H.G., Fernandes, H.C. Deviation time for defect depth estimation in composite materials. In: 5th Brazilian Conference on Composite Materials – BCCM 5 (2020)
6. Schmolz + Bickenbach: "EN AW-6082". datasheet, Mar. 2011
7. Evonik: "Vestamid® HT plus M1033", datasheet, Sep. 2008
8. Torayca: "T300/FT300 commercial documentation", datasheet, May 2012
9. Sika: "Biresin® CR170 mit Biresin® CH150–3 Härter", datasheet, Jun. 2010
10. Maldague, X.: Theory and Praxis of Infrared Technology for Nondestructive Testing. Wiley (2001)
11. Bernhard, F.: Handbuch der technischen Temperaturmessung. Springer (2014)
12. Wu, D., Busse, G.: Lock-in thermography for nondestructive evaluation of materials. J. Appl. Phys. Revue Générale De Thermique **37**(8), 693–703 (1998)
13. Rösner, H., Netzelmann, U., Hoffmann, J., Karpen, W., Kramb, V., Meyendorf, N.: Thermographic Materials Characterization. Springer (2004)
14. Hübschen, G.: Elektromagnetische Ultraschall (EMUS-) Wandler zur Erzeugung horizontal polarisierter Transversalwellen. NDTnet **3**(3) (1998)
15. Castaings, M.: SH ultrasonic guided waves for the evaluation of interfacial adhesion. Ultrasonics **54**(7), 1760–1775 (2014)
16. Su, Z., Ye, L.: Identification of Damage Using Lamb Waves. Springer (2009)

Resistance Welding of FRP to Steel Components in High-Volume-Production

Jens Lotte(✉), Uwe Reisgen, and Alexander Schiebahn

Welding and Joining Institute (ISF), RWTH Aachen University, Pontstraße 49, 52062 Aachen, Germany
{lotte,office,schiebahn}@isf.rwth-aachen.de

Abstract. Parts or Components made of fibre-reinforced plastics (FRP) are often connected to metallic components. However, joining FRP to metals is currently a particular challenge. As a rule, such connections are currently glued and/or mechanically joined. Recent research has focused on innovative hybrid connections, which are reinforced by small-scale 3D structures in the z-direction of the component axis. Although different approaches are being followed for the production of FRP/metal hybrids, there is still a great need for a joining process that meets the technical and economic challenges in this context.

A new innovative joining concept, which is being worked on at the Institute of Welding and Joining Technology at RWTH Aachen University, is based on weldable inserts that are integrated into the manufacturing process of the FRP semi-finished product. As a result, these FRP structures can be joined locally with metallic connecting components using conventional resistance welding processes. The integration of the inserts is reproducible in the RTM process, which is suitable for large-scale production, and requires no post-processing. With the new joining process, high static composite strengths and a quasi-ductile post-processing behaviour can be achieved. Furthermore, the inserts can be freely varied in terms of the number of pins used and thus their arrangement can be adapted to the respective application.

In addition to investigations of the static bond strength in different load directions, this paper also examines the failure behaviour of top-hat profiles and the combinability of the new joining technology with bonding processes.

Keywords: FRP-/metal-hybrids · Lightweight design · Adhesive bond

1 Introduction

In the production of lightweight structures, the use of hybrid structures based on fibre-reinforced plastics (FRP) and metals can increasingly be observed in a growing number of industrial sectors (automotive, aerospace, wind energy, mechanical and plant engineering). The lightweight construction potential of such multi-material structures has been demonstrated in numerous studies [1–3]. Components made of fibre-reinforced plastics (FRP) are often joined with metallic components. However,

K. Dröder and T. Vietor (Eds.): *Technologies for economic and functional lightweight design*, Zukunftstechnologien für den multifunktionalen Leichtbau, pp. 290–299, 2021. https://doi.org/10.1007/978-3-662-62924-6_25

the joining of FRP with metals still poses a particular challenge at present. Today, such connections are usually bonded and/or mechanically joined. Mechanical joining processes, such as self-piercing rivets, usually cause fiber damage in fiber-reinforced plastics, which has a strong negative effect on the direction-dependent material properties [4, 5]. On the other hand, a structural adhesive joint offers only limited ductility and therefore fails brittlely without any early detection of imminent failure. In addition, the adhesive only bonds the joining partners superficially, which is why the forces are not sufficiently introduced via the matrix into the deeper, load-transmitting fiber layers [6, 7].

For the reasons mentioned above, newer research approaches concentrate on innovative hybrid connections, which are reinforced by small-scale 3D structures perpendicular to the joining surface z-direction of the component axis (cf. e.g. [8–13]). In the research project "Verfahrensentwicklung zur Herstellung von hybriden FVK/Stahl Strukturen mittels eines neuartigen Blechverbindungselementes" weldable insert structures are integrated into FRP components with thermoplastic matrices. These can then be used as local welding points. However, this solution can only be implemented in connection with thermoplastic matrices [14]. Since the majority of matrices used in industry are thermoset, other solutions must be found here. Although different approaches are being pursued for the production of plastic/metal hybrids, there is still a great need for a joining process that meets the technical and economic challenges in this context. Among these challenges is the problem of intrinsic bonding of such compounds. This requires the integration of the insert structures into the still uncured semi-finished fibre products. The assembly of the semi-finished plastic products, as well as their production itself, would therefore have to take place simultaneously. However, such a concept is not economically attractive for the end user. For this reason, the research project looked for a possibility to decouple the production of such semi-finished plastic products from their final assembly in terms of time and logistics.

A new, innovative joining concept, which is being developed at the Institute for Welding and Joining Technology at the RWTH Aachen and the Institute for Plastics Processing (IKV) in Industry and Craft at the RWTH, is based on weldable inserts (carrier plates with welded-on pin structures), which are integrated into the manufacturing process of the semi-finished fibre composite product. Thus, after consolidation of the semi-finished plastic product, it can be welded locally with metallic connecting components using conventional resistance welding processes. The inserts are integrated reproducibly and without post-processing in the RTM process, which is suitable for large series production. With the new joining process, high static strengths and quasi-ductile failure post-processing behaviour can be achieved. In addition, the inserts can be freely varied in terms of the number of pins used and thus their arrangement can be adapted to the respective application.

The aim of the investigations was to produce thermoset FRP/metal hybrid structures using established welding methods, economically, quickly and reliably with high mechanical properties and good bond strength between the joining partners, without negatively affecting the mechanical-technological properties of the FRP components.

2 Materials and Methods

The new joining technology approach is based on an innovative, modified arc welding process, which forms metallic pin structures in one operation and welds them onto a metallic surface (Fig. 1, step 1). In this way, weldable inserts are produced, which can be integrated into the plastic components during semi-finished product production (Fig. 1, step 2). After the plastic matrix has cured, the pin heads protrude locally from one side of the plastic surface and the carrier sheet from the other. The pin structures thus allow current to flow transversely to the component plane, thereby enabling such structures to be welded to connection plates (Fig. 1, step 3).

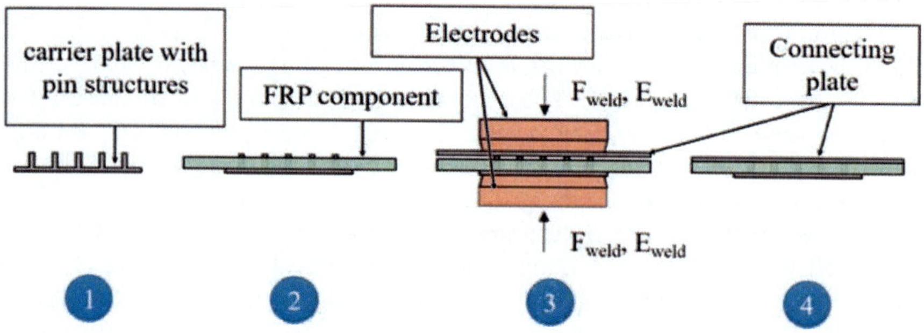

Fig. 1. Process chain for the production of hybrid components made of FVK and metal using innovative joining technology

The structuring of the inserts with the dimensions 30 × 30 × 1 [mm] is based on the so-called CMT process, with the help of which small-scale metallic pin structures (2–5 mm) can be formed directly from the welding wire in one step and without additional prefabricated components. The CMT process extends conventional metal arc welding (MSG welding) by a process control that enables dynamic wire movement by reversing the wire feed direction. For this work a welding system TransPlus Synergic 4000 VR 7000 from Fronius International GmbH, Pettenbach, Au stria, is used. The torch head is integrated into a three-axis CNC system.

The semi-finished fibre products (preforms) used consist of the quasi-unidirectional HP-U600E fabric made from HP textiles, Schapen, Germany, with a total surface weight of 650 g/m^2. Altogether 4 fabric layers are used for each preform, resulting in a total surface weight of 2600 g/m^2. The individual layers are bonded by applying the thermoplastic binder Spunfab® PA 1807 from Spunfab Ltd, Ohio, USA, between the individual layers and subsequent thermoforming. By applying an oscillating movement (amplitude: 1.8 mm, frequency: 300 1/s) and downward pressure on the base plate of the inserts, complete penetration of the pin structures through the preforms without damaging the fibers is achieved.

The infiltration of the matrix resin into the preforms is carried out by a Resin Transfer Moulding (RTM) process, for which an epoxy resin system (Araldite XB 3585 and Aradur 3474 from Huntsman, The Woodlands, Texas, US

A) is used. The resin system is injected by means of pistons via distribution channels into the mould cavity into which the dry preforms have previously been placed, where it cures under heat and pressure. The components are degassed and individually preheated to 50 °C under vacuum for 60 min before mixing. The resin system is infiltrated into a preheated (120 °C) and evacuated square mould measuring $500 \times 500 \times 5$ mm^3. A holding pressure of 3 bar is then applied for the total curing time of 3 min. The welding spots (pin structures) are separated from the matrix material by silicone membranes so that they remain blank and can be welded immediately after the matrix has cured. Further steps are not necessary.

In a final step, the hybrid samples are welded to a metallic connecting component. The welding tests are performed on an MFDC resistance welding machine (type BM mD 24/3 × 260MF, Nimak GmbH, Wissen), whereby a particularly short welding time of 10 ms is selected.

The carrier plates of the inserts and the metallic connection plates of the tests presented here are made of DC01 (material number 1.0330). The wire material used is G2Si1 (material number 1.5125). The destructive mechanical tests are performed on the Zwick/Roell RetroLine 10 tensile testing machine. The free specimen length, i.e. the unclamped specimen length, is 150 mm. The clamped specimen length is 50 mm each. At the start of the test a preload of 100 N is applied and the tests are performed at a test speed of 5 mm/min at room temperature.

3 Experiments

3.1 Mechanical Characterization of the Joint

The geometries of the test specimens and their load direction in the mechanical test are shown in Fig. 2.

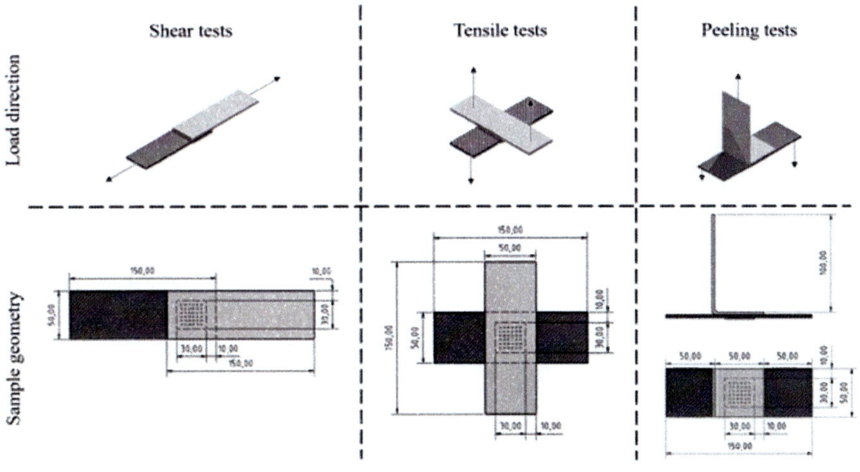

Fig. 2. Geometries of the test specimens and their load direction in the mechanical test

For the mechanical characterization of welded joints, shear tests, tensile tests and peeling tests (following the standards: DIN EN ISO 14273, DIN EN ISO 14272 and DIN ISO 14270) are being carried out. The test specimen geometries deviate from the corresponding standards in favour of better integration of the insert structures. In addition, the number of pins (16 pins and 21 pins) is examined for the bond strength of the hybrid connections. A constant welding force of 156 N per pin is used. This corresponds to a welding force of 2.5 kN or 3.2 kN when welding 16 or 21 pins. The results of the mechanical tests are shown in Fig. 3.

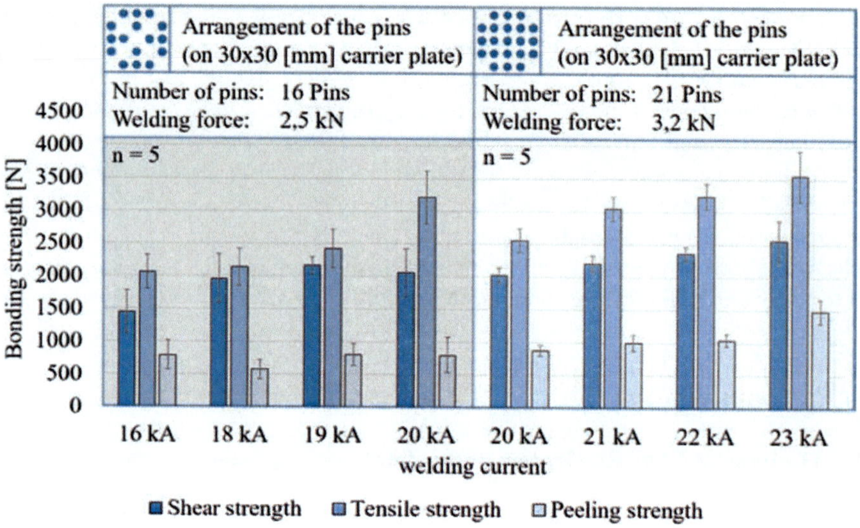

Fig. 3. Comparison of the bond strengths as a function of the load direction and the number of pins

With increasing welding currents, higher composite strengths can be produced. At the same time, however, the tendency to spatter and the melt-down stroke also increase. Therefore, the welding current cannot be increased indefinitely, but must be limited depending on the number of pins. For 21 pins (10 ms welding time) a maximum welding current of 23 kA and for 16 pins a maximum welding current of 20 kA is determined. Here, the assessment with regard to weld spatter that can be tolerated is subjective.

For 21 pins, which are welded simultaneously, an average of 2565 N can be transmitted in shear tests, 3553 N in tensile tests and 1484 N in the peel tests at these maximum currents. For 16 pins, an average of 2165 N in shear tests, 3214 N in tensile tests and 808 N in peel tests can be achieved.

With peeling stress, only individual pin rows are loaded. If they fail, the force applied is transmitted by the next pin row in the direction of the gap. For this reason, only a fraction of the forces that can be transmitted in shear tests or tensile tests can be absorbed in the peeling test.

If high composite forces are generated, the deflection of the sheets in the joining zone also causes peeling stress on the specimens in the head tensile test. The pin rows are therefore separated from "outside to inside" and in this case the peeling becomes the actual cause of failure of these pins. The pin arrangement in the outer pin array of both specimen types (21 and 16 pins) correspond to each other, therefore the transmittable head tensile forces of these two specimen types are also very similar.

3.2 Investigation of Component-Like Structures

For a further evaluation of the new joining technology, tests are carried out on component-like specimens. These consist of a plate made of (glass)-FRP and a hat profile bent from steel. All materials correspond to those of the tests described above. The aim of these tests is the direct comparison of bonded and welded components.

On the one hand, the two components are bonded with a 2 K epoxy resin (3 M™ Scotch-Weld™ DP 410). The two load-transferring surfaces in the legs are limited to a total area of 30 × 400 [mm] by means of a PTFE foil and an adhesive gap of 0.25 mm is set. Before gluing, the surfaces are wiped with isopropanol. On the other hand, a total of 10 weldable inserts are integrated into the legs of the hat profile and each one is individually welded to the steel hat profile. The inserts used were each provided with 21 pin structures (Fig. 4). A welding force of 3.0 kN and a welding current of 23 kA (with a welding time of 10 ms) is used.

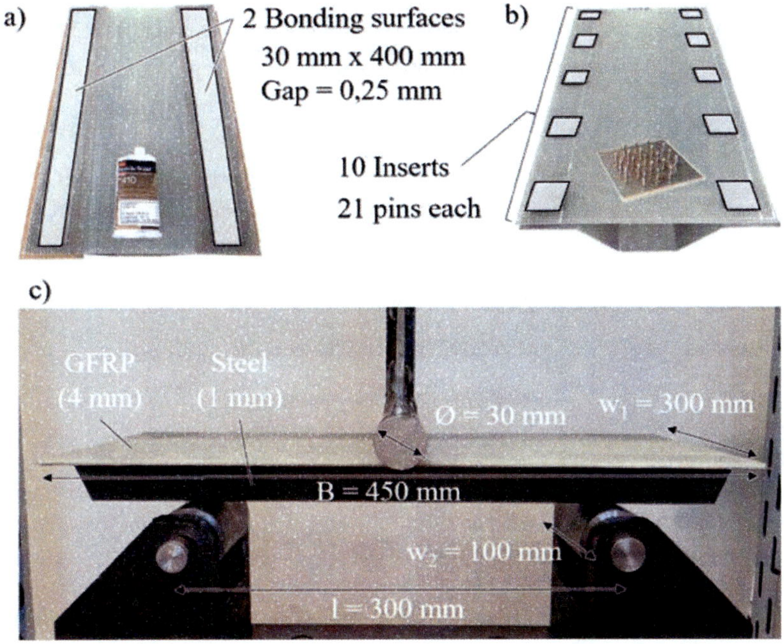

Fig. 4. Combination of the innovative welding process with conventional bonding processes

The results of the 3-point bending tests are shown in Fig. 5. The adhesively bonded profiles fail brittle in the 3-point bending test at an average maximum force of 6056.16 N. The failure occurs adhesively or near the surface on the FRP sides. Here, a maximum crosshead travel of 14.51 mm is achieved. The average maximum force of the welded joints is 5844.84 N, slightly below the adhesive joints. Individual pin structures fail during the test, but a complete failure of the connections cannot be determined during the tests. All tests are therefore considered to be completed at a maximum crosshead travel of 70 mm.

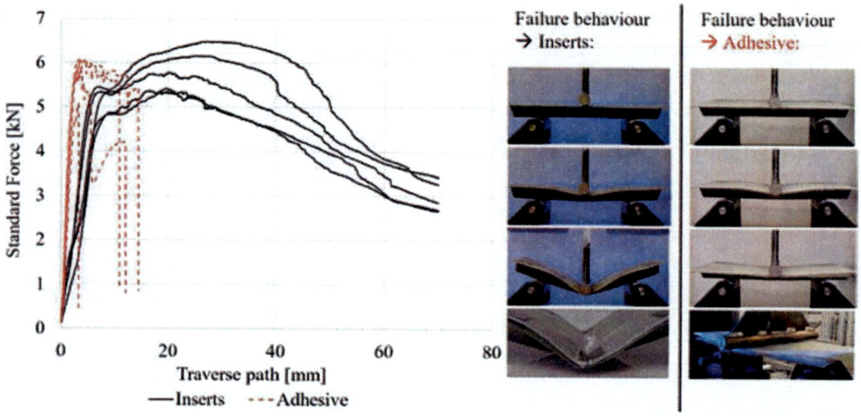

Fig. 5. Combination of the innovative welding process with conventional bonding processes

A quasi-ductile post-breakage characteristic generated by multiple bonding of the pin structures can be observed on component-like components. In comparison to the brittle and very sudden failure of the adhesively bonded joints, the welded joints fail stepwise under strong noise generation. Monitoring of the joint is therefore easily possible during operation.

3.3 Combination of the Innovative Welding Method with Bonding Processes

Finally, the combinability of the new innovative welding process with conventional adhesive bonding processes is investigated. Reference shear test specimens are welded according to the previously described geometries, materials and welding parameters on the one hand by means of 16 pin structures and on the other hand bonded by 85 g of a 2-component epoxy resin adhesive (Scotch-Weld DP 490, 3 M, Neuss). In a last step, the adhesive is applied between the 16 pins in the joining zone, these are welded and the adhesive is subsequently cured (Fig. 4b). All test samples are degreased with isopropanol before the adhesive is applied (Fig. 6).

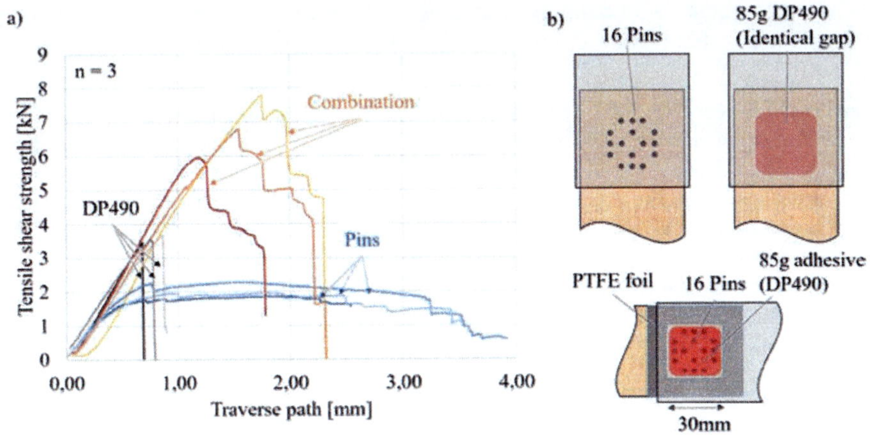

Fig. 6. Combination of the innovative welding process with conventional bonding processes

In the shear tensile test (Fig. 4a), the ductile post-break characteristics of the pins can also be verified for the combination with the bonding process. In addition, the combination of both processes achieves an average shear tensile strength of 6775.6 N. Compared to the average shear tensile strength of purely bonded specimens of 3553.6 N and those of purely welded specimens of 1942.17 N, this exceeds the sum of both individual shear tensile strengths. This effect is possibly caused by the additional stiffening effect of the adhesive layer in the joining zone.

4 Conclusion

It was shown that weldable insert structures can be realized by means of small-scale form-fitting elements. Furthermore, it was shown that these structures can be reproducibly integrated into fibre-reinforced plastic components by means of manufacturing processes suitable for large-scale production (RTM) and that they can be welded to metallic connecting components by means of established resistance welding processes.

The findings were used to manufacture near-series hybrid components using the new joining technology and to reference otherwise identical, conventionally bonded hybrid components. It was proven that the findings of the samples from the laboratory scale can be transferred to real components. Furthermore, welded insert structures can be used to produce composite strengths comparable to those of a bonded composite. In this case, however, the post-fracture behaviour of the welded components is more favourable than that of the bonded components.

The positive features of the new joining process are summarized below:

- Suitable for large series production
- Reproducible and robust process control possible
- Immediate handling strength after the joining process

- Can be combined with bonding processes
- Avoidance of fibre damage due to small-scaled form-fit elements

Acknowledgements. The IGF project "Entwicklung eines alternativen Fügeverfahrens zur wirtschaftlichen und prozesssicheren Herstellung von faserverstärkten Kunststoff-/ Metallhybridstrukturen auf Basis des Widerstandsschweißverfahrens mittels integrierter metallischer Inserts", IGF project no. 19.466 N, of the Research Association for Welding and Allied Processes of the DVS, was funded by the AiF within the framework of the programme for the promotion of joint industrial research (IGF) of the Federal Ministry of Economics and Energy on the basis of a resolution of the German Bundestag.

Our thanks go to all institutions.

References

1. Lässig, R., Eisenhut, M., Mathias, A. et al.: Serienproduktion von hochfesten Faserverbundbauteilen – Perspektiven für den deutschen Maschinen- und Anlagenbau. Studie der Roland Berger Strategy Consultants (2012)
2. Lotus Engineering: An Assessment of Mass Reduction Opportunities for 2017–2020, Model Year Vehicle Program. International council on clean transportation, Washington, DC (2010)
3. Sahr, C., Berger, L., Lesemann, M., et al.: Systematische Werkstoffauswahl für die Karosserie des Superlight-Car, ATZ 112. Springer Fachmedien, Wiesbaden (2010). https://doi.org/10.1007/BF03222166
4. Schürmann, H.: Konstruieren mit Faser-Kunststoff-Verbunden, 2. Aufl. Springer, Berlin (2007). ISBN 978-3-540-72190-1
5. Puck, A.: Festigkeitsanalyse von Faser-Matrix-Laminaten. Hanser, München (1996). ISBN 3-446-18194-6
6. Habenicht, G.: Kleben – Grundlagen, Technologien, Anwendungen, Bd. 6. Springer, Berlin (2009). ISBN 978-3-540-85266
7. Klein, B.: Leichtbau-Konstruktion. Springer Vieweg, Wiesbaden (2013) ISBN 978-3-658-02271-6
8. Oluleke, R., Strong, D., Ciuca, O., Meyer, J., Oliveira, A., Prangnell, P.: Mechanical and microstructural characterization of percussive arc welded hyper-pins for titanium to composite metal joining. Mater. Sci. Forum **765**(1), 771–775 (2013). https://doi.org/10.4028/www.scientific.net/MSF.765.771
9. Blackburn, J., Hilton, P.: Producing surface features with a 200 W Yb-fibre laser and the Surfi-Sculpt® process. Physics Procedia **12**(Part A), 529–536 (2011). https://doi.org/10.1016/j.phpro.2011.03.065
10. Tu, W., Wen, P., Hogg, P., Guild, F.: Optimisation of the protrusion geometry in Comeld (TM) joints. Compos. Sci. Technol. **71**(6), 868–876 (2011). ISBN: 978-988-18210-7-2
11. Dröder, K., Große, T., Brand, M.: HyTensile – Klammerstrukturen für innovative Hybridbauteile, Ingenieur Spiegel, Public Verlagsgesellschaft, 2014, Issue 3 (2014)
12. Ucsnik, S., Kirov, G.: New possibility for the connection of metal sheets and fiber reinforced plastics. Mater. Sci. Forum **690**(1), 465–468 (2011). https://doi.org/10.4028/www.scientific.net/MSF.690.465

13. Ucsnik, S., Pahr, D., Scheerer, M., Zaremba, S.: Experimental investigation of a novel hybrid metal-composite joining technology. Compos. Part A Appl. Sci. Manuf. (2010). https://doi.org/10.1016/j.compositesa.2009.11.003
14. Dröder, K., Sterz, J., Kühn, M., Ballschmiter, G., Jüttner, S., Obruch, O.: Verfahrensentwicklung zur Herstellung von hybriden FVK/Stahl Strukturen mittels eines neuartigen Blechverbindungselementes, EFB-Forschungsbericht Nr. 476 (2017). ISBN 978-3-86776-527-5

Novel Ultrasonic-Based Joining Methods for Metal-Plastic Composites (MPC)

Matthias Riemer[1(✉)], Christian Kraus[2], Mathias Kott[3], Koen Faes[4], Marcin Korzeniowski[5], and Marc Götz[3]

[1] Fraunhofer Institute for Machine Tools and Forming Technology IWU, Reichenhainer Strasse 88, 09126 Chemnitz, Germany
matthias.riemer@iwu.fraunhofer.de

[2] Fraunhofer Institute for Machine Tools and Forming Technology IWU, Nöthnitzer Strasse 44, 01187 Dresden, Germany
christian.kraus@iwu.fraunhofer.de

[3] Fraunhofer Institute for Process Engineering and Packaging IVV Branch Lab Processing Technology, Heidelberger Strasse 20, 01189 Dresden, Germany
mathias.kott@ivv-dd.fraunhofer.de

[4] Belgian Welding Institute, Technologiepark 48, B-9052 Zwijnaarde, Germany
Koen.Faes@bil-ibs.be

[5] Wroclaw University of Science and Technology, Wybrzeze Wyspianskiego 27, 50-370 Wroclaw, Germany
marcin.korzeniowski@pwr.edu.pls

Abstract. The use of metal-plastic composites (MPCs) plays an important role for lightweight products. MPCs provide an innovative substitute for solid metal sheets due to their low weight, excellent stiffness, and due to their thermal and acoustic insulation. A challenge that links all the different applications is the need to efficiently process the MPCs at high quality and join them with other materials, using suitable joining techniques. Due to their special features (layer structure, material mix, etc.), conventional manufacturing processes are therefore only of limited or of no use at all. In particular, the polymer core layer is a barrier for the application of conventional joining methods. This contribution presents a novel joining approach for MPCs. The basic approach is the local melting of the polymer layer by energy of ultrasonic waves, and displacement of the molten plastic material by pressure on the cover sheets. Two different strategies are investigated for this purpose. In the first approach, the joining tool is directly superimposed by ultrasonic waves and the polymer is displaced before and during joining. In the second approach, the polymer layer is displaced in the joining area during a prior sheet forming process step. This paper describes the simulation of the ultrasonic-assisted displacement of the polymeric layer. These simulations are validated using experimental results. An exact analysis of the process parameters and infrared images are used for the validation of the simulation results. The results lead to an optimal design of

K. Dröder and T. Vietor (Eds.): *Technologies for economic and functional lightweight design*, Zukunftstechnologien für den multifunktionalen Leichtbau, pp. 300–312, 2021. https://doi.org/10.1007/978-3-662-62924-6_26

special ultrasonic tools, which enable clinching and resistance spot welding in one step or allow pressing in a forming tool.

Keywords: Metal-plastic composite · Joining · Ultrasonic

1 Introduction

The metal-plastic composite (MPC) materials discussed in this article consist of two thin, solid metal sheets in between a thermoplastic core layer. In general, these hybrid materials offer different advantages, mainly based on the properties and thickness of the core material. Thicker and stiffer cores improve the bending stiffness, while thin and softer ones tend to lower the structure-borne noise, especially large sheet metal structures. Even though these composites show benefits for different applications [1], a challenge of all these applications lies in their requirement to process the MPCs efficiently at high quality and to join them with other materials by using suitable joining techniques. The forming behavior of MPCs was investigated in numerous research projects [2, 3]. Due to the polymer core layer, joining is one of the most critical challenges for processing MPCs [4]. It requires new approaches compared to conventional sheet metal joining processes. State of the art joining of MPCs is shortly presented in the following. The main challenge when using conventional resistance spot welding (RSW) is the polymer core layer's electrical insulation. Different process approaches for RSW of MPCs with a metallic part are presented in [5, 6]. All approaches consist of heating and subsequent displacement of the molten core layer. These approaches mainly differ in the way the polymer core layer is heated. In the known approaches, heating is either performed by conduction through active heating of the welding electrode or by resistance heating with an additional power circuit. The easiest way to heat the polymer layer lies in applying an additional shunting circuit, which enables heating the area in the direct proximity of the electrode tip and removing the polymer layer [7]. The main drawback of these approaches comprises the relatively long heating time, which increases the process time. In [8], the joinability of MPCs with metals was fundamentally investigated using a wide variety of conventional joining methods (e. g. clinching, self-piercing riveting (SPR), and RSW). Joining was carried out without displacing the core layer. It was observed that there were several limitations regarding the joining direction, joined materials, fatigue strength when joining MPCs without displacing the core layer. Currently, no process approach is known that combines the short processing time of concepts without suppressing the core layer with the high joint quality and flexibility of those with a suppressed core layer. This paper presents a novel concept for joining MPCs with metals. The main target of the development is the application of modified and improved conventional joining technologies to MPCs without a significant increase in the processing time or decrease in joint quality.

2 Novel Joining Concept

State of the art shows that the displacement of the thermoplastic core layer is a practical solution to join MPCs with metals with good joint quality. The main challenge is to heat up the core layer quickly and without extensive process modifications. In contrast to conventional solutions, the polymer layer's local melting is realized by ultrasonic (US) waves. Applications of ultrasonic waves with a frequency range of 20–50 kHz for heating thermoplastics have been established for decades and have mainly been used for ultrasonic welding [9]. The main advantage of this technique is that the energy is directly coupled into the polymer layer. For this reason, the heating of the polymer layer until melting can be realized in a very short period of time. Two concepts for the ultrasonic-assisted joining are presented below. In the first approach, the joining tool is directly superimposed by ultrasonic waves, and the polymer layer is displaced during joining (see Fig. 1a). This approach will be used for clinching and resistance spot welding. For processes such as self-piercing riveting or refill friction stir spot welding, the joining tool cannot be used to inject ultrasonic waves in the MPCs. A second approach will be investigated for these joining processes (see Fig. 1b). Here, the polymer layer is displaced during a sheet metal forming process. Forming tools stimulated by ultrasound will be used for this purpose. Afterward, the MPCs can be joined like monolithic materials. These two procedures have in common that there is no need for an additional process step to remove polymer material from the joining zone. Thus, the same process time can be achieved for joining MPCs as for joining monolithic sheet metal materials.

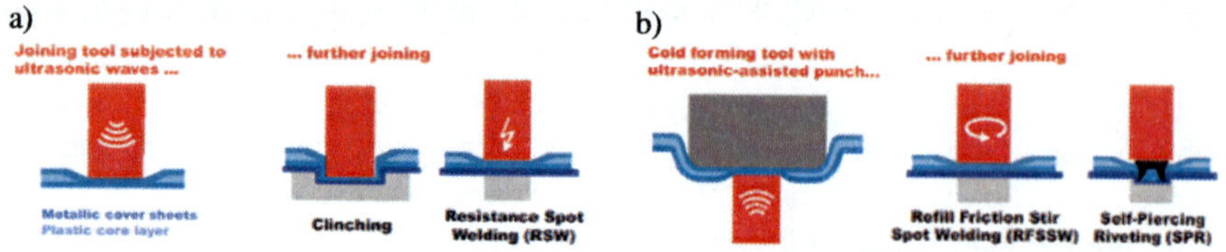

Fig. 1. Schematic representation of the two process approaches for ultrasonic-assisted joining

Regardless of the chosen approach, the first step is to examine the displacement of the core layer in detail, which is realized both experimentally and by using FE simulations. Subsequently, the design of the joining process is realized based on the results of this examination.

3 Ultrasonic Heating and Displacement – Proof of Concept

To prove whether and to what extent the ultrasonic waves can be coupled into the material and can be used to heat up the inner polymer layer, an experimental set-up was developed, which allowed the heating process to be recorded using a thermal camera. A linear tool geometry was chosen. The samples were clamped against a sapphire glass to realize frontal thermographic images. The analysis was performed using a flat sonotrode and a round anvil ($R=2.5$ mm). The result showed that the ultrasonic waves were very well transmitted through the metal skin into the polymer layer, which led to heating in the core layer (see Fig. 2). An exact temperature determination was not necessary (high effort required for calibrating the system) since the melting behavior is reflected in the recorded force-displacement curves of the ultrasonic device.

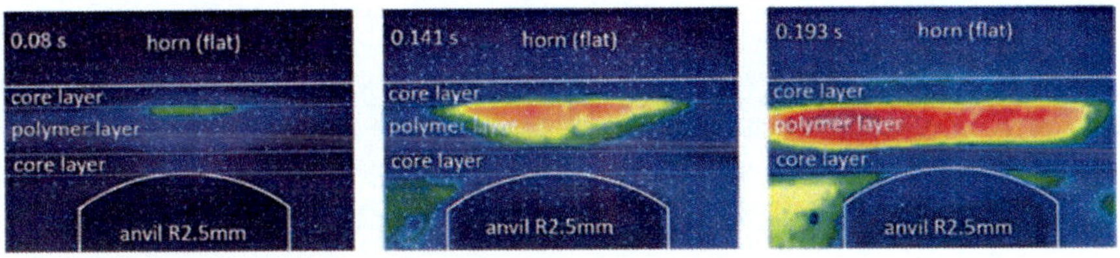

Fig. 2. Time-dependent heating of the polymer layer by ultrasonic waves

Subsequently, heating and displacement tests were carried out with the same arrangement. A radius of 2.5 mm was used for the linear sonotrode and the anvil. According to the used ultrasonic setup parameters, a force of 950 N led to the complete displacement of the polymeric interlayer in less than 1.0 s (see Fig. 3).

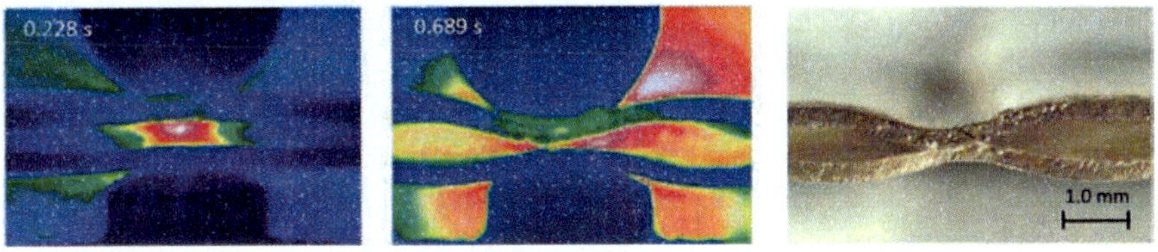

Fig. 3. Heating and displacement of polymer layer by ultrasonic waves

In summary, the heating and displacement process of such hybrid composites were realized using a conventional ultrasonic welding machine. Furthermore it was shown that the process times common in ultrasonic welding of plastics, were maintained. The effectiveness of the ultrasonic heating and displacement process can be influenced by the shape of the tools (energy directors) and by the machine parameters. A FE simulation of the heating and displacement process was carried out (Sect. 5) for the process design. Thus, the materials used were characterized in the first step.

4 Material Characterization

The novel joining concept is investigated using commercially available MPCs with aluminum and a steel cover layer. The material HYLITE® manufactured by 3A Composites is used as MPC with an aluminum cover layer and the ThyssenKrupp Steel material Litecor® as MPC with a steel cover layer (see Table 1).

Table 1. Selected MPC materials [10, 11]

Material	Overall Thickness t [mm]	Core Layer		Cover Layer	
		Material	Thickness t_k [mm]	Material	Thickness t_c [mm]
Hylite®	1.2	Polypropylene	0.8	EN AW-5182	0.2
Litecor®	1.6	PA/PE compound	0.7	CR300IF	0.3

In order to simulate the US-assisted joining process, modeling is required regarding the polymer core being heated by US waves and concerning the displacement of the polymer core layer. Therefore, a LS-Dyna internal elastic-viscoplastic material-model (*MAT_ELASTIC_VISCOPLASTIC_THERMAL [12]) is used for the polymer core layer. The mechanical and thermal properties of the selected materials were determined in order to parametrize the material model for the FE simulation. Besides the use of conventional testing methods (tension testing, dynamic mechanical analysis, differential scanning calorimetry), a particular compression test is used, providing temperature-dependent stress-strain curves under compression loads. Several displacement tests are implemented to determine the material behavior at different temperatures. For this purpose, a set of cylindrical MPC specimens with a diameter of $d_0 = 8mm$ and a height of $h_0 = 4mm$ are loaded at different temperatures. Force-displacement curves are measured during the displacement test. Using the exemplary working temperature of 180 °C, the following describes the procedure of material testing and material fitting for the simulation based on the experimental data. In order to determine the die force-displacement curves, the material test is carried out using the testing machine BT1-FR005TN.A50 by ZwickRoell GmbH & Co. KG. This testing machine performs the material test under a compressive load. The upper pressure plate moves downwards at a constant strain-rate of 0.001 $\frac{1}{s}$. Fig. 4 shows the experimental setup schematically.

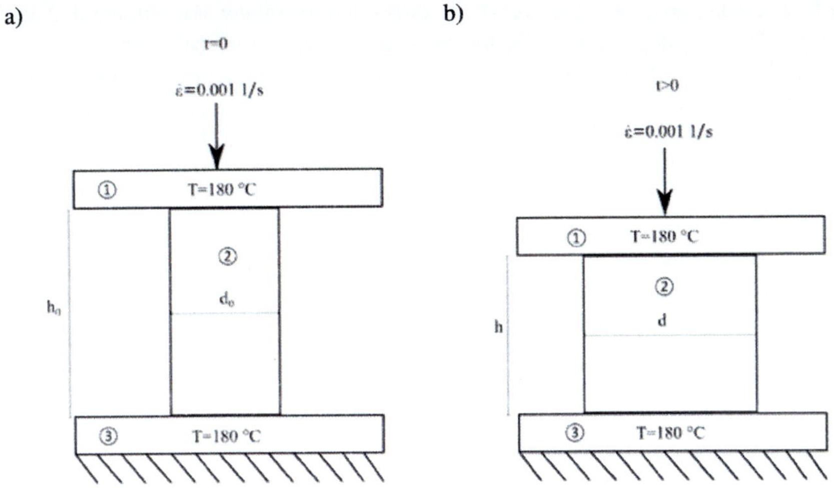

Fig. 4. Schematic setup of the pressure test: 1) movable upper pressure plate, 2) cylindrical specimen, 3) fixed lower pressure plate, a) deformation of specimen at start (t=0), b) deformation of specimen at any time of the pressure test (t>0)

The pressure plates are heated to the desired testing temperature (here $T = 180$ °C) and transfer the heat to the specimen. The force-displacement data obtained from the pressure test are used to fit the material data for the simulation. First, the average of the measured curves from several trials is calculated, resulting in the mean-curve. The adjusted curve is calculated subsequently in order to correct signal offsets. Fig. 5a shows the experimental results and the calculated mean curve and the adjusted curve.

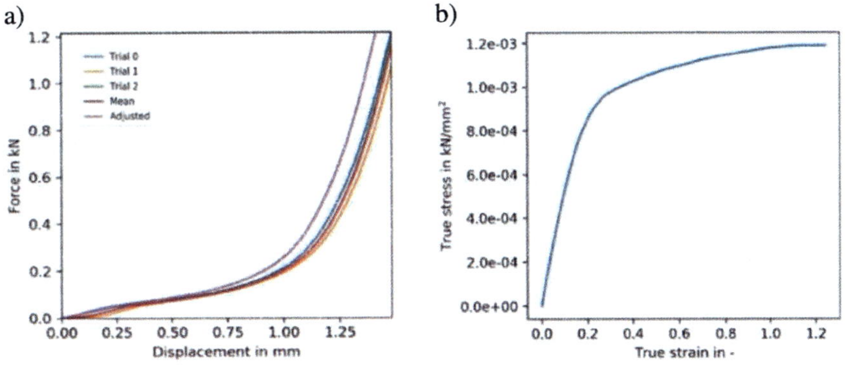

Fig. 5. Experimental and adjusted force-displacement curves (a) and calculated flow curve (b) at $T = 180$°C

In the next step, the flow curve is calculated from the experimental data (see Fig. 5b). Thus, it is assumed that the volume of the cylindrical specimen stays the same during the whole trial. The following equation calculates the time-dependent diameter d (see Fig. 4b):

$$d = d_0 \cdot \sqrt{\frac{h_0}{h_0 - h}} \tag{1}$$

with d_0 and h_0 being the diameter and the height at the beginning of the displacement test and with h being the current height, respectively (see Fig. 4). For the simulation, the true stress-strain curve is fitted by an analytical function and continued with a linear extrapolation (see Fig. 6a). In this case, the G'sell-Jonas law is used for fitting the test data [13].

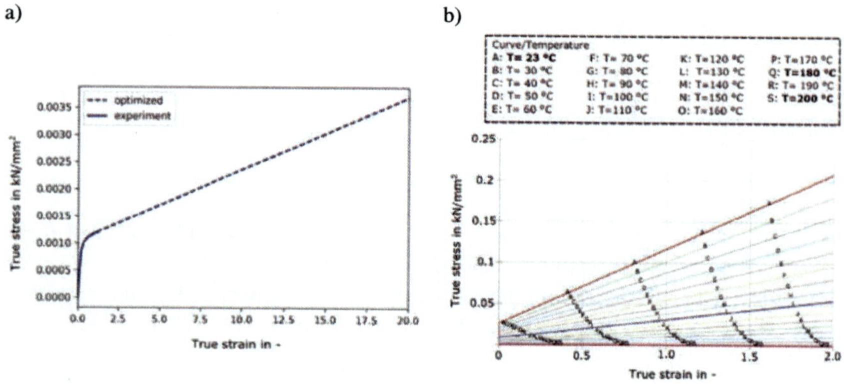

Fig. 6. Fitted data T = 180°C (a) and complete set of flow curves in the relevant temperature range (b)

Interpolation is required among the fitted temperature-dependent strain-stress curves for the complete material characterization. In Fig. 6b, the measured stress-strain curves are marked in bold (curve A, Q, and S), and the interpolated strain-stress curves are illustrated in standard font, respectively.

5 FE Modeling

The US-assisted joining process has to be modeled in order to realize an effective design of tool and process. Modeling of the ultrasonic-assisted joining process is discussed below. The displacement of the core layer and the joining process are modeled separately due to the complexity of the whole process.

5.1 Ultrasonic-Assisted Displacement of the Core Layer

The simulation of the displacement consists of three phases. First, the sonotrode exerts a force on the specimen. Second, a strain is applied due to the amplitude of the ultrasonic waves and third, a heating process takes place due to inner molecular friction. The following equation calculates the heat generation rate $\dot{Q}$ due to intermolecular friction during the ultrasonic sealing process:

$$\dot{Q} = \pi \cdot f \cdot \varepsilon_0^2 \cdot E'' \tag{2}$$

where f is the fixed ultrasonic amplitude, ε_0 is the strain amplitude and E'' is the loss modulus as a material property, respectively. The strain amplitude ε_0 is influenced by the horn amplitude and the sealing force. Simulations of oscillations are not recommended due to the required small time step that results in long durations of simulation. So far, there has been no material model in LS-Dyna that calculates heat generation due to strain. For that reason, the simulation advances within blocks (see Fig. 7). The first block calculates long-term effects such as deformation and the temperature field evolving from the heat generation and the force-controlled displacement of the sonotrode. The simulation of the oscillation of the sonotrode is realized in the second block.

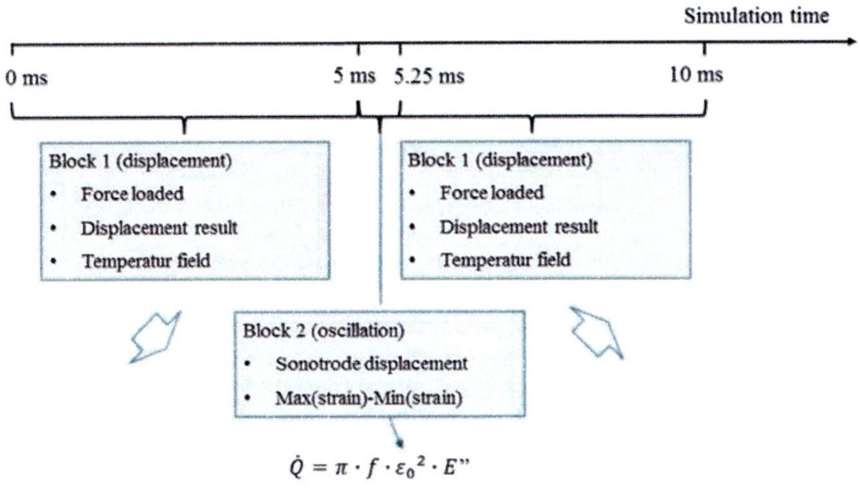

Fig. 7. Schematic workflow of the ultrasonic sealing simulation

For this purpose, the effective strain during one oscillation cycle is calculated by a position-controlled displacement of the sonotrode (see Fig. 7). The description below gives a more detailed insight into the simulation workflow. In 'Block 1' (see Fig. 7), there is a strain caused by the force that the sonotrode exerts on the specimen. Furthermore, the temperature is known in this state of the model. It is justified to assume that the temperature is constant. Subsequently, in 'Block 2', there is an additional strain due to the sonotrode's amplitude. The resulting strain ε_0 from Eq. (2) is calculated based on the difference between the strain ε_a occurring before the vibration of the sonotrode affects the material and the remaining strain ε_b when the movement of the sonotrode has stopped. Fig. 8 shows the corresponding results for Litecor®.

Thus, the strain ε_0 from Eq. (2) is calculated with the difference between the strain before and after the sonotrode displacement:

$$\varepsilon_0 = \varepsilon_a - \varepsilon_b \tag{3}$$

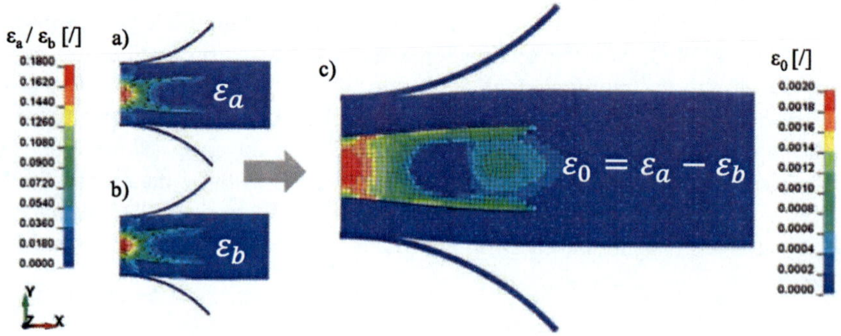

Fig. 8. a) Strain before sonotrode displacement ε_a, b) strain after sonotrode displacement ε_b and c) difference of strain ε_0

The total heat Q applied to the system is calculated by:

$$Q = \dot{Q} \times \Delta t \tag{4}$$

with $\dot{Q}$ being the calculated heat flux using Eq. (2), and Δt being the time of one amplitude, respectively. The resulting heat is mapped to the nodes of the model, and the next simulation step starts again with 'Block 1'. This workflow allows simulating the displacement process of the plastic layer. The simulation process is not able to simulate the complete displacement of the plastic interlayer yet. One to two rows of elements remain intact (see Fig. 9). The output of the displacement simulation is then used as the input data for the joining simulation. However, no material model in any simulation software exists that considers heat generation due to friction. Appyling the shown methodology, it is possible to model US-assisted heating of MPCs and the displacement of the core layer.

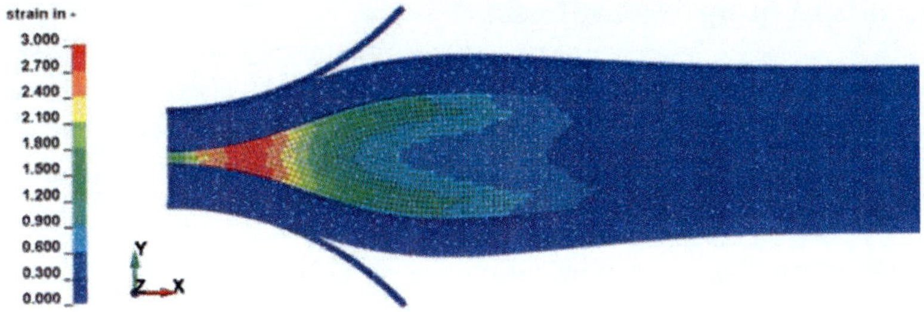

Fig. 9. Displacement process in the last time step (144 ms) of the simulation

5.2 Joining Processes

When the plastic core has been displaced, the two remaining cover sheets of the MPC can be joined to a second joining part, e. g. through clinching. This joining task is similar to a three-sheet connection, in which two sheets have a minimal thickness compared to typical joining tasks. On the one hand, it shall be prevented that the thin sheet layers are locally damaged. Additionally, the process shall be designed so that the best possible joint behavior can be achieved under load conditions. This goal shall be achieved by forming simulations.

Compared to the usual setup for clinching with larger sheet thicknesses, smaller process windows are expected in the present application. These process windows shall be determined by parameter studies regarding the influence of the clinching tool parameters on the joint formation. Other geometric parameters must be considered besides the main dimensions of the clinching punch and die. The radius at the punch edge is of particular importance. If it is too small, damage to the cover sheets is expected, while too large a radius prevents a form fit in the joint (Fig. 10). In the shown example of a mixed joint on MPC with aluminum, it is clear that a radius greater than 0.3 mm is required to avoid excessive thinning of the cover sheets. It is also apparent that an increased radius is accompanied by decreasing interlock values.

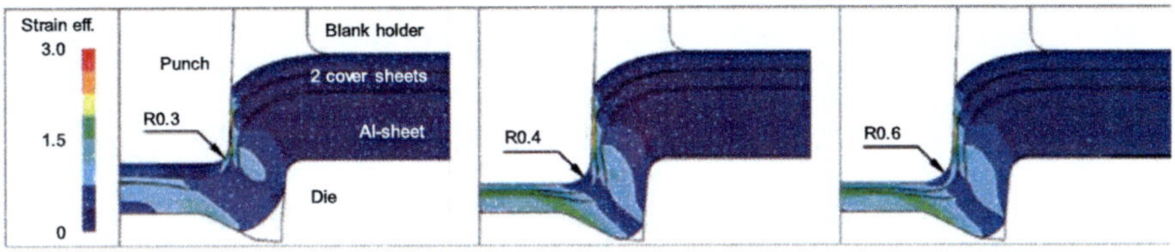

Fig. 10. Impact of tool geometry on clinch joint; Litecor® cover sheets, t=0.3 mm+EN-AW 5182, t=1.1 mm

6 Results of Suppressing Tests

A zone with a 16 mm diameter was chosen as a case study, considering the intended joining processes to be investigated in the project. The tests for heating and displacement of the polymer layer were carried out on the ultrasonic welding machine “MS Ultrasonic—SonxTop Newton”. In the presented setup, the material HYLITE® was used with an overall thickness $t = 1.2$ mm.

In ultrasonic processes, the effectiveness of the energy transfer primarily depends on the design of the contact area between the sonotrode and the workpiece. When joining polymer films, the tools (sonotrode and anvil, or anvil alone) are equipped with energy direction transmitters, which enable punctual or linear contact. In order to analyze the influence of different energy directors, three different anvil contours were chosen (spherical, ring-shaped, wave-shaped) with a flat round sonotrode $D = 16$ mm. The polymer layer’s displacement is terminated when the two outer skins of the MPC are in contact. This termination could not be achieved with the wave-shaped anvil, since the pressure in the melted polymer was too high, which led to the cracking of one sheet metal. A sonotrode with a flat surface was used for comparison (Fig. 11).

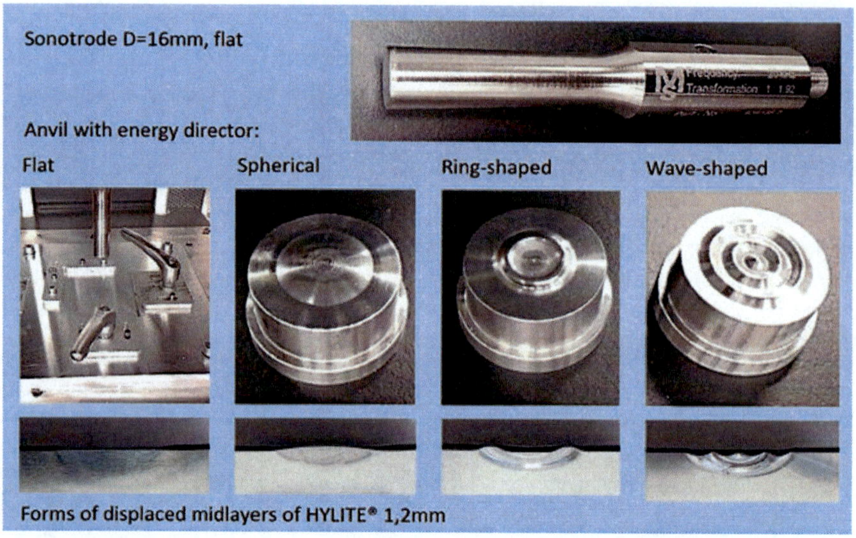

Fig. 11. Results of displacement with different energy directors

The result shows that the efficiency of heating increases significantly due to the energy directors. The heating time is reduced from approx. 3.5 s to approx. 1 s. Moreover, the force required to press the two outer skins together can be reduced from approx. 4000 to approx. 2000 N. The shape of the energy direction sensors plays a subordinate role and can be specified by the required tool geometry within certain limits, e. g. for clinching or spot welding. During further investigations, the specimens with the displaced core layer will be joined by clinching, RSW, SPR, and refill friction stir spot welding.

7 Conclusion

The presented work describes a novel joining method for MPCs. Compared to state of the art solutions, the polymer core layer is displaced by the application of US vibrations. The aim is to significantly reduce the process time (especially the heating time) compared to conventional solutions where heating is realized, for example, by conduction. The working principle is investigated utilizing US heating and displacement tests, with the temperature distribution in the sandwich being recorded by a thermal camera. It has been proven that ultrasound can be used to generate heat directly in the polymer core layer.

Furthermore, the resulting processes have many degrees of freedom regarding tool geometry and process parameters. Both the US-assisted displacement process and the subsequent joining process are simulated for efficient process design. For this purpose, the used materials are characterized mechanically. This paper presents a specific testing and evaluation procedure for determining temperature-dependent flow curves under compression. Based on these results, a workflow is developed and carried out for modeling the heat generation due to intermolecular friction. The results of these simulations are used as input data for the joining simulation to determine the tool geometry's impact on the quality of the clinched joint. Finally, displacement tests are performed using various tool geometries. In the next step, the specimens with the displaced core layer will be joined by clinching, RSW, SPR, and refill friction stir spot welding.

Acknowledgements. The presented studies are funded within the project 'Ultrasonic supported processing of hybrid materials' of industrial research under grant no. 241 EBR/P 1374. The project is running in the Collective Research Network 'Cornet' under the coordination of 'Forschungsvereinigung Stahlanwendung e. V. (FOSTA)'.

References

1. Greshham, J., Cantwell, W., Cardew-Hall, M., Compston, P., Kalayanasundaram, S.: Drawing behaviour of metal composite sandwich structures. Compos. Struct. **1–4**(75), 305–312 (2006)
2. Kroll, L.,Wührl, M., Landgrebe, D., Riemer, M., Marburg, S., Geweth, C.: Reduktion der Schallabstrahlung durch gezielte Schubdämpfung in hybriden Metall-Kunststoff-Verbunden. EFB-Forschungsbericht 446, Hannover (2016)
3. Harhash, M.: Forming Behaviour of Multilayer Metal/Polymer/Metal Systems. Technische Universität Clausthal (2017)
4. Gude, M., Lieberwirth, H., Meschut, G., Zäh, M.F.: FOREL-Studie—Chancen und Herausforderungen im ressourceneffizienten Leichtbau für die Elektromobilität (2015)

5. Chergui, A., Niesen, A.: Mehrstufiges Widerstandsschweißen von Sandwichblechen. DE102014109505A1, 14 Jan 2014
6. Voigt, A., Graul, M.: Fügen von blechartigen Bauelementen mit Zwischenschicht aus thermoplastischen Kunststoff. DE102011109708A1, 7 Feb 2013
7. Tanco, J.S., Nielsen, C.V., Chergui, A., Zhang, W., Bay, N.: Weld nugget formation in resistance spot welding of new lighttweight sandwich material. Int. J. Adv. Manuf. Technol. **80**, 1137–1147 (2015)
8. Meschut, G., Schmal, C.: Mechanisches Fügen und Hybridfügen von Metall-Kunststoff-Hybriden mit Metallen. EFB-Forschungsbericht 481, Hannover (2018)
9. DIN ISO 857: 1977–09: Schweißen; Schweißen von Kunststoffen, Verfahren
10. Hylite Homepage. https://media.alucobond.com/pdf/hylite/HYLITE_Flyer_DE.pdf. Last accessed 28 Feb 2020
11. DSpace Homepage. https://dspace.cvut.cz/bitstream/handle/10467/79567/F2-BP-2018-Vlk-Vladimir-priloha-201500716_VSM_Technische_Informationen_LITECOR.pdf?sequence=-1&isAllowed=y. Last accessed 28 Feb 2020
12. LSTC – Livermore Software Technology Corporation: LS-Dyna® Keyword user's manual – Volume 2 Material Models (2018)
13. Becker, F., Kraatz, A., Wieser, J.: Determination of the Behaviour of Thermoplastics at High Strain-Rates Using the Invariant Theory. 7. LS-DYNA Anwenderforum, Bamberg (2008)

Experimental Parameter Identification for the Bending Based Preforming of Thermoplastic UD-Tape

Daniel Kupzik(✉), Alexej Bachtin, Sven Coutandin, and Jürgen Fleischer

Karlsruhe Institute of Technology, wbk Institute of Production Science, Kaiserstraße 12, 76131 Karlsruhe, Germany
daniel.kupzik@kit.edu

Abstract. The combination of different types of plastics can be advantageous to obtain a good component performance at a reasonable price. For this, a compromise between mechanical properties and cheap manufacturing has to be found. In the research project GRK2078, the manufacturing of hybrid components consisting of long fiber reinforced thermoplastics (LFT) and unidirectional endless fiber reinforced material (UD-tapes) is researched. The LFT is the main constituent of the components. Local UD-tape reinforcements are added in areas with high load for a large effect or low geometric complexity to minimize preforming effort. To keep the cost low, a novel preforming process for the UD-tapes is in development at wbk Institute of Production Science. The aim is to enable the shape flexible, tool less forming of UD-tapes. To obtain this, the process is based on sequential bending of the UD-tape on several positions along its longitudinal axis during handling operations. To conduct the process, a supply unit and an industrial robot with a gripper with integrated heating devices are used. In this paper, the experimental examination of the process and the identification of suitable parameters on the shape of the preform are presented. The main influencing factors are the type of UD-tape, the heating of the tape and the movement of the robot. The processes are conducted with PA and PP carbon fiber tapes. The process quality with contact and radiation heating are compared and the respective heating duration is identified. For the robot movement, a kinematic description of the process is derived and compared to a circular bending movement. With the identified parameters, the process can be conducted reliably. The resulting accuracy limit and the process time with these parameters are described in this paper.

Keywords: Preforming · Handling · UD-tape · Bending · Kinematics

K. Dröder and T. Vietor (Eds.): *Technologies for economic and functional lightweight design*, Zukunftstechnologien für den multifunktionalen Leichtbau, pp. 313–325, 2021. https://doi.org/10.1007/978-3-662-62924-6_27

1 Introduction

In the manufacturing of CFRP components, preforming is an essential step for the mechanical performance and has a huge influence on the manufacturing cost. In the following section, a short overview over a variety of preforming processes is given. Afterwards, requirement and peculiarities of the manufacturing of locally reinforced, hybrid components are described. From these, a novel process, based on the local bending, is derived. The main section of this paper is focused on the description of the necessary bending movement as well as the identification of suitable process parameters. In the end, an outlook onto further fields of work is given.

In CFRP manufacturing, preforming can be an important step for achieving good quality in complex parts. During preforming, the raw material will be brought to a near net shape geometry which can be formed into the desired component during molding. To enable a reliable molding process, the preform will have to lie stable in the mold before closing and it either needs to have the right shape before molding or has to reliably be formed to the right shape in the mold without wrinkles and other defect. Components with low three dimensional complexity like plates (e.g. car roof or hood) may be molded from a flat layup without a preforming step while more three dimensional components like (e.g. the firewall or inner panels of a car) often need a more cautious preparation of the stack. In thermoplastic FRP manufacturing, preforming may [1] or may not [2] be used depending on the workpiece complexity. In manufacturing, separate stations can be used for preforming like the stamp based preforming in RTM [3, 4]. This way, the process can be controlled sensitively but additional process steps and a cost increase are caused. Furthermore, stamp draping requires component specific tools. To avoid these disadvantages, several approaches of handling integrated preforming are researched. In [5], a compliant end effector which can be used to drape dry textiles onto a component specific counterpart is presented. Due to the large contact area in the end effector and counterpart model, results cannot be directly transferred to tacky pre-impregnated materials. In [6], a unit for the preforming of stacks is presented which is based on a number of flexibly movable gripping units. Although being component flexible, this approach is limited to low degrees of preforming due to the complex kinematics of the preforming of finished stacks. Additionally, the grippers would have to be adapted for the handling of preimpregnated material. In [7] a handling device for the simultaneous grasping, heating and preforming of thermoplastic organic sheets is described. The device is grasping the organic sheets with vortex gripper, heating them by applying hot air and then bending them along a hinge within the end effector. In this approach, preforming is limited to a single bend in the stack while the degree and position of the bend are flexible. However, this may be sufficient and very efficient for some approaches. An end effector for the preforming of longish UD-tape strips was implemented in [8]. In this approach, the gripper units are moved by pneumatic cylinders between two component specific positions. This allows a high degree of preforming while forming a whole stack of tape strips. The disadvantage is that the end effector can only be used for one component. In [9], it is demonstrated, that preforms can be manufactured sequentially and assembled after storing them in a storage system in some applications. In all these processes, requirements of the final part determine the effort which is necessary for the preforming. If e.g. a large surface stack has to be preformed, more degrees of freedom of the process have to be controlled compared to the manufacturing of a linear reinforcement structure. In the latter case, sharp corners and the most difficult component regions can be avoided. A second simplification can be the use of subpreforming strategies, where smaller preforms are made and later assembled.

In the current research at wbk Institute of Production Science within the GRK2078 project, the last three approaches [7–9] are combined to find a new process for the relatively simple prefoming tasks for reinforcement structures. In the novel process, flexibility is guaranteed by determining the subpreform's shape by iteratively combining local bends with variable positions and degrees of bending. This process is made efficient and the potential for complex geometries is increased by processing longish tape strips. The subpreforms can afterwards be assembled to a complex reinforcement geometry which is co molded with LFT or unreinforced thermoplastics.

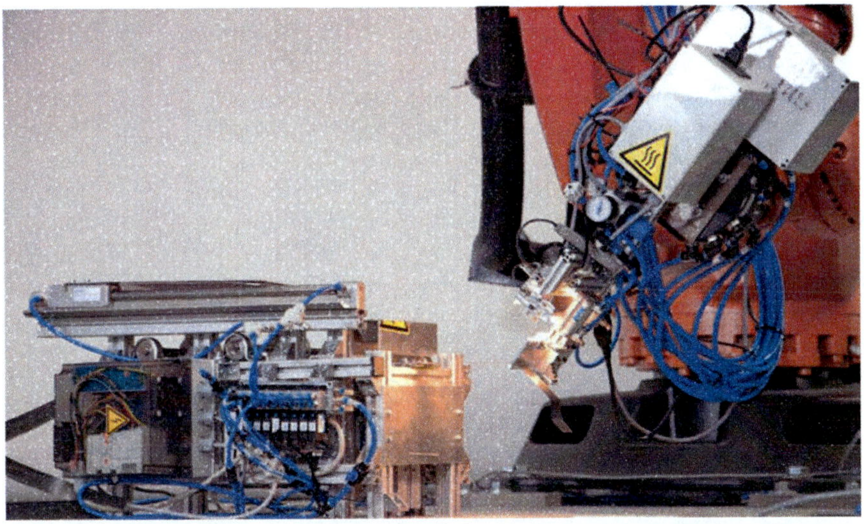

Fig. 1. Bending of a single Tape strip: The strip is pushed out of the supply unit (left). Then it is heated and bent by the robot repeatedly. Finally, it is cut by the supply unit.

The iterative bending process is implemented in the experimental robot cell shown in Fig. 1. The cell consists of a supply unit for the controlled supply of a certain length of 30 mm wide tape strip, an end-effector mounted to an industrial robot and an early prototype of a joining station (from left to right). In this paper, the bending of a single preform is described. For this, four steps are necessary (see [10] for a video):

1. The supply unit will move the tape strip to a defined position.
2. The end effector will move to the gripping position on the tape. The tape is clamped by the gripper and heated by the heating (see Fig. 2) unit which is mounted to the end effector.
3. The robot will conduct a bending movement and wait for the molten area on the tape to cool. If several bends are desired, the process will jump back to step 2.
4. In the end, the tape is gripped close to the supply unit, cut off from the roll and handed over to a storage or the joining station.

As the geometry is defined by the bends on the tape strip, certain shape limitations occur. To overcome this disadvantage, an automated optimization of the position and angles of the bends for certain workpieces was developed [11]. In the initial experimentation on the process conduction, two challenges became apparent. The first is the choice of a bending movement and the second the choice of appropriate heating parameters for achieving a high accuracy and low process time.

Initially, a pure rotation around a fixed axis was chosen as bending movement. This proved to be insufficient, as the correct length of the fibers had to be maintained which is geometrically impossible with this approach if the bending radius is larger than zero. Depending on the bending angles, the distance between supply unit and clamp would either increase or decrease. If the distance in the end of the movement was too large, there would be tension on the tape which would prevent the bending line from being in the molten area. If the distance was too small, buckling would occur. Therefore, a kinematic description was derived which is described in Sect. 2.

The second challenge is the correct heating of the tape. Too little heat would not melt the tape sufficiently and therefore increase process forces while too much heat would unnecessarily increase process time and might damage the matrix material. Two heating units were built which can be mounted to the end effector alternately (see Fig. 2). The parameters which are used within the kinematic description and the heating parameters are examined in Sect. 3.

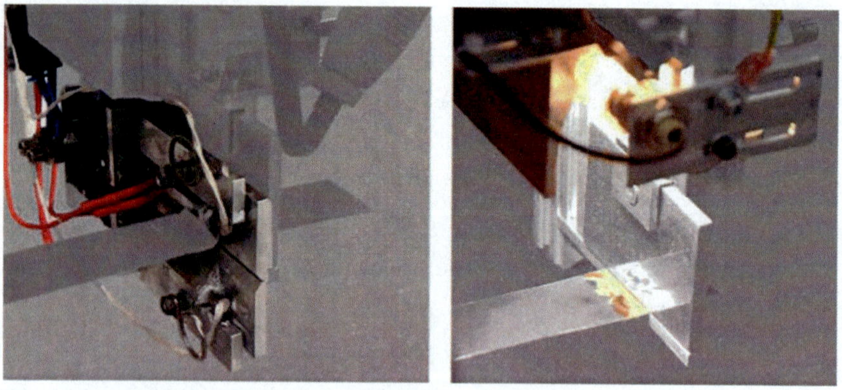

Fig. 2. Contact heating (left): left of the clamping unit, a gripper with heated aluminum jaws is mounted. It can be closed to melt the tape locally. Heating by IR-radiation (right): Above the gripper, a lightbulb with a mirror is mounted above the gripper to heat the tape locally.

2 Kinematic Description of the Forming Movement

The initial experimentation with a fixed predefined tool-center-point (TCP) as rotational axis of the end effector showed that the assumption of a pure rotation is an oversimplification and leads to manufacturing errors. The new kinematics of the forming movement introduced for this reason is essentially based on the unfolding of the bending radius R_0 along the neutral axis. The constant arc length b_0 and the ideal circular arc of the bending radius serve as assumptions for the mathematical equations. Fig. 3 schematically shows the three states of the forming in black (unfolded geometry/undeformed tape), orange (current bending geometry) and green (nominal bending geometry).

By introducing the parameters and assumptions according to Fig. 3, the expression of the geometric relationship and its transformation according to the current radius is:

$$b_0 = R_0 \cdot q_{30} = R(t) \cdot q_3(t) = const., \tag{1}$$

$$R(t) = R_0 \frac{q_{30}}{q_3(t)}. \tag{2}$$

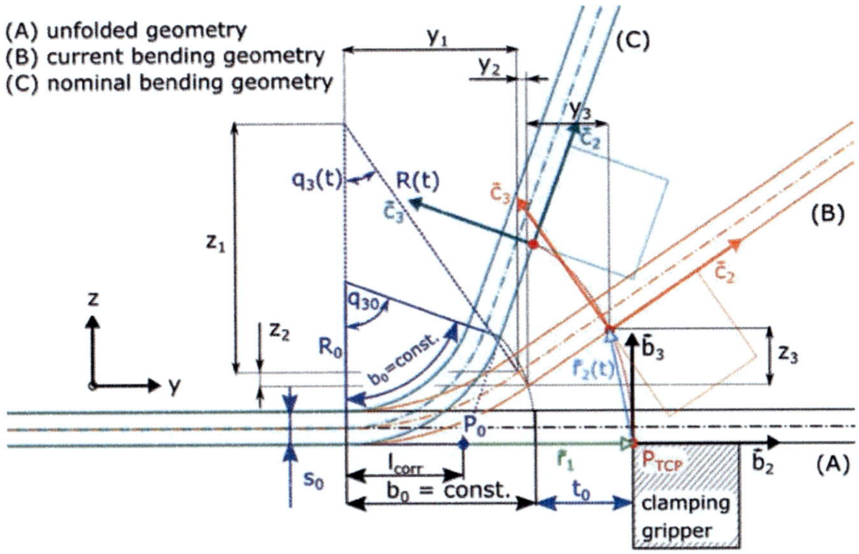

Fig. 3. Graphical representation of the forming kinematics. (b_0: arc length, R_0: nominal bending radius, $R(t)$: actual bending radius at the time t, q_{30}: nominal bending angle, $q_3(t)$: actual bending angle at the time t, s_0: material thickness, t_0: tool-center-point (TCP) offset, P_0: point of intersection, $\vec{r}_1$: approach vector, $\vec{r}_2$: bending vector)

Related to the same figure, further trigonometric equations can also be set up, which are necessary for setting up the vectors to describe the kinematics. The directional dependency is ensured by setting the nominal angle in amount.

$$y_1(t) = R(t) \cdot sin(q_3(t)) = \frac{|q_{30}|}{q_3(t)} \cdot R_0 \cdot sin(q_3(t)) = \frac{b_0}{q_3(t)} \cdot sin(q_3(t)) \quad (3)$$

$$y_2(t) = \frac{s_0}{2} \cdot sin(q_3(t)) \quad (4)$$

$$y_3(t) = t_0 \cdot cos(q_3(t)) \quad (5)$$

$$z_1(t) = R(t) \cdot cos(q_3(t)) = \frac{|q_{30}|}{q_3(t)} \cdot R_0 \cdot cos(q_3(t)) = \frac{b_0}{q_3(t)} \cdot cos(q_3(t)) \quad (6)$$

$$z_2(t) = \frac{s_0}{2} \cdot cos(q_3(t)) \quad (7)$$

$$z_3(t) = t_0 \cdot sin(q_3(t)) \quad (8)$$

The point of intersection P_0 is suitable as a reference point for describing the vectors and, at the same time, for maintaining the position of the finished formed surface. The coupling of the equations between the approach vector $\vec{r}_1$ and P_0 must take place over the length l_{corr}, which varies according to the target geometry. The correction is necessary because the shape optimization [11] is calculated with perfectly sharp bends.

$$l_{corr} = \left(R_0 + \frac{s_0}{2}\right) \cdot tan\left(\frac{q_{30}}{2}\right) = \left(\frac{b_0}{q_{30}} + \frac{s_0}{2}\right) \cdot tan\left(\frac{q_{30}}{2}\right),$$
$$with \quad l_{corr} = \begin{cases} 0, & |\ q_{30} = 0 \\ \left(\frac{b_0}{q_{30}} + \frac{s_0}{2}\right) \cdot tan\left(\frac{q_{30}}{2}\right), & |\ else. \end{cases} \quad (9)$$

Using this relationship, it is now possible to establish the approach vector $\vec{r}_1$ and then the bending vector $\vec{r}_2$. The rotation of the end effector is determined by the nominal angle q_{30}.

$$\vec{r}_1^{\,B} = (t_0 + b_0 - l_{corr}) \cdot \vec{b}_2 = \left(t_0 + b_0 - \left(R_0 + \frac{s_0}{2}\right) \cdot tan\left(\frac{q_{30}}{2}\right)\right) \cdot \vec{b}_2$$
$$= \left(t_0 + b_0 - \left(\frac{b_0}{q_{30}} + \frac{s_0}{2}\right) \cdot tan\left(\frac{q_{30}}{2}\right)\right) \cdot \vec{b}_2,$$
$$with \quad \vec{r}_1^{\,B} = \begin{cases} \vec{0}, & |\ q_{30} = 0 \\ \left(t_0 + b_0 - \left(\frac{b_0}{q_{30}} + \frac{s_0}{2}\right) \cdot tan\left(\frac{q_{30}}{2}\right)\right) \cdot \vec{b}_2, & |\ else. \end{cases} \quad (10)$$

$$\vec{r}_2^B(t) = \begin{bmatrix} 0 \\ \Delta y(t) \\ \Delta z(t) \end{bmatrix}_{\{\vec{b}_1,\vec{b}_2,\vec{b}_3\}} = \begin{bmatrix} 0 \\ -b_0 - t_0 + y_1(t) + y_2(t) + y_3(t) \\ \frac{s_0}{2} + R(t) - z_1(t) - z_2(t) + z_3(t) \end{bmatrix}_{\{\vec{b}_1,\vec{b}_2,\vec{b}_3\}},$$

$$\vec{r}_2^B(t) = \begin{bmatrix} 0 \\ -b_0 - t_0 + \left(\frac{b_0}{q_3(t)} + \frac{s_0}{2}\right) \cdot sin(q_3(t)) + t_0 \cdot cos(q_3(t)) \\ \frac{s_0}{2} + \frac{b_0}{q_3(t)} - \left(\frac{b_0}{q_3(t)} + \frac{s_0}{2}\right) \cdot cos(q_3(t)) + t_0 \cdot sin(q_3(t)) \end{bmatrix}_{\{\vec{b}_1,\vec{b}_2,\vec{b}_3\}}, \quad (11)$$

$$with \quad \vec{r}_2^B(t) = \begin{cases} \vec{0}, & | \; q_3 = 0 \\ \vec{r}_2^B(t), & | \; else. \end{cases}$$

In this view, the vectors describe only the two-dimensional case in relation to the base system $\left\{\vec{b}_1, \vec{b}_2, \vec{b}_3\right\}$. The robot-based preforming process also permits more complex geometries in the forming of the UD-tapes, as shown in the Fig. 4 in two rotation steps. The additional angle q_1 describes the rotation of the end effector around the a_3 -axis. This also allows the production of askew bending edges. To realize this forming movement, the established vectors are transferred into the three-dimensional case via the transformation relationship

$$\vec{r}_i^A = \left(\vec{r}_i^B\right)^T {}^B\tilde{\boldsymbol{\omega}}^A \tilde{\boldsymbol{a}} \; with \; {}^B\tilde{\boldsymbol{\omega}}^A = \begin{bmatrix} cos(q_1) & sin(q_1) & 0 \\ -sin(q_1) & \cos(q_1) & 0 \\ 0 & 0 & 1 \end{bmatrix}, \tilde{\boldsymbol{a}} = \begin{bmatrix} \vec{a}_1 \\ \vec{a}_2 \\ \vec{a}_3 \end{bmatrix}, \tilde{\boldsymbol{b}} = \begin{bmatrix} \vec{b}_1 \\ \vec{b}_2 \\ \vec{b}_3 \end{bmatrix} \quad (12)$$

With this relationship, the defined vectors $\vec{r}_1^B$ und $\vec{r}_2^B$ in the fixed base system $\{\vec{a}_1, \vec{a}_2, \vec{a}_3\}$ are in general case

$$\vec{r}_1^A = \begin{bmatrix} -\left(t_0 + b_0 - \left(\frac{b_0}{q_{30}} + \frac{s_0}{2}\right) \cdot tan\left(\frac{q_{30}}{2}\right)\right) \cdot sin(q_1) \\ \left(t_0 + b_0 - \left(\frac{b_0}{q_{30}} + \frac{s_0}{2}\right) \cdot tan\left(\frac{q_{30}}{2}\right)\right) \cdot cos(q_1) \\ 0 \end{bmatrix}_{\{\vec{a}_i\}}, \quad (12)$$

$$with \quad \vec{r}_1^A = \begin{cases} \vec{0}, & | \; q_{30} = 0 \\ \vec{r}_1^A, & | \; else. \end{cases}$$

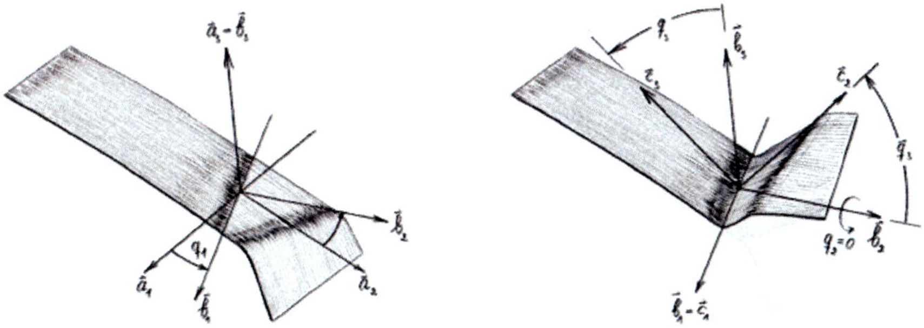

Fig. 4. Rotation of the end effector around the a_3-axis (left) and the following bending (right).

$$\vec{r}_2^A(t) = \begin{bmatrix} \left(t_0 + b_0 - \left(\frac{b_0}{q_3(t)} + \frac{s_0}{2}\right) \cdot sin(q_3(t)) - t_0 \cdot cos(q_3(t))\right) \cdot sin(q_1) \\ -\left(t_0 + b_0 - \left(\frac{b_0}{q_3(t)} + \frac{s_0}{2}\right) \cdot sin(q_3(t)) - t_0 \cdot cos(q_3(t))\right) \cdot cos(q_1) \\ \frac{s_0}{2} + \frac{b_0}{q_3(t)} - \left(\frac{b_0}{q_3(t)} + \frac{s_0}{2}\right) \cdot cos(q_3(t)) + t_0 \cdot sin(q_3(t)) \end{bmatrix}_{\{\vec{a}_i\}},$$

$$with \quad \vec{r}_2^A(t) = \begin{cases} \vec{0}, & \mid q_3 = 0 \\ \vec{r}_2^A(t), & \mid else. \end{cases} \tag{13}$$

These formulas describe the kinematic description based on the movement presented in Fig. 3. They are implemented in the robot control to maintain the correct length of the fibers during the movement for all relevant angles.

3 Process Parameters and Influencing Factors

In the preforming process presented here, two different heating processes can be used, which are necessary to initiate the forming process. These are the local heating by means of radiation heating or by means of contact heating. The investigation of the process, respecting the newly introduced kinematics for describing the bending movement, leads to the process parameters presented in Table 1. In the two right columns it is checked for which heating method the parameters are relevant. The deviations of the nominal approach angle dq_1 and the nominal bending angle dq_3 are mainly examined as target values. The interactions with increasing UD-tape length L also form a target of investigation.

Table 1. Definition and assignment of process parameters

	Process parameters		Unit	IR heating	Contact heating
1	approach angle	$[q_1]$	°	✓	✓
2	bending angle	$[q_3]$	°	✓	✓
3	TCP offset	$[t_0]$	mm	✓	✓
4	arc length	$[b_0]$	mm	✓	✓
5	preheating time	$[t_{PH}]$	s	✓	✓
6	heating time	$[t_H]$	s	✓	
7	cooling time	$[t_C]$	s	✓	✓
8	heating temperature	$[T]$	°C		✓
9	prestressing	$[P]$	mm		✓
10	distance of the bend	$[L]$	mm	✓	✓

For the evaluation of the angular deviations dq_1 and dq_3 different measuring methods are used in this study. The first is a 3D-angle-measuring-device developed within the framework of the research. The device is manufactured using the 3D printing method Fused Deposition Modeling (FDM) and allows on-site measurement of both angles in a simple way. This device was used for all measurement of single bends in this thesis. The realization of the device is shown in the picture below.

Fig. 5. The design and 3D printing of the 3D-angle-measuring-device.

Additionally, industrial 3D scanning technology (ATOS Triple Scan Technology) is used for measuring larger tape strips with several bends. The samples are scanned individually using the stereo camera principle [12]. This allows the measurement of the samples under constant conditions, which is especially useful for longer samples due to the disturbing effect of gravity. Measuring against the CAD model is also possible using the best-fit method.

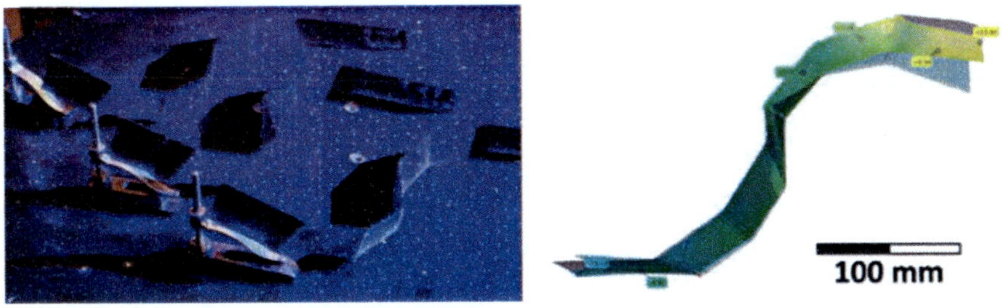

Fig. 6. Arrangement of the samples and the result in the GOM-software [12].

Currently, an internally developed stereo camera-based measurement technology is also being integrated into the experimental robot cell. The angles are measured inline. This device shall be used for future measurement to simplify experimentation and for an automated process control. The research on the automated measurement and correction of the parameters using the camera system shall be published separately. The comparison showed that the results of all three measurement systems in their applicable range lies within an error range of 2°. Currently, the only outlier is the internally developed stereo camera system which cannot measure angles between -20° and 20° or above 50° and therefore has a low reliability. Therefore, the device shown in Fig. 5 was used for the evaluation of the experimentation.

Due to the large number of adjustable process parameters, the analysis of the manufacturing with IR heating is based on fractional factorial experimental design. The resulting alias structure of the preferred experimental designs allows only a reliable statement about the main effects. The interactions cannot be clearly assigned. In addition to the derived process parameters, two different UD-tape variants are also considered. These are carbon fiber (CF) reinforced thermoplastics with polyamide (PA) (Cetex TC910 PA6, Carbon) and polypropylene (PP) (Cetex TC960 PP, Carbon) matrix.

Figure 7 shows the average effects of both UD-tape variants on the deviation of the bending angle dq_3 when adjusting the main process parameters. Across all process parameters, the mean bending angle is 4° smaller than the nominal value of q_3 . In general, it is recommended to differentiate between matrix material groups. The effect can be up to 3°, depending on the choice of material. The most significant mean effects are seen when modifying TCP offset t_0 and the cooling duration t_C. The adjustment of the t_0 value from 0 mm to 10 mm leads to a mean reduction of the deviation by almost 3°. Increasing the cooling time from 4 s to 8 s also contributes to an improvement and leads to an effect of about 1.5°. The arc length b_0 and the heating duration t_H shows hardly any influence here. The effect of the preheating duration t_{PH} is also only weakly pronounced and contributes to a small deterioration of less than 1° when increasing from 0 s to 4 s. It may be necessary to differentiate between the materials here.

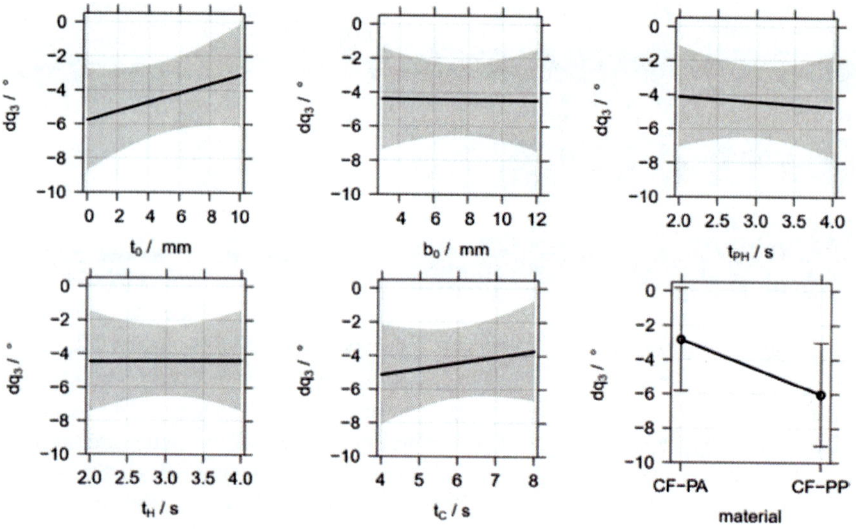

Fig. 7. IR heating – Main effects on bending angle deviation dq_3 in degree.

While the arc length has no major effect on the angle q_3, a controlled change in radius can be achieved by the interaction of t_0 and b_0 and the melting width controlled by the heating time. The Fig. 8 below illustrates the aspect of radius enlargement and highlights the possibilities of the newly introduced bending kinematics.

Fig. 8. IR heating – Deliberate radius enlargement from 3 mm to 7 mm due to adjustments in the bending kinematics' parameters.

Fractional factorial experimental designs are also the first choice for the investigation of effects using contact heating. An additional series of experiments with full factorial designs allows a detailed analysis of the interactions when varying the process parameters. A particularly important factor is the heating temperature T, which is set by a temperature controller. Depending on the choice of material and the composition of the matrix, different temperatures are required. If the temperatures are too low, the melting region is too small and the bending surface tends to spring back after forming. Based on the results of the experiments, this mean tendency is visible in the Fig. 9.

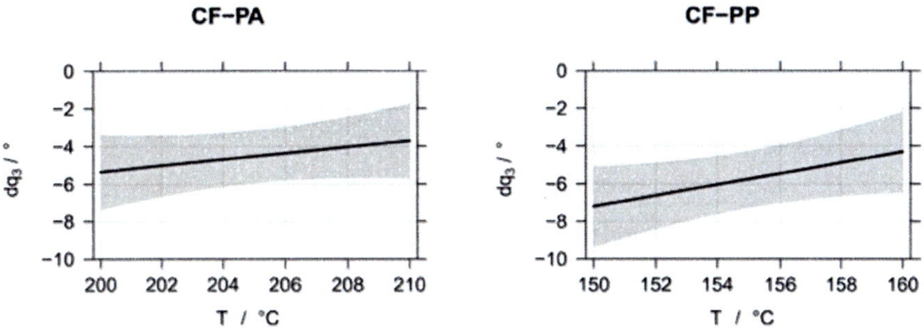

Fig. 9. Contact heating – Effect of temperature *T* on the deviation of the bending angle dq_3.

Using contact heating, the left diagram in the Fig. 10 also shows a significant reduction in the bending angle deviation dq_3 when the cooling time is increased to 8s. Reducing the preheating time t_{PH} to 2 s can slightly support this effect. The diagram on the right indicates an interaction between the prestressing P and the adjustment of the tcp offset t_0. Only if the prestress remains inactive (P = 0 mm), a significant effect of 5° can be seen when modifying t_0. This can be caused by the design of the contact heating. The contact heating protruding 6 mm in front of the tool-center-point can create a leverage effect at $t_0 = 0mm$. During bending, a kind of "dynamic prestress" occurs, which has a positive effect on the result.

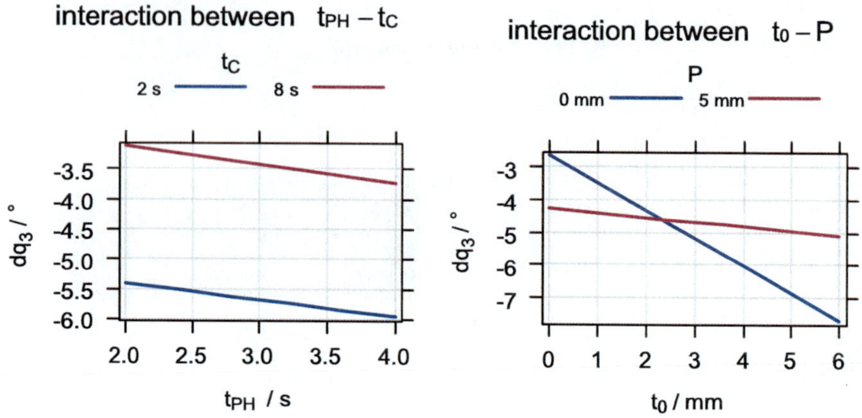

Fig. 10. Contact heating – Interaction between t_0-P in [mm] and $t_{PH} - t_C$ in [s].

The bending sequence used in these series of experiments as shown in the Fig. 6 also allows an insight into the behavior of the length-dependent angular deviations dq_3. The UD-tape was bent every 50 mm along its length with q_3 varying from $q_3 = -45°$ to $+45°$ with variable approach angle $q_1 = -25°/0°/+25°$. The effects when the parameters are adjusted in the front area of L = 50 mm show a relatively small effect in the Fig. 11. At a tape length of 300 mm these can make up to 12° difference. Therefore, the geometry of the preform is also an important influencing factor.

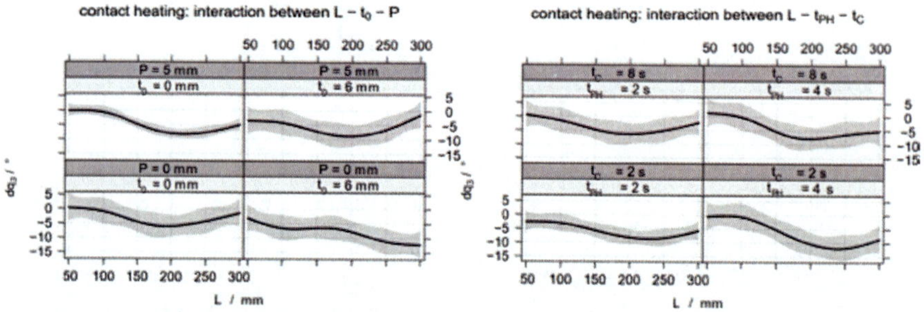

Fig. 11. Contact heating – Length-dependent interaction between t_0-P in [mm] and $t_{PH} - t_C$ in [s].

4 Summary and Outlook

In this paper, a novel process for the preforming of UD-tape strips has been introduced. To avoid manufacturing errors, a kinematic description which maintains a correct unfolded length of the fibers in the tape was found and implemented in the robot cell. The second necessary ingredient for a correct bend geometry are good parameters for this description and sufficient heating before bending as well as cooling after bending. Fractional factorial experimental designs were conducted to identify the main effects of the parameters and reasonable values of the parameters. An initial solution for bending errors of less than 3° could be found. In future work, large experimental plans will be conducted to fine tune the values of the parameters and to characterize process variations. The stereo camera measurement system which has recently been implemented will help to automatically execute those experimental plans. Also, it will be used to characterize and correct systematic manufacturing errors.

Acknowledgement. The research documented in this manuscript has been funded by the German Research Foundation (DFG) within the International Research Training Group "Integrated engineering of continuous-discontinuous long fiber reinforced polymer structures" (GRK 2078). The support by the German Research Foundation (DFG) is gratefully acknowledged.

References

1. Altstädt, V., Spörrer, A., Mühlbacher, M., Michel, P., Seidel, S.: Großserientauglicher Hochleistungsleichtbau. *Kunststoffe* **5** (2012)
2. Kropka, M., Muehlbacher, M., Neumeyer, T., Altstaedt, V.: From UD-tape to final part – a comprehensive approach towards thermoplastic composites. Procedia CIRP **66**, 96–100 (2017). https://doi.org/10.1016/j.procir.2017.03.371
3. Behrens, B.-A., et al.: Automated stamp forming of continuous fiber reinforced thermoplastics for complex shell geometries. Lightweight Des. **66**, 113–118 (2017)
4. Coutandin, S., Brandt, D., Heinemann, P., Ruhland, P., Fleischer, J.: Influence of punch sequence and prediction of wrinkling in textile forming with a multi-punch tool. Prod. Eng. Res. Devel. **1**(1), 5 (2018). https://doi.org/10.1007/s11740-018-0845-9
5. Kunz, H., Raatz, A., Dilger, K., Dietrich, F., Schnurr, R., Dröder, K.: Form-flexible handling technology for automated preforming. *ICCM19* **2013** (2013)
6. Förster, F., Ballier, F., Coutandin, S., Defranceski, A., Fleischer, J.: Manufacturing of textile preforms with an intelligent draping and gripping system. Procedia CIRP **66**, 39–44 (2017). https://doi.org/10.1016/j.procir.2017.03.370
7. Bruns, C., Raatz, A.: Simultaneous grasping and heating technology for automated handling and preforming of continuous fiber reinforced thermoplastics. Procedia CIRP **66**, 119–124 (2017). https://doi.org/10.1016/j.procir.2017.03.286
8. Moll, P., Ohlberg, L., Salzer, S., Coutandin, S., Fleischer, J.: Integrated gripping-system for heating and preforming of thermoplastic unidirectional tape laminates. Procedia CIRP **85**, 266–271 (2019). https://doi.org/10.1016/j.procir.2019.10.006
9. Schuster, A., et al.: Smart Manufacturing of Thermoplastic CFRP Skins. Procedia Manuf. **17**, 935–943 (2018). https://doi.org/10.1016/j.promfg.2018.10.147
10. Kupzik, D.: Flexible preforming of CFRP by robot based bending. https://youtu.be/HjTxdq9gwqM. Accessed 2 May 2019
11. Kupzik, D., Biergans, L., Coutandin, S., Fleischer, J.: Kinematic description and shape optimization of UD-tape reinforcements manufactured with a novel preforming process. In: Procedia CIRP, pp. 78–83. Accessed 13 Jan 2020
12. GOM, *Webpage.* https://www.gom.com/metrology-systems/atos/atos-triple-scan.html (2020). Accessed 14 Feb 2020

Design and Simulation

Development of a Hybrid Crash-Relevant Car Body Component with Load-Adapted Thickness Properties: Design, Manufacturing and Testing

Alan A. Camberg[1(✉)], Thomas Tröster[1], and Clemens Latuske[2]

[1] Automotive Lightweight Design, Paderborn University, Mersinweg 7, 33100 Paderborn, Germany
{alan.camberg,thomas.troester}@uni-paderborn.de
[2] Thyssenkrupp Steel Europe AG, Kaiser-Wilhelm-Straße 100, 47166 Duisburg, Germany
clemens.latuske@thyssenkrupp.com

Abstract. Semi-finished sheet products with load- or forming-adapted properties are classified as tailored blanks. By locally adjusting sheet thickness or material properties, the overall performance of the component can be improved while reducing the weight of the part. State-of-the-art tailored blanks are realized by rolling, welding or tailored heat treatment of monolithic materials and consider a change in properties with respect to the sheet plane. A further weight reduction could be achieved by combining the idea of tailored blanks with a multi-material design approach along the sheet thickness. For this purpose, a top-down material design is proposed to allow a demand-oriented hybrid tailored stacked blank design. Within this contribution an optimization-based top-down design methodology is applied on a crash relevant car body part. Based on benchmark crash simulations of a reference BIW structure, a critical body component is determined. The identified demonstrator component is later subdivided into multiple layers and submitted to an optimization loop in which the developed methodology varies the material parameters for each single layer. The result is a tailored stacked hybrid blank consisting of steel and FRP layers. In order to meet formability restrictions of the novel semi-finished product, the part under investigation is redesigned and compared with the reference BIW structure. Finally, the hybrid component is manufactured and tested on a dynamic crash device. Compared to a monolithic DP800 component, a mass reduction of 22% was achieved.

Keywords: Tailored stacked hybrid blanks · Material concepts · Simulation · Crash · Fiber-metal-laminates

K. Dröder and T. Vietor (Eds.): *Technologies for economic and functional lightweight design*, Zukunftstechnologien für den multifunktionalen Leichtbau, pp. 329–341, 2021. https://doi.org/10.1007/978-3-662-62924-6_28

1 Introduction

New drivetrain concepts can enable and accelerate the reduction of CO_2 emissions in the transportation sector worldwide, so long as the electricity is provided by renewable energy sources [1]. However, electric drive train concepts are observed to be a favorable strategy to meet emission regulations. In the last two decades this domain was traditionally dominated by lightweight concepts, since the vehicles weight has a proportional influence on three of four terms of the driving resistant forces [2]. Though, the vehicle's CO_2-footprint will be in the future, regardless of the drivetrain concept, still dependent on the vehicle's weight [3], and so lightweight and material selection will still play a substantial role in the car design. Furthermore, a vehicle in motion is exposed to more than just driving resistant forces. Dynamic forces, as for example during lane change, cornering, or, as an extreme example, in the event of an accident, are acting on the vehicle structure as well. In all these scenarios, weight-specific material properties, and therefore lightweight, play a significant role for the vehicle handling qualities, passive safety and its ecological footprint.

The demand for lighter and high strength materials lead to an increased portfolio development of conventional materials over the past years. Press hardened steels, dual and complex phase steels as well as high strength and ductility aluminum alloys enabled weight reductions in new car generations, despite of higher safety requirements or additional weight due to customer driven infotainment demands. Even individual CFRP concepts, as e.g. the BMW i3, came into series. Nevertheless, every single of those monolithic materials show limitations in terms of lightweight due to material specific drawbacks [4].

Motivated by a deliberate compensation of disadvantages of individual materials, the approach to integrate different materials into a functional vehicle component gained increased interest in both academia and industry [5]. Moreover, first multi-material vehicle components have been realized in series production [6–9]. Especially hybrid components made of metal and fiber-reinforced plastics (FRP) promise high weight-specific component properties at reasonable extra costs. Based on a local adaptation of the strength of thin-walled structures through FRP reinforcements, the focus of the considerations is primarily on safety-relevant structural body parts [10]. With the market launch of the new BMW 7 Series in 2015, several hybrid component concepts were brought to series production, enabling mass savings of up to 40 kg per body-in-white [11]. The strategy of hybrid lightweight design is continued by BMW also in the body structure of the new 8 Series [8]. Thus, hybrid lightweight design seems to be a promising lightweighting approach not only from the point of view of research, but also from that of industry.

Multi-layer materials that combine materials with different properties to a semi-finished product represent another form of multi-material design. The approaches here are partially very different. Metal–metal hybrids as for example tribond® [12] or Novelis Fusion™ [13] provide a better formability and crash performance due to an adapted ductility in the outer clad layers. Materials such as bondal® or litecor® [14] consist of a polymer core enclosed by steel clad layers and thus represent classic sandwich sheets with high specific stiffness and damping. Another subgroup is represented by so-called fiber-metal laminates (FML), i.e. combinations of thin metal sheets and FRP. Probably the best-known representative of this material class is GLARE, which is used for large fuselage areas of the Airbus A380. GLARE consists of several alternating layers of aluminum and glass-fiber reinforced plastic (GFRP) and shows extraordinary crack-resistant properties thanks to its multi-layer and multi-material structure [15]. Other examples of fiber-metal laminates have so far been limited to research projects, as for example CAPAAL and CAPET [16] or developments from the LEIKA project [17]. However, all multi-layer composites mentioned above have a common disadvantage – they were developed as semi-finished products with generalized properties. Consequently, their lightweight design potential is limited as well, as they are not optimized for the requirements of a specific application.

Not so with tailored blanks, as semi-finished products with locally varying material grade, thickness or surface quality, they offer the capability of spatially selective adaptation to the requirements of the finished component. In this way, tailored blanks lead to a weight reduction of the parts by subsequently improving the crashworthiness and formability [18]. By combining the approaches of tailored blanks with multi-layer materials, the lightweight design potential of parts manufactured from requirement-adapted semi-finished products can be further increased, as shown in [4, 19, 20].

Within this contribution, a top-down material design methodology for a crash-relevant and energy absorbing car body component is introduced. With the aim to design a semi-finished product with tailored through-thickness characteristics, the layer properties are modeled by parametrized bi-linear material models in which material behavior between ideally elastic to ideally plastic can be reproduced by the algorithm. In a further step, the component's geometry is redesigned to account for the material specific manufacturing limitations. Finally, the car body component is manufactured from the newly designed tailored stacked blank and tested on a dynamic crash device.

2 Proposed Optimization Method for Tailored Hybrid Stacks

To pursue a demand-oriented material design, a top-down material development process is applied. Starting with full vehicle crash benchmark simulations, a crash-relevant car body component is selected for the optimization. To reduce the computational cost of a single optimization run, the layer property optimization is carried out by using a surrogate model. The results of the most promising design candidates are then evaluated by means of restrictions given by the benchmark simulations. In the next step, the idealized material parameters of the winning candidate are compared with a material database in order to match them with automotive relevant material pendants. In a second optimization loop, the thicknesses and material orientations of previously selected real materials are optimized under consideration of manufacturing limitations (Fig. 1).

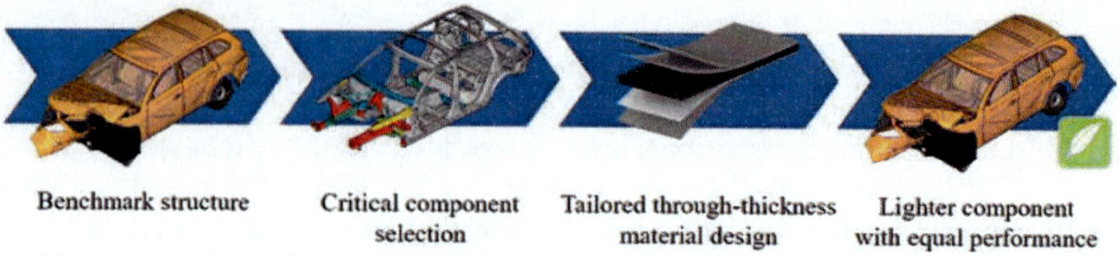

Fig. 1. Top-down approach for tailored through-thickness material properties.

2.1 Parametrized Bi-linear Multi-Material Approach

Among a variety of approaches for structural optimization, the "Solid Isotropic Material with Penalization" (SIMP) [21, 22] had gained a wide popularity in commercial FE-software. In order to take into account more than just one material, the SIMP method was successfully adapted for multi-material optimizations by several approaches [23–25]. In [20], a revisited method was proposed to allow the optimization algorithm to select from a range of common automotive lightweight materials. However, SIMP and its extensions consider only the elastic regime of the material and are therefore only applicable to linear problems.

To account for material non-linearity, it is proposed to replace the functional dependence between stiffness and density by a parameterized isotropic bi-linear material model. The design space of the material parameters is, in analogy to [20], within boundaries of common automotive materials. Depending on the selected parameters, the bi-linear model can replicate an ideally elastic, an ideally elastic–plastic and any material behavior in between. If the parameters permit plastic flow, the yield locus is based on an isotropic J_2-plasticity. However, density is assumed to be invariant on material parameters. The constitutive equations and exemplary material characteristics are given in Fig. 2. The design space of parameters used is given in Table 1.

Kinematic: $\varepsilon = \varepsilon^p + \varepsilon^e$

Stress: $\sigma = E\varepsilon^e = E(\varepsilon - \varepsilon^p)$

Flow rule: $\dot{\varepsilon}^p = \dot{\lambda}\dfrac{\partial f}{\partial \sigma}$

Isotropic hardening: $\dot{R} = H_1\dot{\lambda}$

Flow function: $f(\sigma, \lambda) = |\sigma| - (Y_0 + R)$

Tangent modulus: $E_t = \dfrac{EH_1}{E + H_1}$

Ideally elastic rule: $E \leq E_t \Rightarrow Y_0 \to \infty$

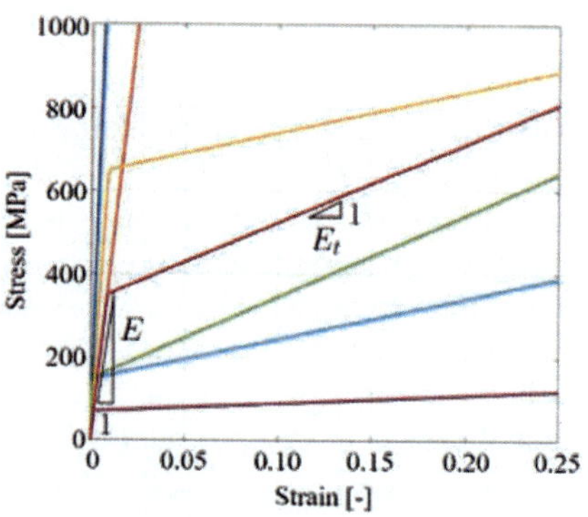

Fig. 2. Parametrized bi-linear isotropic multi-material: Constitutive equations and exemplary material behavior for different parameters of E, Y_0 and H_1.

The parametrized bi-linear material model can be applied on structural problems to find an optimal material property distribution within the design space and can be understood as an “inverse homogenization scheme” [24]. Within this contribution the approach is directly applied on multi-layered shell elements. The material parameterization is invoked at the level of shell element integration points clustered into plies. This, in turn, enables the representation of layers with different properties within the investigated component and the design of an application-adapted layered hybrid material.

Table 1. Set of parameters used in the bi-linear multi-material approach

Parameter	Parameter Range [MPa]
E	40E+03, 70E+03, 140E+03, 210E+03,
Y_0	70, 100, 150, 200, 250, 300, 350, 400, 450, 500, 550, 600, 650, 700
H_1	2E+02, 1E+03, 2E+03, 4E+03, 6E+03, 10E+03, 20E+03, 25E+03, 50E+03, 10E+04

2.2 Optimization Methodology

The proposed parametrized isotropic bi-linear material method uses idealized material data which have no associated information about the material mass and orientation. As a result, the mass of the component cannot be optimized directly, but only the structural cross-sectional properties. To account for this limitation, a two-stage optimization is introduced. In the first loop the monolithic reference part is subdivided into N^l integration point layers k, while the overall wall thickness t of the part remains unchanged. In the first run the absorbed energy E_{abs}, as an integral of the load–displacement curve F-s obtained from the crushing simulation, is to be maximized by varying the material parameters of each layer.

Once an optimum is found, the solution space is reduced by constrains given by the reference structure, i.e. the maximum force and maximum intrusion. The still idealized material properties of each layer are then compared with a material database and replaced by concrete pendants by taking into account real material properties, as for example non-linear hardening or anisotropy. Within the last optimization run the layers are optimized in terms of layer thicknesses t^l and material orientations θ^l. The objective is to maximize the specific energy absorption *SEA* defined as the energy absorbed per unit mass of material [26], where M_{crush} is the crushed mass and L is the structure length with respect to the crushing direction; φ_M is the mass fraction; M and M_0 is the mass of the current and the reference design, respectively. The boundaries for t^l and θ^l are set in a way to meet manufacturing limitations. Figure 3 depicts a flowchart of the defined multi-material optimization task.

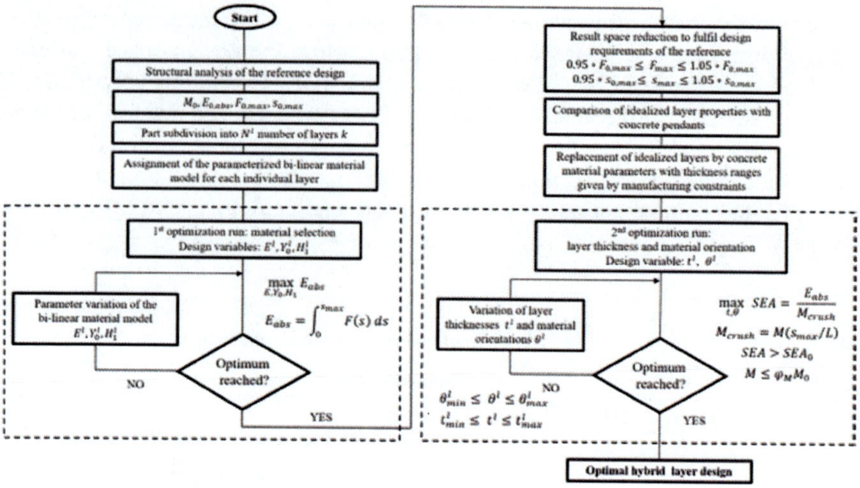

Fig. 3. Flowchart of the developed method for an optimization-based hybrid stacked blank design.

3 Reference Structure Benchmark

The thyssenkrupp InCar® plus, an OEM-independent full vehicle model with a steel-intensive state-of-the-art car body structure [27], is used as a benchmark for the material development. The relevant load cases are chosen under the consideration of the crash scenario probability in real traffic events. As reported in [28], 65,7% of all traffic accidents are front crashes. Therefore, worst-case configurations of European and US rating agency front crash load cases are selected for benchmarking, i.e. the Euro NCAP Moderate Overlap Deformable Barrier Frontal Test (64 km/h), the US NCAP Full-Width Rigid Wall Frontal Test (56 km/h) and the IIHS Small Overlap Rigid Barrier Frontal Test (64 km/h).

To classify the BIW parts into load case relevant groups and evaluate their impact on the overall crash performance of the vehicle, a strain energy based method introduced in [19] is used. The methodological component selection ensures that performance improvements will be realized on parts that show a high relevance for the global BIW properties and so, additional material costs can be justified by better overall vehicle characteristics. As a result of the described procedures the front long member is selected for a tailored material design trail. The reference front long member is made of DP-K®440Y780T tailored blank with a gauge size of 1.8 mm and 2.0 mm, respectively [4, 19].

4 Component Optimizations

4.1 Layer Design

To reduce the computational cost of a single optimization run, a surrogate model of the selected component is derived from the full vehicle model. The part is divided into 5 layers, where each of the integration point layers is coupled to an independently parameterized bi-linear material model, as described above. In result, 15 parameters are to be optimized. To perform sampling and optimization the capabilities of LS-OPT are used. Due to high non-linearity, numerical noise and non-convex objective functions with numerousness extrema of the optimized crashworthiness issue, a Response Surface Model with domain reduction (SRSM) is used for optimization. The SRMS metamodel is built by quadratic polynomials. For point selection, a D-Optimal design with 205 simulation points per iteration and a space filling method scheme are applied. The optimization is performed with a hybrid Adaptive Simulated Annealing (ASA) algorithm. The approximate global optimum is searched by ASA, while the accurate optimum search is performed with a Leaping Frog algorithm (LFOP). For detailed information about the algorithms please refer to [29]. In sum, the optimization task was performed in 73 iterations and took 14.965 single simulations. The surrogate model of the long member as well as the optimization results are presented in Fig. 4.

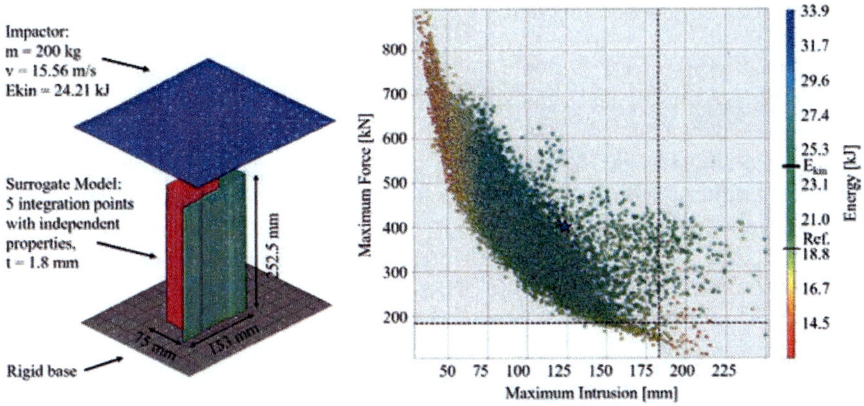

Fig. 4. Left: Surrogate long member model used for the optimization of the material distribution, Right: Diagram with optimized solutions; both adapted from [30].

The global optimum shows a supposedly energy absorption of E_{abs} of 33.9 kJ (marked as a star in Fig. 4), what exceeds the kinetic energy E_{kin} of the impactor. This is due to a, unfortunately, faulty implemented numerical integration scheme of the absorbed energy. When integrating the F-s curve, only the increment of the displacement of the impactor is taken into account without considering the direction of translation. Therefore, in solutions with a high elastic fraction, the rebound of the impactor is erroneously attributed to the energy absorption. Since, however, only solutions within a certain force maximum and penetration range are of interest with regard to the requirements on the full vehicle, only the ungreyed solution space in the diagram in Fig. 4 is considered anyway. Within these limits, only 7 solutions show higher energy absorptions than the reference design, and fortunately none of these solutions are affected by the rebound effect. In the relevant time interval of 0 ms to 14 ms, the best solution shows an energy absorption of 20.8 kJ, which is 9.5% more than with the reference material. The layer properties as well as the selected real material pendants are presented in Fig. 5.

After a comparison with a material database, a DP-K®290Y490T steel sheet is selected for the clad layers (IP layer 1 and 5). For the core material a unidirectional GFRP is selected, since it shows a good match with the optimized material properties of the inner layers (IP layer 2–4).

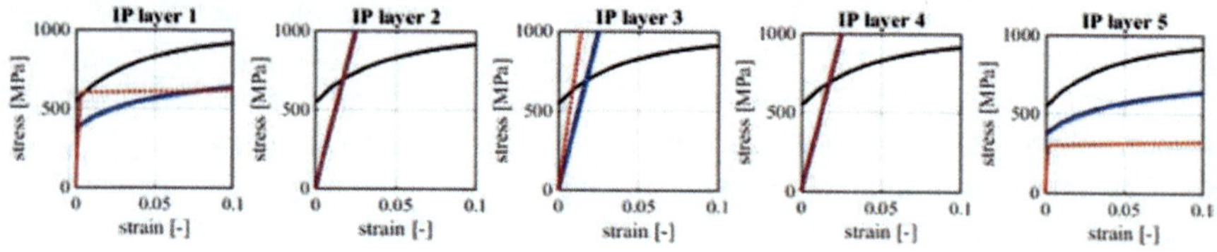

Fig. 5. Integration point (IP) layer properties: Black – reference, Red – idealized optimal properties, Blue – matched real material pendants.

The optimization of layer thicknesses and orientations is performed in analogy with the setup of the material selection run. The design space for the outer steel sheet thicknesses is limited from 0.2 mm to 1.0 mm, the GFRP core is divided into 4 separate layers with a thickness design space between 0.2 mm and 2.0 mm and possible discrete material orientations of 0° and 90°. The optimum of 2651 simulation runs is a sandwich-like fiber-metal-laminate with 0.5 mm outer steel sheet layers and a 2.1 mm GFRP core with a 60/40 fraction of 0° and 90° fiber orientation, respectively. The specific energy absorption was increased by 62.7% from 19 kJ/kg to 30.9 kJ/kg compared to the reference design.

4.2 Component Design

In order to adapt the component design to manufacturing restrictions of the new hybrid blank, the longitudinal member is redesigned in an iterative process by evaluating and adapting the new design in crash simulations described in Sect. 3. With the aim to allow a near net shape blank geometry and decrease the joining complexity in the car body assembly, the final design of the hybrid long member consist of two U-profiles as presented in Fig. 6. The benchmark requirements for passive safety were mainly met with the new component design, although the global deformation mode of the hybrid part shows a significant difference to the reference design.

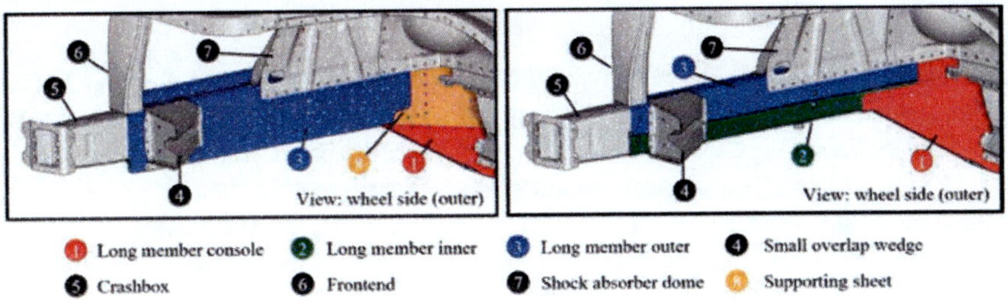

Fig. 6. Left: Reference long member design, Right: New long member design; adapted from [31].

5 Component Manufacturing and Testing

An experimental series of tests is carried out in order to evaluate the numerical results. The tailored stacked hybrid blank is manufactured according to the optimized design and material availability. The final stack consists of 0.55 mm outer DP-K®290Y490T steel sheet layers, a 2.3 mm Quadrant QTex 50/50 glass fiber reinforced thermoplastic PP composite laminate core and two 0.1 mm Nolax Cox391 bonding agent layers.

5.1 Manufacturing

The manufacturing process is adapted from [32] and can be divided into two sequences. In the first sequence (OP10 and OP20), the monolithic materials are cut to net shape blanks and stacked together. In the next step the stack is simultaneously heated and bonded to a hybrid blank within a press heating station with heated contact plates. After cooling down, the semi-finished product can be stored until further processing. In the second sequence (OP30 and OP40), the hybrid blank is heated in a furnace to a temperature above the melting point of the thermoplastic core and then transferred to a cold tool, where it is formed into its final shape in a single-stage forming process with simultaneous quenching, see Fig. 7. Afterwards, the two formed halves of the component are joined by blind steel rivets and a structural adhesive.

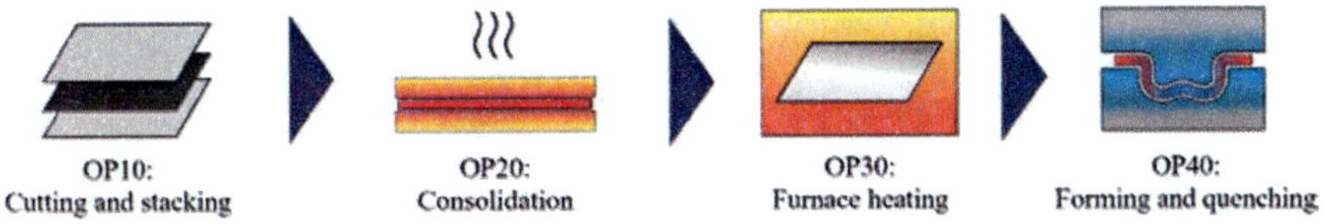

Fig. 7. Tailored hybrid stacked blank and component manufacturing process. Adapted from [32].

5.2 Crashworthiness

The crashworthiness of the new hybrid long member was investigated on a crash test apparatus of the LiA Crash & Impact Lab. The crash sledge impact energy was selected to be in a range from 10.5 kJ up to 14 kJ given by an impact mass of 205 kg up to 288 kg and an initial impact velocity of 9.8 m/s up to 10.8 m/s. To impose a progressive buckling, four trigger holes were machined at the circumference of the front edge of the long member. The crash tests have shown that an increased *SEA* between 25.5 kJ/kg and 34.1 kJ/kg can be realized with the new tailored material. However, the achieved *SEA* shows a strong dependence on the initial instability of the structure and the resulting local buckling characteristics (Fig. 8).

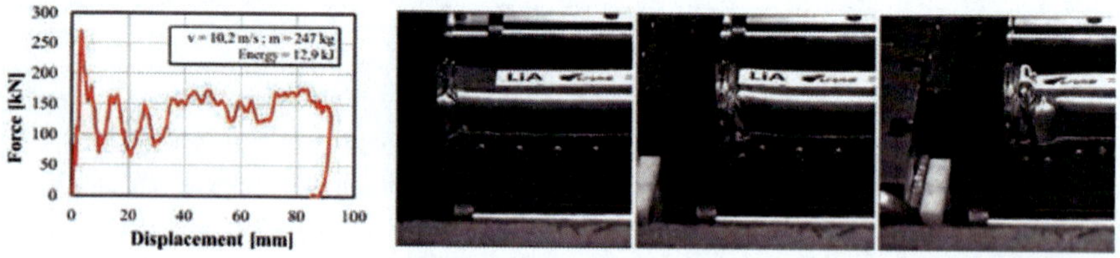

Fig. 8. Axial crushing of the hybrid long member made of the designed tailored stacked blank [33].

Six unique crushing mechanisms of the hybrid laminate could be identified from macroscopic fractographs of the deformed long member, i.e.: 1) Plastic deformation of the metal part, 2) brittle fracture of the FRP part, 3) lamina bending, 4) local lamina buckling 5) delamination of Mode I/II and 6) formation of a debris wedge in the crushing front, see Fig. 9. As reported in [34], the friction associated with the latter mechanism can contribute to more than one third of the energy absorption in FRP structures. This seems to be a clear advantage of the hybrid material developed compared to, e.g. asymmetric hybrid energy absorbing structures, where the formation of a debris wedge has not been observed [35, 36].

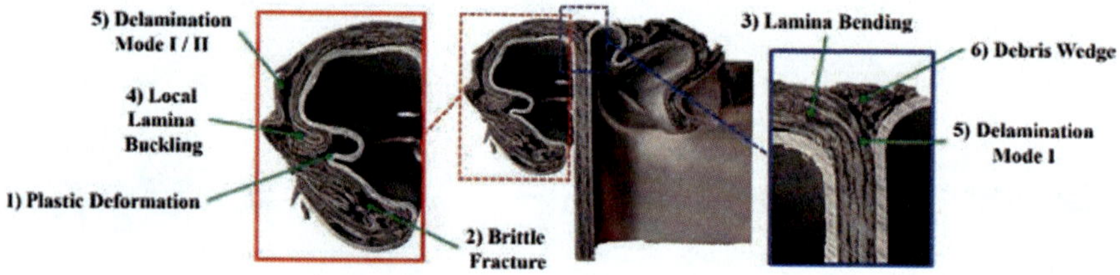

Fig. 9. Identified crushing mechanisms of the tailored hybrid stacked blank. Adapted from [33].

Finally, the experimental test data of the worst-case hybrid long member (25.5 kJ/kg) where compared with simulations of monolithic steel designs with the same component geometry. By an equal maximum force and maximum intrusion, the component mass can be reduced by 22% compared to a DP-K®440Y780T (19.4 kJ/kg) and up to 30% compared with a DP-K®290Y490T (18.5 kJ/kg).

6 Discussion

The here presented parametrized isotropic bi-linear material approach, even though if with some room for improvement (e.g. integration scheme of E_{abs}), delivers interesting results for material optimization in crashworthiness problems. However, as already stated in [24], the largely unrestricted design space carries the risk of ending up with artificial and non-physical material parameters which cannot be reproduced with technically relevant materials. Furthermore, crashworthiness is a highly non-linear problem in terms of deformation and material behavior, where the crashworthiness strongly depends on material strength and fracture characteristics. Obviously, that phenomena cannot be reproduced by a generalized bi-linear material model. Therefore, a direct optimization with discrete real material parameters seems to be a more practicable and cost-effective way to find an optimum in through-thickness material distribution for engineering problems. Nevertheless, although the selected technical materials differ in their properties from the ideal material parameters, an improvement of the specific energy absorption could be achieved. A relatively strong dependency of the *SEA* on the initial buckling of the structure in the experiments suggests, that a deliberate investigation on robust trigger geometries and mechanisms is needed.

7 Conclusions

It has been shown that the crashworthiness of automotive structures can be increased by optimizing the through-thickness material properties. For this purpose, a parameterized bi-linear material approach was introduced to determine the ideal layer property distribution. By approximating the idealized properties by means of Advanced High Strength Steel cover layers and a GFRP core, the specific energy absorption of the investigated front longitudinal member was successfully increased by up to 80% while reducing the part weight by 22%. However, the presented beta version of the optimization scheme has still several weaknesses, which will be improved in future work.

Acknowledgements. The authors gratefully acknowledge the funding of the projects "LHybS" (www.tecup.de/lhybs/) and "HyOpt" (www.hyopt.de) by the European Regional development Fund and the State of North Rhine-Westphalia. Sincere thanks also due to all industrial project partners, especially to the thyssenkrupp Steel Europe AG for providing the InCar® plus model. Special thanks are due to Patrick Besting M.Sc., Christian Bielak M.Sc., Felix Beule M.Sc. and Niklas Schulenberg B.Sc. for their contributions to the presented results during their thesis work at the Chair of Automotive Lightweight Design, Paderborn University.

References

1. Jochem, P., Babrowski, S., Fichtner, W.: Assessing CO_2 emissions of electric vehicles in Germany in 2030. Trans. Res. Part A Policy Pract. **78**, 68–83 (2015)
2. Gänsicke, T., Goede, M.: Die Technische Motivation. In: Friedrich, H. (ed.) Leichtbau in der Fahrzeugtechnik. Springer Vieweg, Wiesbaden (2013)
3. Cabrera, S.A., Norman, J.B., Allwood, J.M.: The impact of reducing car weight on global emissions: the future fleet in Great Britain. Philos. Trans. R. Soc. A **375**(2095), 1–16 (2017)
4. Camberg, A.A., Engelkemeier, K., Dietrich, J., Heggemann, T.: Top-down design of tailored fibre-metal laminates. Lightweight Des. Worldwide **11**(2), 24–28 (2018)
5. Bader, B., Türck, E., Vietor, T.: Multi material design. A current overview of the used potential in automotive industries. In: Dröder, K., Vietor, T. (eds.) Technologies for Economical and Functional Lightweight Design. Springer, Berlin (2019)
6. Jäschke, A., Dajek, U.: Dachrahmen in Hybridbauweise. Sonderdruck Aus VDI-Tagungsband Nr. **4260**, 25–45 (2004)
7. Frei, P.: The hybrid B-Pillar in the new BMW 7-Series. An example for the manufacturing implementation of an innovative Steel-CRFP lightweight design concept. In: Automotive Circle, Materials in Car Body Engineering Conference, Bad Nauheim (2016)
8. Henseler, U., Eras, A.: Introducing the new 8 series coupe. In: Euro Car Body Conference, Bad Nauheim (2018)
9. Wawers, U., Stein, S.: The development and Production of the new Porsche 911 body structure: the new 911 Carrera. In: Eckstein, L. (ed.) Proceedings of Aachener Body Engineering Days 2019. Institut für Kraftfahrzeuge, RWTH Aachen, Aachen (2019)
10. Ickert, L.H.: FKV-Metall-Hybridbauweise für die automobile Großserie. Dissertation, RWTH Aachen University, Institut für Kraftfahrzeuge, Aachen (2014)
11. N.N.: Technical Training – Product Information – G12 Introduction. BMW AG (2015)
12. Pieronek, D., Keßler, L., Richter, H, Myslowicki, S.: Virtuelle Produktentwicklung und Crashauslegung von Stahl-Werkstoffverbundsystemen. 14. Deutsches LS-DYNA Forum, Bamberg (2016)
13. Bischoff, T.F., Hudson, L.G., Wagstaff, R.B.: Novelis fusion TM, a novel process for the future. In: Grandfield, J.F., Eskin, D.G. (eds.) Essential Readings in Light Metals. Springer, Cham (2016)
14. Pieronek, D., Böger, T., Röttger, R.P.: Modelling approach for steel sandwich materials in automotive crash simulations. In: 12th German LS-DYNA Forum, Ulm (2012)
15. Vlot, A.: Glare, History of the Development of a New Aircraft Material. Kluwer Academic Publishers, Delft (2001)
16. Wielage, B., Nestler, D., Steger, H., Kroll, L., Tröltzsch, J., Nendel, S.: CAPAAL and CAPET – new materials of high-strength, high-stiff hybrid laminates. In: Fathi, M., Holland, A., Ansari, F., Weber, C. (eds.) Integrated Systems, Design and Technology 2010, pp. 23–35. Springer, Berlin (2011)
17. Wiedemann, S., Eckstein, L., Modler, N., Kawalla, R., Meschut, G., et al.: LEIKA – Effiziente Mischbauweisen für Leichtbau-Karosserien. Platform Forel, Dresden (2017)
18. Merklein, M., Johannes, M., Lechner, M., Kuppert, A.: A review on tailored blanks – production, applications and evaluation. J. Mater. Process. Technol. **214**, 151–164 (2014)
19. Camberg, A.A., Tröster, T.: Optimization-based material design of tailored stacked hybrids for further improvement in lightweight car body structures. In: Hausmann, J.M., Siebert, M., von Hehl, A. (eds.) Proceedings of 3rd Hybrid Materials and Structures 2018, pp. 240–245. DGM, Berlin (2018)

20. Camberg, A.A., Stratmann, I., Tröster, T.: Tailored stacked hybrids – an optimization-based approach in material design for further improvement in lightweight car body structures. In: Dröder, K., Vietor, T. (eds.) Technologies for Economical and Functional Lightweight Design, pp. 119–131. Springer, Berlin (2019)
21. Bandsøe, M.P.: Optimal shape design as a material distribution problem. Struct. Optim. **1**(4), 1–16 (1989)
22. Zhou, M., Rozvany, G.I.N.: The COC algorithm, part II: topological, geometry and generalized shape optimization. Comput. Methods Appl. Mech. Eng. **89**(1), 309–336 (1991)
23. Li, C., Kim, I.Y.: Multi-material topology optimization for automotive design problems. Proc. IMechE Part D J. Aut. Eng. **232**(14), 1950–1969 (2017)
24. Bandsøe, M.P., Sigmund, O.: Topology Optimization: Theory, Methods and Applications, 2nd edn. Springer Vieweg, Berlin (2013)
25. Zuo, W., Saitou, K.: Multi-material topology optimization using ordered SIMP interpolation. Struct. Multidiscip. Optim. **55**, 477–491 (2016)
26. George, C.J., Fellers, J.F., Simunovic, S., Starbuck, J.M.: Energy absorption in polymer composites for automotive crashworthiness. J. Compos. Mater. **36**(7), 813–850 (2002)
27. N.N.: Benchmark 2.0: Die aktualisierte Referenzstruktur. Thyssenkrupp InCar plus. ATZextra **19**(10), 84–88 (2014)
28. Kramer, F.: Integrale Sicherheit von Kraftfahrzeugen, 4th edn. Springer, Wiesbaden (2013)
29. Stander, N., Roux, W., Basduhar, A., Eggleston, T., Goel, T., Craig, K.: LS-OPT user's manual. A design optimization and probabilistic analysis tool for the engineering analyst. 5.2 Ver. LSTC, Livemore (2015)
30. Besting, P. (Supervisor: Camberg, A.A.): CAE-basierte Entwicklung einer Längsträgerstruktur mit idealen Schichteigenschaften in Bauteildickenrichtung hinsichtlich Energieabsorption und Leichtbau, M.S. thesis, Paderborn University, Paderborn (2017)
31. Bielak, C.R. (Supervisor: Camberg, A.A.): Konzeptbewertung hybrider Längsträgerstrukturen in der FE-Gesamtfahrzeugsimulation, semester thesis, Paderborn University, Paderborn (2018)
32. Behrens, B.-A., Hübner, S., Neumann, A.: Forming sheets of metal and fibre-reinforced plastics to hybrid parts in one deep drawing process. Procedia Eng. **81**, 1608–1613 (2014)
33. Schulenberg, N. (Supervisor: Camberg, A.A.): Herstellung und experimentelle Untersuchung eines hybriden PKW Längsträgers mit angepassten Schichteigenschaften in Werkstoffdickenrichtung, semester thesis, Paderborn University, Paderborn (2019)
34. McGregor, C., Vaziri, R., Xiao, X.: Finite element modelling of the progressive crushing of braided composite tubes under axial impact. Int. J. Impact. Eng. **37**(6), 662–672 (2010)
35. Song, H.W., Wan, Z.M., Xie, Z.M., Du, X.W.: Axial impact behavior and energy absorption efficiency of composite wrapped metal tubes. Int. J. Impact. Eng. **24**(4), 385–401 (2000)
36. Reuter, C., Tröster, T.: Crashworthiness and numerical simulation of hybrid aluminium-CFRP tubes under axial impact. Thin-Walled Struct. **117**, 1–9 (2017)

Operationalization of Manufacturing Restrictions for Hybrid Tailored Forming Components

Tim Brockmöller(✉), Renan Siqueira, Iryna Mozgova, and Roland Lachmayer

Institute of Product Development, Leibniz University Hannover, Hannover, Germany
{brockmoeller,siqueira,mozgova, lachmayer}@ipeg.uni-hannover.de

Abstract. In the Collaborative Research Centre (CRC) 1153, the process chain for the manufacture of hybrid high-performance components by Tailored Forming is being investigated. This involves the production of hybrid solid components. These have properties that are locally adapted to the respective load case, by combining different materials. The application potential of these multi-material components arises mainly where conventional mono-material components reach their technological development limits.

However, the Tailored Forming process chain is more complex than conventional forming processes, as the combination of materials results in a larger solution space when designing components, because the distribution of materials must also be taken into account. Tailored Forming is also a new manufacturing technology whose fundamentals are currently being researched. So, there is only a small number of manufactured samples whose properties have not yet been adequately analyzed. As a result, few design guidelines are known to date according to which Tailored Forming components can be designed.

In order to make statements about an optimized shape, the topology optimization method Interfacial Zone Evolutionary Optimization (IZEO) has been developed, which can be used to determine the material distribution for Tailored Forming components. Therefore, the manufacturing restrictions of Tailored Forming are numerically translated into geometric constraints that restrain the evolution throughout the optimization process. The solution achieved delivers a gross distribution of materials in the specified domain, while obeying the manufacturing constraints imposed. Further design development includes a fine construction of the model, where specific analysis can be executed about the joining zone of the different materials. With this process, the solution space can be efficiently explored and manufacturable designs can be produced, from which first guidelines for the construction of Tailored Forming components are derived.

Keywords: Tailored forming · Multi-material · Topology optimization · IZEO

K. Dröder and T. Vietor (Eds.): *Technologies for economic and functional lightweight design*, Zukunftstechnologien für den multifunktionalen Leichtbau, pp. 342–352, 2021. https://doi.org/10.1007/978-3-662-62924-6_29

1 Introduction

Within the scope of the Collaborative Research Center 1153 a process chain for the production of multi-material forming components is being investigated. The aim of the Tailored Forming technology, is to manufacture hybrid high-performance components with locally adapted properties by using the material required for the prevailing operating conditions at the respective point of the component. In comparison with conventional mono-material components, especially where the technological development limits have been reached, Tailored Forming components offer weight advantages and at the same time high resistance to wear, thermal influences or corrosion. This results in a better material utilization [1, 2].

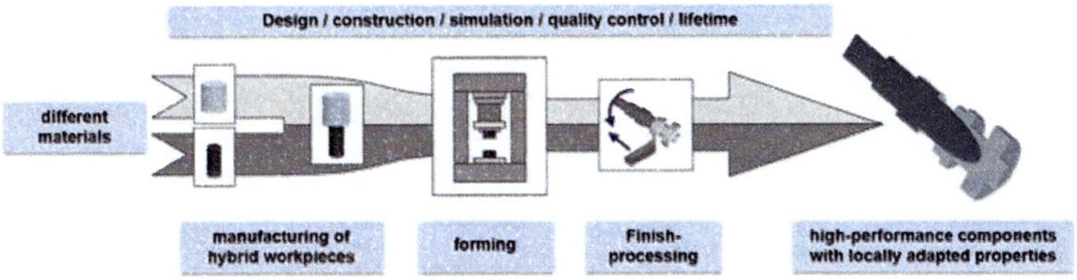

Fig. 1. Tailored Forming process chain, according to [1]

The Tailored Forming process chain consists of several process steps, which are shown in Fig. 1. The process always begins with the manufacturing of a semi-finished workpiece. The composite of different materials can be achieved by friction welding, welding cladding, ultrasonic-assisted laser beam welding or rod extrusion. The next step is forming, in which both materials are formed simultaneously. Die forging, extrusion and cross wedge rolling are used for forming. Machining is used to achieve the final geometry [1]. In order to provide the necessary forming temperatures and to be able to set desired material properties (e.g. hardened surfaces), various heating strategies are used in the process chain [3].

The challenge here is that there are hardly any design guidelines for Tailored Forming components, as the technology is currently undergoing basic research. This is particularly problematic for the forming processes, as the forming tools required complex effort to manufacture [4] and may not function properly because they cannot be designed correctly. Nevertheless, in order to be able to realize a material distribution optimized for the application case under consideration of the manufacturing restrictions, the topology optimization method Interfacial Zone Evolutionary Optimization (IZEO) has been developed.

The paper introduces the IZEO method in Chap. 2. In Chap. 3 the simplified version of a suspension arm is introduced as an example component. Furthermore the production processes used to manufacture the suspension arm and the associated manufacturing restrictions are presented. Subsequently, the implementation of IZEO is shown in Chap. 4 including the load case and the boundary conditions. In Chap. 5, generated suspension arm geometries with and without manufacturing restrictions are compared. Chapter 6 gives a summary of the paper and an outlook on following research projects.

2 Interfacial Zone Evolutionary Optimization—IZEO

Whenever working with mono or multi-material designs, the search for the best material utilization is a central research question. The materials must be distributed in a defined design domain in an optimal way, which is the main focus of topology optimization methods [5]. Moreover, manufacturing restrictions must be considered in this optimization process, so that the generated solution is suitable for the used manufacturing process. IZEO, firstly introduced in [6] and further extended in [7], is an evolutionary topology optimization method that was developed with the intention of dealing with multi-materials and manufacturing constraints in a unified method, suitable for the conceptual design phase of Tailored Forming components.

The working principle of IZEO is presented in Fig. 2. The model consists of a domain, which is the maximal space that the component can occupy and the referent load and boundary conditions. Initially, the whole domain is made of the stronger material (material 1) and on every iteration a small portion of the weaker material (material 2) is added. This is the same working principle of other evolutionary methods, such as Bi-directional Evolutionary Structural Optimization (BESO) [8, 9], but here the evolution is limited to a surface, which allows the implementation of manufacturing restrictions, as explained in [7]. Since similar topology optimization methods are not capable of distributing several materials in the domain in addition to optimizing the geometry and simultaneously complying with applicable Tailored Forming manufacturing restrictions, it is difficult to validate the results obtained with IZEO using other methods. In this case, it is necessary to investigate the properties of real manufactured components in order to compare the results. As shown in [7], the result of an IZEO without the consideration of manufacturing restrictions is similar to the BESO approach.

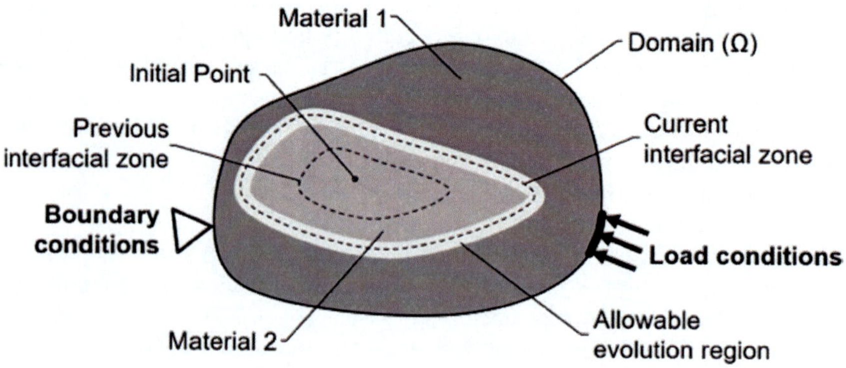

Fig. 2. Representation model of the interfacial evolutionary process of IZEO [7]

3 Suspension Arm and Required Manufacturing Processes

The component under consideration is a miniature version of a hybrid suspension arm from automotive engineering. The suspension arm here is a geometrically simplified version of a wheel guiding arm that lies transverse to the wheel plane and is mainly subjected to horizontal forces [10].

In the CRC 1153 the suspension arm is produced by the manufacturing processes rod extrusion for semi-finished products and die forging. The rod extrusion process is the so-called Lateral Angular Co-Extrusion (LACE) process in which an extruded steel profile is joined with aluminum. For this purpose the aluminum is plastically deformed and flows around a steel profile. The joined materials are pressed together through a die and thus obtain the desired semi-finished workpiece geometry [11]. For the suspension arm an L-section steel profile is used. The multi-material strand is then divided into pieces of defined thickness. Each of them serves as a semi-finished workpiece for die forging. The right side of Fig. 3 illustrates the schematic representation of the LACE process. The extrusion directions are indicated by the arrows. On the left side, the geometry of the semi-finished hybrid suspension arm is depicted after to the LACE process.

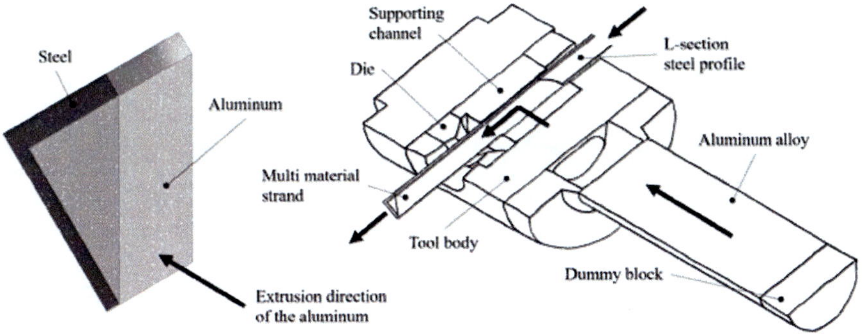

Fig. 3. Geometry of a semi-finished suspension arm (left) and a schematic representation of the LACE process (right), according to [12]

Closed dies are used for die forging. Due to the manufacturing processes used and the associated manufacturing restrictions, the available solution space is limited, since it is no longer possible to produce any geometry [13]. In general, for both manufacturing processes, no undercuts are possible in the direction of extrusion during the rod extrusion process and in the direction of tool movement during die forging with closed dies [4]. The rod extrusion process ensures that the distribution of the materials in the cross-section to the direction of extrusion is always identical.

4 Applying IZEO for the Suspension Arm

IZEO was already used for a concept of a multi-material rocker arm [14], where only a 2-dimensional version of the method was implemented. However, the current challenge of the suspension arm presents is a 3-dimensional nature, since the forces and the degrees-of-freedom are present in a 3D space. Therefore, the design domain is defined as seen in Fig. 4, with the correspondent load cases and boundary conditions. The two forces present are different load cases, which means they are not simulated at the same time. The two load cases are simulated in 2 parallel simulations simultaneously during the optimization process, and both influences are considered in the evolutionary design process. As a fundamental constraint, the regions where the forces and the boundary conditions are set must be necessarily made out of steel and will not have their material changed during the evolution. This guarantees sufficient hardness and strength at the bearing points.

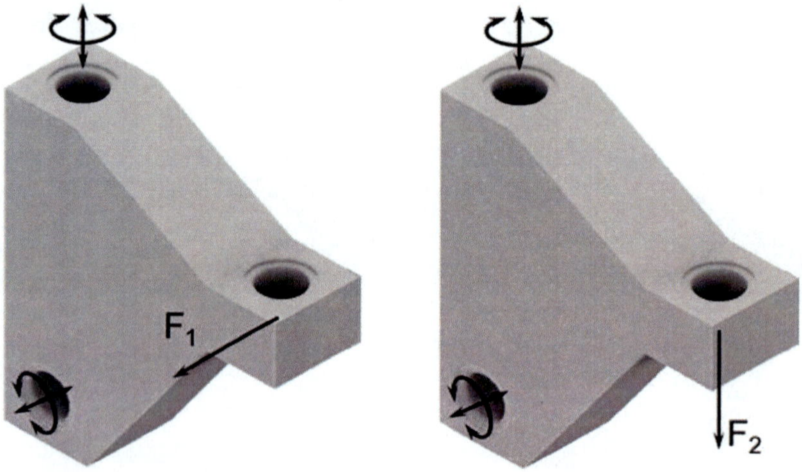

Fig. 4. Design domain for the problem of the suspension arm with load and boundary conditions for two simultaneous load cases

The first result generated is to a case where no manufacturing restriction is present. The main input parameters were:

- Materials: steel and aluminum (elasticity modulus: 210 GPa und 70 GPa, respectively)
- Objective function: stiffness maximization
- Volume constraint: 50% of the initial

- Material proportions: 50% steel, 50% aluminum in volume
- The bearing arrangements are floating
- The ratio of the forces F1 and F2 is 1:1
- The forces are static and each represent one of two simultaneous load cases

In this case a qualitative optimization of the geometry and material distribution was sought. The reason for this is that it is not known what the limits of the available production machines are and how large the suspension arm can be produced. Therefore an approach was chosen where the optimized geometry can be scaled, assuming linear material properties.

Besides the parameters given above, other optimization control parameters inherent to IZEO are also present and may also influence the result, but its calibration goes out of the scope of this paper and will not be further explained. The final result is to be seen in Fig. 5. The shown domain of the suspension arm model is slightly changed compared to Fig. 4 and originates from previous tests. Nevertheless, without the presence of manufacturing restrictions, the algorithm tries to set the most of the material to the outer part of the design space, in order to increase the inertia against flexion forces. The aluminum is than distributed in regions with less stress, also not obeying any kind of manufacturing restriction. This produces a suspension arm that is basically hollow on the inside and cannot be produced by the Tailored Forming process chain.

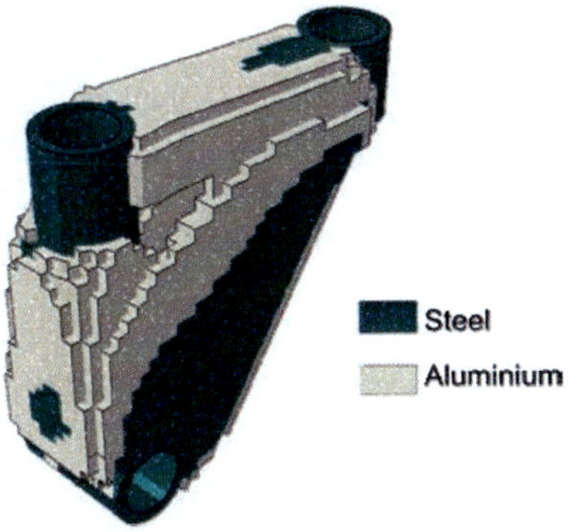

Fig. 5. Design results for a multi-material suspension arm without manufacturing restrictions using IZEO

For the next case, some other constraints must be considered, such as an extrusion constraint to describe the LACE extrusion process and a directional constraint to describe the die forging process. These constraints are represented next in Fig. 6.

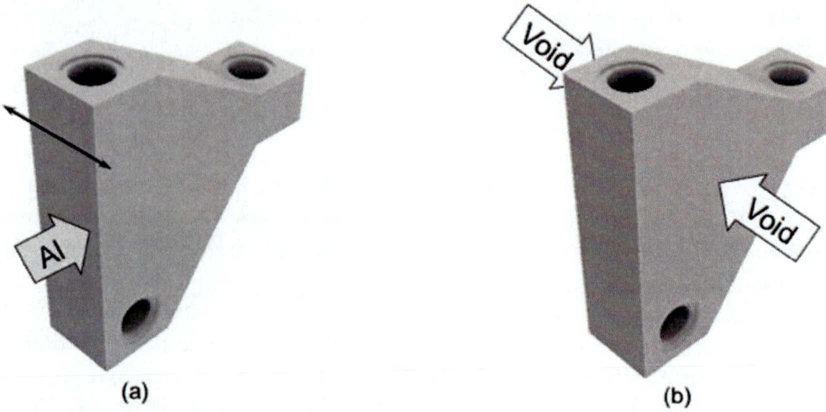

Fig. 6. Manufacturing restrictions representation showing the directions allowed to add materials, according to [6]. (a) Extrusion; (b) Die forging.

One of the main advantages of IZEO for multi-materials is the possibility of implementing different manufacturing restrictions for each phase of the Tailored Forming process chain. The bi-material phase and the third material phase. The bi-material phase is the joining process between steel and aluminum, which is the LACE extrusion process. This process creates, an extrusion constraint in a given direction of the model (represented by the double-pointed arrow in Fig. 6a). At the same time, the aluminum can be added in the structure from one side only, according to the manufacturing restriction, creating an additional manufacturing constraint. The third material phase is related to the form of the whole structure, which is in this example generated by the die forging process. This corresponds to the third material (void) in the optimization process, which is here limited to grow only from the same direction as the forming tool movement of the die forging process. Different combinations can be used for these manufacturing restrictions, also with different control parameters. One of these results is seen in Fig. 7, where the manufacturing restrictions for extrusion were implemented. In this result isosurface was used as a post-processing method in order to get a smooth result.

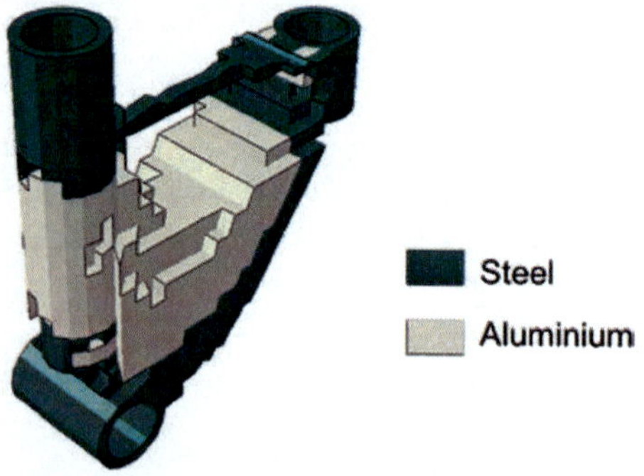

Fig. 7. Design results for a multi-material suspension arm with manufacturing restrictions

Here, both manufacturing restrictions were combined and considered to generate the solution. As seen, considering such basic manufacturing restrictions in the topology optimization process generates a solution that is totally different from the first one (Fig. 5) without restrictions. This enables the design engineer to generate concepts that are more meaningful to the next phases of development. A manufacturable construction derived from the result seen in Fig. 5 would require big changes in the distribution of the materials in the design domain, which would, consequently, cause a loss of the optimized characteristics provided by the algorithm. However, this means that the computing time for the suspension arm with manufacturing restrictions is 8.5 times higher than for the one without restrictions (722373 s to 84825 s).

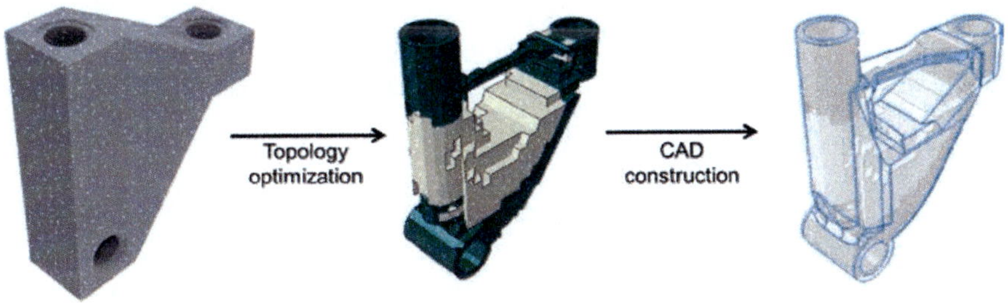

Fig. 8. Development process for the design of the suspension arm

In the next phases of development, following the guidelines according to [15], a more detailed design must be generated. The output of topology optimization, although containing important information about material distribution, is a concept. However, it works as base for a CAD construction of the same model, as presented in Fig. 8. This is a process that must be performed manually by the design engineer, in order to simplify the structure and possible create a meaningful parametric model.

5 Comparison of Hybrid Suspension Arms with and Without Manufacturing Restrictions

In the following, the CAD models derived from the hybrid IZEO models are compared. The suspension arm from Fig. 5 without manufacturing restrictions (a) and the suspension arm from Fig. 7 with manufacturing restrictions (b) are analyzed for their deformation under load. The load described in Fig. 4 is used as the load case. Figure 9 shows the result of the FE-simulation. Due to the creation of the CAD model and the resulting adaptation of the geometry, the suspension arms deviate from the material ratio of 1:1 specified by IZEO. Table 1. Shows the differences between the individual CAD models.

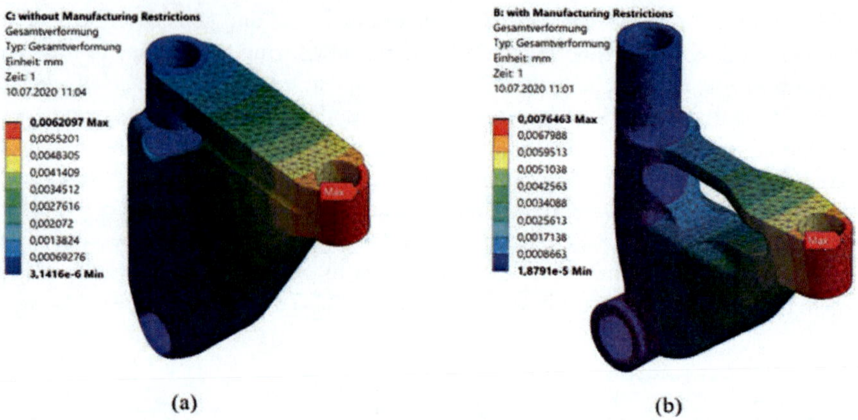

Fig. 9. Comparison of the deformation of hybrid suspension arms without (a) and with (b) manufacturing restrictions

It can be seen that without manufacturing restrictions the suspension arm is heavier and takes up a larger volume. The deviation is between 5% and 7%. As expected, the deformation due to the forces acting on the suspension arm with manufacturing restrictions is much greater (approx. 23%), since the available solution space is further limited by the restrictions to be observed. As a result, the suspension arm without manufacturing restrictions has a higher stiffness compared to its weight. However, this cannot be produced with the Tailored Forming process chain, as there are undercuts in this design and the suspension arm is basically hollow. To be able to produce it, the manufacturing process must be adapted. For example, an additive multi-material process can be used (as described in [16]).

Table 1. Relative comparison of both suspension arms

Suspension arm properties	Ratio a:b
Volume (St)	1:0,94
Mass (St)	1:0,94
Volume (Al)	1:0,97
Mass (Al)	1:0,97
Total Volume	1:0,95
Total Mass	1:0,95
Max. Deformation	1:1,23

6 Conclusion

IZEO offers the user the possibility to optimize the geometry and material distribution of hybrid components according to the loads, while at the same time taking into account applicable Tailored Forming manufacturing restrictions. Thus IZEO offers an advantage over other topology optimization methods such as SIMP or BESO, which do not offer these possibilities. The solution achieved delivers a gross distribution of materials in the specified domain, while obeying the manufacturing constraints imposed. Further design development includes a fine construction of the model, where specific analysis can be executed about the joining zone of the different materials. With this process, the solution space can be efficiently explored and manufacturable designs can be produced, from which first guidelines for the construction of Tailored Forming components are derived. Future works will involve a further exploration of the design solution space with IZEO, considering different material proportions and manufacturing restrictions, and a final CAD design of the component according to other specifications from the manufacturing side. In this way, IZEO is used as part of the Tailored Forming engineering environment that is being researched in the CRC 1153 [17].

To validate the suspension arm geometries forming simulations for the LACE process and die forging are currently being carried out in the CRC 1153. Here, the geometry of the suspension arm is optimized for the subsequent forming process. The results generated in this process will be implemented in the IZEO model of the suspension arm with regard to the manufacturing restrictions to be observed, so that the topology optimization can achieve more meaningful results. In addition, it is planned to analyze the properties of the manufactured suspension arm by performing test bench trials in order to compare the IZEO results with the real component behavior and to adjust the topology optimization if necessary.

Acknowledgements. The results presented in this paper were obtained within the Collaborative Research Centre 1153 (Project number: 252662854) "Process chain to produce hybrid high performance components by Tailored Forming" in the subproject C2. The authors would like to thank the German Research Foundation (DFG) for the financial and organizational support of this project.

References

1. Behrens, B.-A., Bouguecha, A., Frischkorn, C., Huskic, A., Stakhieva, A., Duran, D.: Tailored forming technology for three dimensional components: approaches to heating and forming. In: 5th International Conference on Thermomechanical Processing, Italy (2016)
2. Behrens, B.-A., Breidenstein, B., Duran, D., Herbst, S., Lachmayer, R., Löhnert, S., Matthias, T., Mozgova, I., Nürnberger, F., Prasanthan, V., Siqueira, R., Töller, F., Wriggers, P.: Simulation-aided process chain design for the manufacturing of hybrid shafts. HTM – J. Heat Treatm. Mat. **74**(2), 115–135 (2019)
3. Herbst, S., Maier, H.J., Nürnberger, F.: Strategies for the heat treatment of steel-aluminium hybrid components. HTM – J. Heat Treatm. Mat. **73**(5), 268–282 (2018)
4. Doege, E., Behrens, B.-A.: Handbuch Umformtechnik – Grundlagen, Technologien, Maschinen. Springer, Berlin (2016)
5. Spillers, W.R., McBain, K.M.: Structural Optimization. Springer Science & Business, Springer (2009)

6. Siqueira, R., Mozgova, I., Lachmayer, R.: Development of a topology optimization method for tailored forming multi-material design. 24th ABCM International Congress of Mechanical Engineering, Brazil (2017)
7. Siqueira, R., Mozgova, I., Lachmayer, R.: An interfacial zone evolutionary optimization method with manufacturing constraints for hybrid components. J. Comput. Des. Eng. **6**(3), 387–397 (2019)
8. Huang, X., Xie, Y.: Bi-directional evolutionary topology optimization of continuum structures with one or multiple materials. Comput. Mech. **43**(3), 393–401 (2009)
9. Xie, Y.M., Steven, G.P.: Evolutionary Structural Optimization. Technology & Engineering. Springer, London (1997)
10. Heißing, B., Ersoy, M., Gies, S.: Fahrwerkhandbuch – Grundlagen, Fahrdynamik, Komponenten, Systeme, Mechatronik, Perspektiven. Springer Vieweg, Wiesbaden (2013)
11. Thürer, S.E., Uhe J., Golovko, O., Bonk, C., Bouguecha, A., Klose, C., Behrens, B.-A., Maier, H.J.: Co-extrusion of semi-finished aluminium-steel compounds. AIP Conference Proceedings **1896**(140002) (2017)
12. Behrens, B.-A., Klose, C., Chugreev, A., Heimes, N., Thürer, S.E., Uhe, J.: A Numerical study on co-extrusion to produce coaxial aluminium-steel compounds with longitudinal weld seams. Metals **8**(717), 1–4 (2018)
13. Gembarski, P.C., Sauthoff, B., Brockmöller, T., Lachmayer, R.: Operationalization of manufacturing restrictions for CAD and KBE-systems. In: Proceedings of the DESIGN 2016—14th International Design Conference, vol. 3, Design Support Tools, the Design Society, Glasgow, pp. 621–630 (2016)
14. Siqueira, R., Lachmayer, R.: A Manufactured Constrained Design Methodology Application for a Tailored Forming Hybrid Component. Tagung Faszination Hybrider Leichtbau, Braunschweig (2018)
15. Pahl, G., Beitz, W., Feldhusen, J., Grote, K.-H.: Engineering Design – A Systematic Approach. Springer, London (2007)
16. Vaezi, M., Chianrabutra, S., Mellor, B., Yang, S.: Multiple material additive manufacturing – part 1: a review. Virtual Phys. Prototyp. **8**(1), 19–50 (2013)
17. Brockmöller, T., Gembarski, P.C., Mozgova, I., Lachmayer, R.: Design catalogue in a CAE environment for the illustration of tailored forming. In: 59th Ilmenau Scientific Colloquium, pp. 11–15, Ilmenau (2017)

A New Numerical Method for Potential Analysis and Design of Hybrid Components from Full Vehicle Simulations: Implementation and Component Design

Thomas Tröster[1(✉)], Alan A. Camberg[1], Nils Wingenbach[1], Christian Hielscher[2], and Julian Grenz[2]

[1] Automotive Lightweight Design, Paderborn University, Mersinweg 7, 33100 Paderborn, Germany
{thomas.troester,alan.camberg, nils.wingenbach}@uni-paderborn.de
[2] BENTELER Automobiltechnik GmbH, 1 An der Talle 27–3, 33102 Paderborn, Germany
{christian.hielscher,julian.grenz}@benteler.com

Abstract. To exploit the lightweight potential of common materials like steel, aluminum or especially fiber-reinforcement plastics (FRP), a load-compliant material concept is mandatory. In keeping with the motto "the best material for the best application", a new approach for a tailored material distribution between FRP and metals is proposed. By defining a scalar value referred as "uniaxiality" and a uniaxiality weighted sensitivity, a numerical method is implemented to identify body-in-white (BIW) components with a high amount of anisotropic loading. The uniaxiality value is gathered from full vehicle crash simulations and is superpositioned over all load cases to access a generalized information which component's area is suitable for isotropic materials like metals and which one for anisotropic materials like FRP. In the next step, a functional component group consisting of an A-pillar and a roof frame section is exemplary engineered by using the findings of the numerical potential analysis. The developed extreme lightweight concept made of aluminum extrusion, unidirectional carbon fiber-reinforcement plastics (CFRP) tapes, warm stamped aluminum and press-hardened steel demonstrates outstanding performance that has been proven in full vehicle crash simulations and experimental tests. Furthermore, the concept is evaluated in terms of the components CO_2 footprint and costs. Based on these data a scalable component concept is possible to meet customer specific requirements between the design objectives: lightweight, costs and environmental impact.

Keywords: Potential analysis · Hybrid components · Finite element analysis · Crash simulation · Multi-material design

K. Dröder and T. Vietor (Eds.): *Technologies for economic and functional lightweight design*, Zukunftstechnologien für den multifunktionalen Leichtbau, pp. 353–365, 2021. https://doi.org/10.1007/978-3-662-62924-6_30

1 Introduction

Hybrid material concepts, as an approach to combine the advantages of two different materials by compensating their disadvantages at the same time, have attracted increased interest in both academia and industry [1]. In recent years, several hybrid components concepts have been introduced into automotive series production. The front roof frame of the Audi A6 (C6, model year 2004) is one of the first examples of how weight reduction can be realized by combining different materials while maintaining cost neutrality [2]. Another early example of hybrid lightweight design is the sidewall frame of a small-series hydrogen derivate of the BMW 7-series (E68, model year 2005). To fulfil the strict safety requirements of a hydrogen-powered vehicle, the steel sidewall frame was reinforced by CFRP patches [3]. The hybrid reinforcement concept allowed a cost-effective solution for small series with high flexibility and quick integration into the existing production process. In addition, a weight saving of 52% compared to a steel solution was achieved [4]. With the new BMW 7 Series (G11, model year 2015), several hybrid components consisting of steel shell structures and CFRP reinforcement patches have been introduced into series production, enabling mass savings of up to 40 kg per BIW [5, 6]. A hybrid lightweight design approach is also featured in the body structure of the new BMW 8 Series (G15, model year 2018) [7]. The latest convertible model of the iconic sports car Porsche 911 (992, model year 2019) uses hybrid parts as well [8]. Here, the hybrid A-pillar is manufactured using a technology developed within the Q-Pro project [9]. Based on the examples given, it can be concluded that hybrid lightweight design is an interesting and promising lightweighting strategy not only from a research point of view, but also from that of industry.

However, the design of lightweight and efficient multi-material concepts requires a detailed information about the stress and strain distributions within a component. This applies in particular to the use of FRP, since the mechanical properties of fiber-based materials are highly dependent on the loading direction. This fact is crucial for applications with different load cases, such as a car body structure that must provide high stiffness under operational loads and a high level of passive safety in any crash scenario. These challenging requirements lead often to conservative designs that do not exploit the full lightweight potential of the materials used. Such as in the hybrid B-pillar of the BMW 7 Series (G11), where the CFRP patch has a quasi-isotropic layup [6].

To increase the utilization rate of FRP in complex car body structures, Durst [10] proposed a method that identifies the suitability of FRP for car body components by evaluating the stress state over different load cases. The method is based on the product of three factors: the principal stress ratio, the variation in the load direction and a load case-weighting factor given by the sum of the absolute values of the major and minor principal stress. The result is a scalar value per every BIW component that characterizes the averaged anisotropy state of the component. Albers et al. [11] presented an enrichment of the method from [10] into the 3D stress space. Furthermore, not only the averaged component's anisotropy value but also the local (element) specific anisotropy value can be evaluated by a contour plot. Klein et al. [12] presented a further enhancement of the method from [10] by introducing a new approach for the load case superposition and by considering several possible fiber orientations. Moreover, the anisotropy value is additionally weighted by stiffness and strength factors. A significant enrichment of Durst's method was proposed by Dlugosch [13]. The introduction of a so-called *principal stress factor* enables a stress tensor addition over several load cases through a modification of the initial stress tensors. An undeniable advantage of this method is the implicit consideration of loading directions and absolute stress values. A further criterion introduced by Dlugosch is the *uniformity* as a component specific degree of homogeneity in the loading directions. Later, both component-specific values are combined to an *anisotropy coefficient* which gives the components suitability for FRP reinforcements. Moreover, the vector of the resulting loading orientation can be plotted element-wise directly on the component's geometry enabling the deduction of an optimized ply design directly from the results of the anisotropy analysis. Eventually, Grote [14] proposed a further improvement to the method from [10]. By introducing a "force flow" weighted orientation factor, material thickness is considered and high loads have a greater impact on the resulting orientation of the superpositioned anisotropy value. Unlike in [10, 11, 12], Dlugosch [13] and Grote [14] applied their methods not only to stiffness but also to crash problems.

With the aim to design components with a highly effective and targeted use of FRP materials, a new design methodology for hybrid components is proposed within this paper. A uniaxiality analysis in analogy to [13] is implemented to analyze the accumulated principal stresses. However, instead of using a generalized *uniformity* value to characterize the FRP-suitability of the whole component, the uniaxiality values are used to weight local sensitivities and identify component areas with a high potential for local FRP reinforcements. The methodology is applied on a full vehicle model in several crash load cases. In a further step, a chosen component group with a high fraction of uniaxial loading is redesigned by taking into account local sensitivities and loading anisotropies. Finally, the a newly designed hybrid car body component is manufactured and tested on a dynamic crash device.

2 Proposed Approach

To pursue an affordable and demand-oriented lightweight design of hybrid BIW components, a stress-based potential analysis in analogy to the *principal stress factor* from [13] is implemented and enhanced by a uniaxiality weighted sensitivity analysis. Starting with benchmark simulations of a full vehicle model, a cumulative anisotropy analysis of each component is performed and a crash-relevant body assembly with a high overall uniaxiality is selected for a hybrid design trial. The single components are optimized with the proposed method and adapted for multi-material design. The core idea of the method is to use unidirectional fiber materials in areas with high anisotropic loads and to carry the isotropic part of the load through a metallic part. In addition, the CO_2 footprint and costs are also evaluated during the design phase of the component. In the final design step, the crashworthiness of the newly designed hybrid assembly is determined in full vehicle simulations and compared with the benchmark structure. The objective is to reduce the weight of the components while maintaining or even increasing the crash performance of the vehicle. The design procedure conducted in this work is shown in Fig. 1.

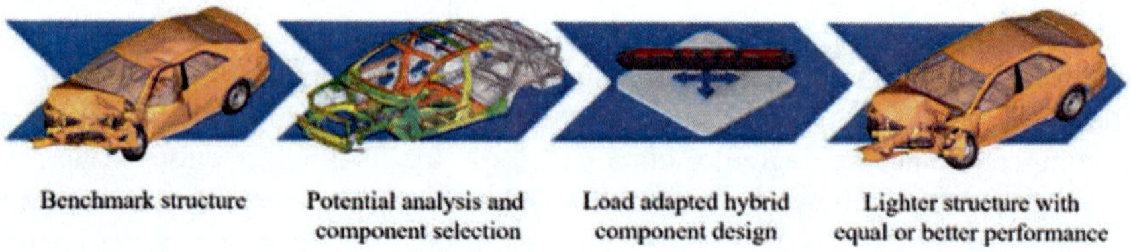

Fig. 1. Proposed approach for analyzing and improving mechanical structures by lightweight hybrid components.

2.1 Uniaxiality Analysis

In analogy to Dlugosch's *principal stress factor* [13], the here presented approach is based on the definition of a scalar value that characterizes the so-called *uniaxiality U* of the stress tensor. As given in Eq. (1), the uniaxiality describes the magnitude of the ratio between the absolute value of the second and first principal normal stress. This characteristic value takes amounts between 0 and 1, where $U = 1$ describes a purely uniaxial (anisotropic) and $U = 0$ a purely equi-biaxial (isotropic) stress state, see Fig. 2. As in [13], the load case superposition is based on the addition of stress components. A specific modification of stresses ensures that the summation does not lead to the annulment of values with different signs. At the same time, it is ensured that the original load direction and the absolute stress values are maintained. The advantage of this procedure is that information about the direction of the resulting stresses remain available and that the individual time steps and load cases are weighted by the magnitude of the stress values. The resulting uniaxiality is determined in the last step from the accumulated stress tensor. The detailed sequence of the algorithm is given in Fig. 3.

$$U = 1 - \frac{|\sigma_2|}{|\sigma_1|}, \quad |\sigma_1| \geq |\sigma_2| (1)$$

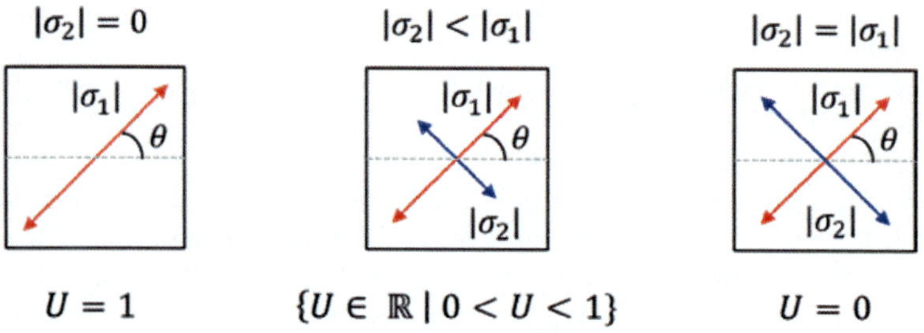

Fig. 2. Uniaxiality values of chosen stress states.

In the here used approach, first the element's stress tensor $\tilde{\sigma}^e$ is evaluated at the most loaded integration point (IP) of a reduced-integrated element type (element type 2 from the LS-DYNA standard element library) with five through-thickness IP and stored for each individual timestep i of the load case j (1). Subsequently, the angle θ^e_{ij} of the principal axis system in relation to the element coordinate system is determined (2) and the absolute values of the principal normal stresses are computed (3). By assuming that tensile and compressive stresses are equivalent, it is ensured that stresses of different signs do not lead to a cancellation of the values during summation. In the next step, a modified element stress tensor $\tilde{\sigma}^{e*}_{ij}$ is computed from the absolute values of the principal normal stresses with the original angle θ^e_{ij} of the principal normal stresses (4). In the next step, all $\tilde{\sigma}^{e*}_{ij}$ values are accumulated to a resulting stress tensor $\tilde{\sigma}^{e**}$ (5). The tensorial value guarantees a weighting of the individual timestep and load case entities during summation and provides information about the resulting orientation of the superimposed stress tensors $\tilde{\sigma}^{e*}_{ij}$. The scalar value of the element uniaxiality U^e is determined in the very last step from the absolute values of the principal normal stresses of the accumulated stress tensor $\tilde{\sigma}^{e**}$ (6).

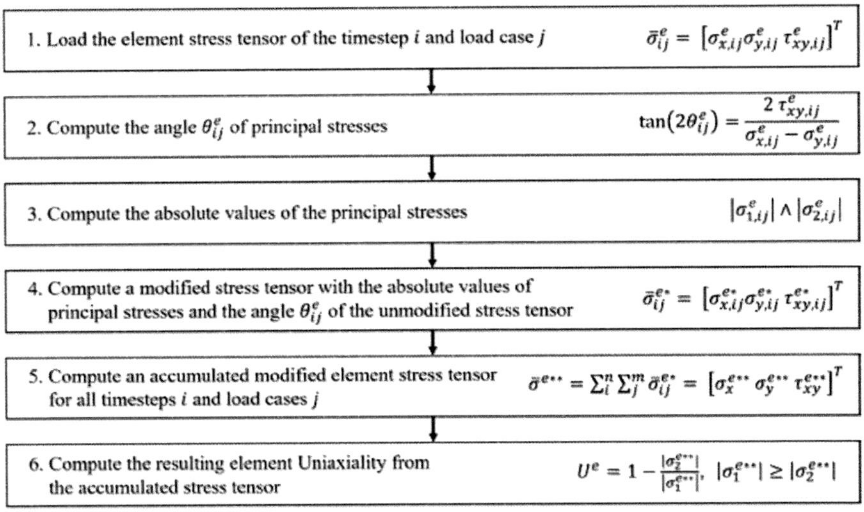

Fig. 3. Scheme of the implemented algorithm for an accumulated Uniaxiality analysis.

In the next step a component specific uniaxiality U^K value is calculated by averaging the element values over the component domains K. With the component specific uniaxiality value U^K, a fast identification of body parts (components) with a high overall uniaxiality fraction is possible. That enables an efficient determination of parts with a high potential for hybrid lightweight design.

2.2 Uniaxiality-Weighted Sensitivity Analysis

Once the uniaxiality analysis has been completed and a car body component with a high total anisotropic loading has been selected for hybrid redesign, a so-called *uniaxiality-weighted sensitivity analysis* of the component is carried out. In order to ensure that the expensive reinforcement fibers are used in areas with pronounced uniaxiality and at same time have a large effect on the structural performance of the component, a procedure depicted in Fig. 4 is introduced.

First, the baseline component K_b is divided into p reasonable subdomains $K_{b,c}$ that gauge sizes T_c are parametrized within a design space given by manufacturing and package restrictions. In the next step, a structural sensitivity analysis of the relevant system response with respect to each design variable is carried out. In parallel, a uniaxiality analysis of each component subdomain of the baseline design is computed. By multiplying the subdomain sensitivity with the subdomain uniaxiality value, a uniaxiality-weighted sensitivity is obtained. Finally, the weighted sensitivities are sorted in descending order to give an indication of which cross-sectional area of the component can be effectively reinforced by fibers in terms of structural performance and costs. With these data and a linear mixing rule of the material fractions within the respective subdomains, a minimization of the component mass M_b and the production greenhouse gas (GHG) emissions E_b can be performed. The production GHG emissions are calculated by multiplying the resulting material fractions φ_c by the material specific emission intensities I. The linear mixing rule is applied in such a way that a structural equivalence to the baseline design is given in terms of stiffness and strength.

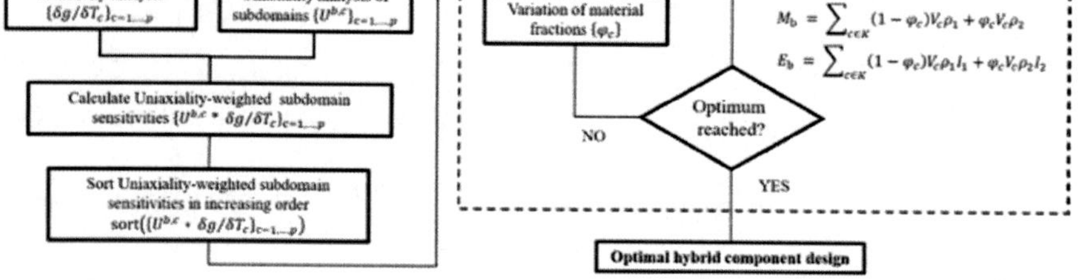

Fig. 4. Flowchart of the uniaxiality-weighted sensitivity analysis and the subsequent optimization loop for the design of sustainable lightweight hybrid components.

3 Accumulated Uniaxiality Analysis

The full vehicle model of the Toyota Camry 2012 [15] provided by the Center for Collision Safety and Analysis, George Mason University, is used as a benchmark structure. The relevant load cases are chosen under the consideration of the crash scenario probability in real traffic events. As reported in [16], frontal impact crashes dominate the total number of impacts (65.7%), followed by side impacts (27.3%). Consequently, worst-case configurations of European and US rating agency front and side crash load cases are selected for benchmarking, i.e. the Euro NCAP Moderate Overlap Deformable Barrier Frontal Test (64 km/h), the US NCAP Full-Width Rigid Wall Frontal Test (56 km/h), the IIHS Small Overlap Rigid Barrier Frontal Test (64 km/h), the US NCAP Side Moving Deformable Barrier Test (62 km/h) and US NCAP Side Pole Test (32 km/h). In order to consider all load case scenarios acting on the front and side area of the passenger compartment, the IIHS Roof Crush Test is incorporated into the benchmarking as well.

The uniaxiality analysis algorithm presented in Sect. 2.1 is implemented as a Tcl/Tk script into the Altair HyperView post-processor software. In order to ensure that the evaluation only takes place on structure-relevant components, a set of body-in-white components is created before the evaluation routine is started. A further measure to reduce the amount of data evaluated is realized by defining a threshold value for critical stress values (e.g. $\sigma_{yield}/2$). As soon as the FE solver output data of all load cases are available, all relevant elements run through the routine depicted in Fig. 3. The result is a contour plot as shown in Fig. 5, in which the uniaxiality can be displayed in the post-processor either as averaged component values or by element. In addition, the resulting direction of the accumulated element stresses can be displayed directly on the elements (not depicted here). With this information, the resulting fiber direction can be determined at the same time without additional computational effort.

Remarkably, several components in Fig. 5 show a uniaxiality value greater than 0.7, what means that the major principal stress is more than 3 times higher than the minor principal stress. As expected, bending dominated components, such as door beams, roof cross members or A-pillars, have a higher uniaxiality value than structures with multiple local buckling modes, such as front longitudinal members or crash boxes. When evaluating the element-wise uniaxiality, numerous areas of almost pure uniaxial loading can be identified.

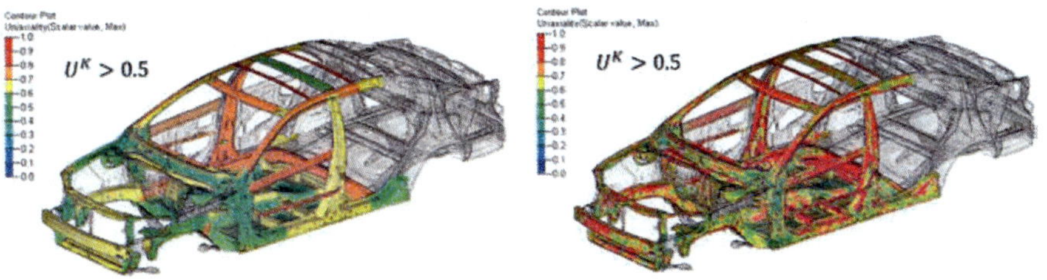

Fig. 5. Result of the uniaxiality analysis over several crash load cases for selected BIW components, threshold: $U^K > 0.5$; Left: Averaged component uniaxiality U^K, Right: Element uniaxiality U^e.

4 Component Design

4.1 Component Selection

The uniaxiality analysis serves as a starting point for the redesign of a selected component with a high hybridization potential. From that, the A-pillar and the front roof frame were selected as a demonstrator group. These components show a high uniaxiality and represent a part of the rigid passenger compartment. Both are excellent prerequisites for the efficient use of local reinforcements made of orthotropic and high-strength carbon fiber composites. In addition, the A-pillar offers not only the potential for increasing passive safety but also the possibility of extending the field of vision, which represents a secondary effect of higher weight-specific properties of this component. Furthermore, the A-pillar gained a lot of attention as a possible hybrid component in the past. Already in 2002, Liedke [17] examined the substitutability of a hybrid lightweight design approach for an A-pillar of a convertible derivate. Using two press-hardened steel shells and a load-adapted aramid-carbon fiber laminate reinforcement, a mass reduction of 12.4% was achieved while maintaining the same crash behavior. Today, the current BMW 7 Series [6] and the current Porsche 911 convertible (−2.7 kg / per vehicle) [8] use hybrid A-pillars in series production. Hence, it is particularly interesting to what design concept and results in weight reduction the development method presented here will lead to.

4.2 Component Design

At the beginning of the component development, a topology study was carried out. Based on the ideal component topology, several structural concept designs were derived, which were evaluated and compared using simplified finite element models. The most promising concept—a continuous and extruded aluminum profile with unidirectional reinforcement tapes—was refined in detail in further development stages. The design method presented in Sect. 2.2 was decisive for the location and the cross-section design of the reinforcing carbon fiber tapes. Unfortunately, the cross-sectional areas with the highest potential fall in the area of joining flanges and are therefore not suitable for fiber reinforcement, as this would lead to a complicated and costly joining situation. However, the selected positions of the carbon fiber unidirectional (UD) reinforcement tapes are located as close as possible to the joining flanges and correspond to a complementary upper and lower belt of the AA6082 T6 aluminum profile.

The lower A-pillar base was designed as a press-hardened steel shell due to a predominant multi-axial loading. The joining between the aluminum and steel parts of the A-pillar could be realized by conventional spot welding thanks to BENTELER-patented SWOPtec joining elements. The Flash Forming Process (FFP) [18], another technology patented by BENTELER, made it possible to manufacture the front roof frame from an AA5182 aluminum sheet and achieve strengths of up to 400 MPa.

Despite a high degree of uniaxiality, the front roof frame could not be designed as a hybrid with any significant mass advantage. The reason for this was the structural instability caused by local fiber reinforcements and the associated abrupt changes in stiffness. The connector node between the A-pillar profile and the roof frame is made of an AA6082 T6 aluminum extrusion, whose cross-section was designed by topology optimization. The final assembly design and the hybrid A-pillar cross-section is depicted in Fig. 6.

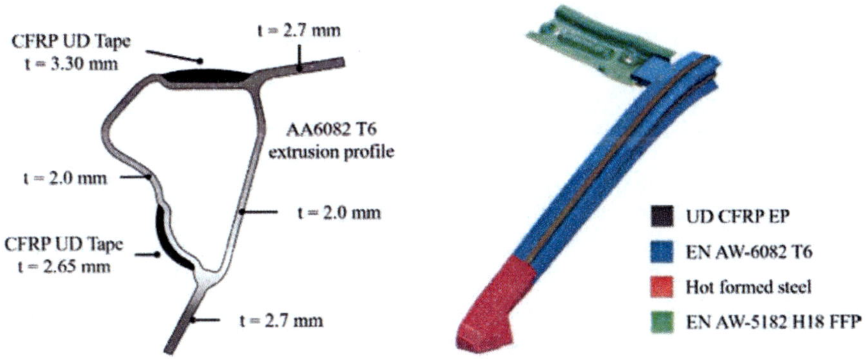

Fig. 6. Left: Hybrid A-pillar cross-section, Right: Final multi-material assembly design.

4.3 CO_2 Footprint and Cost Analysis

Lightweight design not only needs to deliver parts with lower mass, but also ensure moderate costs and a lower CO_2 footprint over the product life cycle. For this reason, a simplified eco-audit and a cost analysis of the hybrid A-pillar were carried out. The cost analysis assumed a production of 60,000 vehicles per year, which corresponds to a usual number of units in the premium segment, for which the use of CFRP is realistic. Even though CFRP was used very targeted, it was not possible to achieve cost neutrality compared to performance equivalent variants in aluminum (+14% weight) and press-hardened steel (+46% weight). The aluminum variant offers a cost-saving potential of 36% and the press-hardened steel variant is even 54% more cost-effective than the hybrid lightweight variant. When evaluating the CO_2 footprint, which was done using the CES Selector software, the hybrid A-pillar shows the lowest end-of-life CO_2 footprint of all concepts. Remarkably, the hybrid concept has a lower CO_2 footprint than the aluminum variant already from the beginning of the use phase, see Fig. 7.

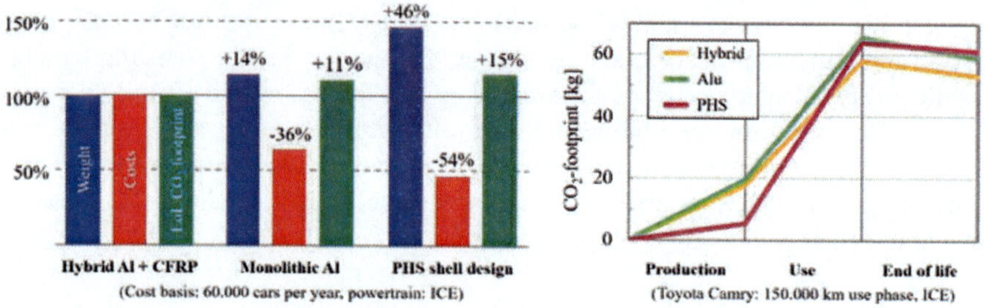

Fig. 7. Cost and simplified Life Cycle Assessment analysis of the hybrid A-pillar and performance equivalent aluminum and steel variants.

4.4 Full Vehicle Validation

Finally, the hybrid A-pillar design concept is validated in full vehicle crash simulations. An equivalent or better structural performance was achieved compared to the benchmark. In particular, the frontal load cases show an increase in passive safety, since the hybrid A-pillar, unlike the baseline design, does not buckle and provides a rigid passenger compartment. A comparison for the Euro NCAP Moderate Overlap Deformable Barrier Frontal Test (64 km/h) between the baseline and the hybrid design is presented in Fig. 8. The hybrid A-pillar provides structural integrity, reduced intrusions and lower maximum decelerations (−12.5%).

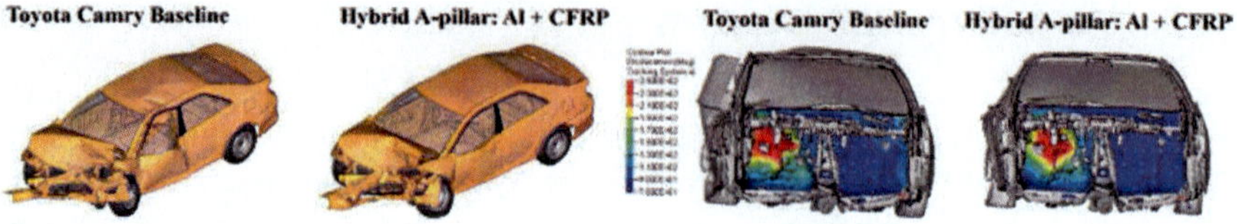

Fig. 8. Euro NCAP MODB Frontal Test (64 km/h): Baseline vs. Hybrid design. Left: Firewall intrusion in mm.

5 Component Manufacturing and Testing

The developed hybrid A-pillar cross-section profile was extruded from AA6082 and stretch-bended by BENTELER Aluminum Systems Norway AS, Raufoss. The unidirectional CFRP tapes (SGL SIGRAPREG® C U300-0/NF E420/38%) were manufactured using a tailored tape placement process at the Institute for Lightweight Design with Hybrid Systems (ILH), Paderborn. The simultaneous curing and bonding of the tapes to the extruded aluminum profile were realized by means of a prepreg press technology using an additional adhesive layer (3M SAT 1010). For comparison purposes, an unreinforced aluminum profile and an aluminum profile with an alternative reinforcement made of a 1.8 mm DP800 steel were manufactured as well. The DP800 reinforcement strips where bonded to the aluminum profile by steel blind rivets (GESIPA G-Bulb 6,4 mm) and a structural adhesive (Dow Automotive BETAMATE 2096).

The crashworthiness of the developed A-pillar cross section (see Fig. 6) was investigated using a representative profile length of 400 mm in a three-point bending test on a drop tower device of the LiA Crash & Impact Lab. The support distance was set to 275 mm, the support and impactor radii were 25 mm. The impact energy was 2.0 kJ at an initial impact speed of 4.5 m/s. The results of the impact tests are summarized in Table 1. The force-displacement curves and post-crash pictures of the deformed profiles are given in Fig. 9. It is shown that the performance of the CFRP and steel reinforced profiles is comparable. Both profiles show a similar increase in maximum force and absorbed energy. However, the achieved mass specific energy absorption (SEA) shows a strong dependence on the reinforcement material. While the SEA increases by 32% for the CFRP reinforcement, it decreases by 5% for steel reinforcement compared to the unreinforced aluminum profile. The test results emphasize the performance of the hybrid aluminum CFRP profile.

Table 1. Experimental results of the energy absorption characteristics of the developed hybrid A-pillar.

Specimen	Weight [kg]	F_{max} [kN]	E_{abs} [kJ]	SEA [kJ/kg]
AA6082 T6	0.556	20.16	1.29	2.32
AA6082 T6+CFRP	0.627 (+12.8%)	29.10 (+44%)	1.92 (+49%)	3.07 (+32%)
AA6082 T6+DP800	0.847 (+52.3%)	28.78 (+43%)	1.87 (+44%)	2.20 (−5%)

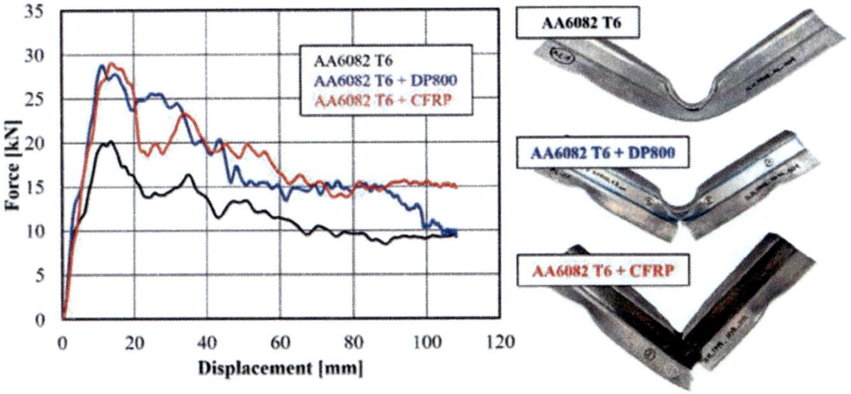

Fig. 9. Force-displacement curves and post-crash pictures of different A-pillar profiles.

6 Conclusions

Based on the findings of the accumulated uniaxiality analysis and the uniaxiality-weighted sensitivity, the design of components with targeted use of unidirectional fiber reinforcement in combination with classical isotropic metallic materials is possible. As shown by the example of the A-pillar, by substituting a material fraction of just two cross-sectional areas by a structural-equivalent amount of unidirectional carbon fibers, the components weight can be reduced by 14%. By means of a simplified life cycle assessment, it was possible to show that the hybrid component has even a better CO_2 footprint already in the production phase than a monolithic aluminum variant. Nevertheless, it was found that the cost of the hybrid component significantly exceeds that of conventional design concepts. This is mainly due to high material and additional processing costs. It must therefore be considered on individual use-case basis whether lower weight and environmental impact is justified by additional costs. However, with smaller production volumes the cost advantages of conventional concepts are reduced, so that a hybrid component concept can be economically reasonable if the requirements for lightweight are particularly high and the number of units is small. That could be particularly the case for derivate-dependent adjustment in crash performance due to e.g. a heavier powertrain. Furthermore, if the results are transferred to electric vehicles, the growing share of renewable energy sources will increase the importance of the production phase in terms of CO_2 emissions in the future, so that the development of sustainable material and component concepts will become increasingly important.

Finally, the hybrid A-pillar developed using the proposed numerical design method was able to prove its outstanding performance in full vehicle crash simulations and experimental impact tests. This confirmed the importance of smart product design methods and shows that the lightweight potential of hybrid components is far from being fully exploited.

Acknowledgements. The authors gratefully acknowledge the funding of the project by BENTELER Automobiltechnik GmbH. Sincere thanks also due to the Paderborn Center for Parallel Computing (pc^2) for the provided computing time. Special thanks are due to Caterina Linnig, Jan Striewe, Bamned Sanitther, Gero Müllers and last but not least Dr. Jörn Tölle for their support and valuable discussions.

References

1. Bader, B., Türck, E., Vietor, T.: Multi material design. A current overview of the used potential in automotive industries. In: Dröder, K., Vietor, T. (eds.) Technologies for Economical and Functional Lightweight Design, pp. 3–13. Springer, Berlin (2019)
2. Jäschke, A., Dajek, U.: Dachrahmen in Hybridbauweise. Sonderdruck aus VDI-Tagungsband Nr. **4260**, 25–45 (2004)
3. Müller, C., Fürst, S., von Klitzing, W.: Hydrogen safety: new challenges based on BMW hydrogen 7. In: Proceedings of 2nd International Conference on Hydrogen Safety 2007, San Sebastian (2007)

4. Derks, M.: CFK-Technologien im Automobilbau. DLR Wissenschaftstag 2007, Braunschweig. BMW Group, München (2007)
5. N.N.: Technical Training – Product Information – G12 Introduction. BMW AG (2015)
6. Frei, P.: The hybrid B-Pillar in the new BMW 7-Series. An Example for the manufacturing implementation of an innovative Steel-CRFP lightweight design concept. In: Automotive Circle, Materials in Car Body Engineering Conference, Bad Nauheim (2016)
7. Henseler, U., Eras, A.: Introducing the new 8 series coupe. In: Euro Car Body Conference, Bad Nauheim (2018)
8. Wawers, U., Stein, S.: The development and production of the new Porsche 911 body structure: the new 911 Carrera. In: Eckstein, L. (ed.) Proceedings of Aachener Body Engineering Days 2019. Institut für Kraftfahrzeuge, RWTH Aachen, Aachen (2019)
9. Haider, D.R., Krahl, M., Koshukow, W., Wolf, M., Liebsch, A., Kupfer, R., Gude, M.: Adhesion studies of thermoplastic fibre-plastic composite hybrid components part 2: thermoplastic-Metal-Composites. In: Hausmann, J.M., Siebert, M., von Hehl, A. (eds.) Proceedings of 3rd Hybrid Materials and Structures 2018, pp. 68–73. DGM, Berlin (2018)
10. Durst, K.G.: Beitrag zur systematischen Bewertung der Eignung anisotroper Faserverbundwerkstoffe im Fahrzeugbau. Cuvillier, Stuttgart (2008)
11. Albers, A., Schmid, A., Zhang, Q., Grönheim, O., Schipperges, J.: Ein CAE-basiertes Konzept zur automatischen Identifikation von Leichtbauteilen im Gesamtsystem. NAFEMS Magazin **23**(3/2012), 54–63 (2012)
12. Klein, D., Witzgall, C., Wartzack, S.: Bewertung und Optimierung der Faserverbundeignung von Leichtbaustrukturen in den frühen Phasen der Produktentwicklung. NAFEMS Magazin **32**(4/2014), 55–66 (2014)
13. Dlugosch, M.: Zur Methodenentwicklung im Entwurf automobiler Strukturkonzepte in FVK-Metall Hybridbauweise unter Crashbelastung. Schriftenreihe Forschungsergebnisse aus der Kurzzeitdynamik, Bd. 36. Fraunhofer Verlag (2018)
14. Grote, M.: Entwicklung einer Methode zur anforderungsgerechten Werkstoffauswahl und Strukturauslegung von PKW-Karosserien im Multi-Material-Design. Siegener Schriftenreihe Automobiltechnik, Bd. 2. Universitätsverlag Siegen, Siegen (2018)
15. Marzougui, D., Brown, D., Park, H.K., Kan, C.D., Opiela, K.S.: Development & validation of a finite element model for a mid-sized passenger Sedan. In: Proceedings of the 13th International LS-DYNA Users Conference, Dearborn, MI (2014)
16. Kramer, F.: Integrale Sicherheit von Kraftfahrzeugen, 4th edn. Springer, Wiesbaden (2013)
17. Liedtke, B.: Faserverbundkunststoff/Metall-Hybridstrukturen im Pkw-Rohkarosseriebau. Fortschritt-Berichte VDI: Reihe 12, Verkehrstechnik, Fahrzeugtechnik 509. VDI-Verlag, Düsseldorf (2002)
18. Camberg, A.A., Bohner, F., Tölle, J., Schneidt, A., Meiners, S., Tröster, T.: Formability enhancement of EN AW-5182 H18 aluminum alloy sheet metal parts in a flash forming process: testing, calibration and evaluation of fracture models. IOP Mat. Sci. Eng. **418**, 012018 (2018)

Graph Based Algorithms to Enhance Mid-Surface Design Fidelity of Finite Element Models of Extrusion Profiles

Johannes Sperber[1(✉)], Enrique Benavides Banda[1], Christopher Ortmann[1], and Axel Schumacher[2]

[1] Volkswagen AG, Wolfsburg, Germany
{johannes.sperber1,christopher.ortmann}@volkswagen.de
enrique.benavides@vw.com.mx
[2] Chair for Optimization of Mechanical Structures, University of Wuppertal, Wuppertal, Germany

Abstract. Extruded profile structures are widely used as efficient and lightweight crash energy absorbers in vehicle structures throughout the automotive industry. For the development and evaluation of new concepts and designs, finite element simulations are used to decrease costs and shorten development periods. In order to get good accordance between physical responses of extrusion profiles and their finite element representations under crash loads, a certain degree of modeling detail is needed. These details can be radii and material accumulations which are usually provided in volumetric CAD (Computer Aided Design) models. Leaving out these details in the profile cross section geometry can lead to differences in the buckling wavelength of walls and therefore in the deformation behavior.

However, CAD designers are usually not incorporated into every design iteration of the CAE (computer aided engineering) departments and therefore these details might be missed out when design changes are rapidly imposed on a finite element shell mesh level. Furthermore, structural optimization methods applied to improve the profile's mechanical behavior might not even allow for CAD designers to be involved. In this work two efficient yet simple to configure graph based algorithms to automatically apply rounded corners and to consider material accumulations at intersection points of walls of the profile are provided. These algorithms work with a graph based representation of the profile's cross section geometry in order to perform geometric calculations. To describe the profile's geometry the graph syntax of the optimization method "Graph and Heuristic based Topology Optimization" is used. Based on the graph a three-dimensional finite element representation of the profile is generated by an extrusion along a defined spline. This finite element model is the basis for further mesh modifications and conclusively for the simulation itself. The algorithms have proven their ability to increase the quality of the finite element representation of extrusion profiles and therefore increase the accordance between physical tests and simulation results in terms of geometry and result accuracy.

K. Dröder and T. Vietor (Eds.): *Technologies for economic and functional lightweight design*, Zukunftstechnologien für den multifunktionalen Leichtbau, pp. 366–382, 2021. https://doi.org/10.1007/978-3-662-62924-6_31

Keywords: Extrusion profiles · Design process automation · Finite element modelling · Crash simulation · Mathematical graphs

1 Introduction

Throughout the automotive industry, extruded aluminum profile structures are broadly used in vehicle bodies. Especially when it comes to lightweight design and the absorption of crash energy, aluminum extrusion profiles have proven their abilities [1, 2]. In the vehicle development process, finite element (FE) simulations are generally used to decrease costs and shorten development periods. To optimize a structure and to evaluate that the vehicles meet all required safety standards, finite element simulations with explicit time integration are used.

While the task of automatically simplifying 3D CAD (Computer Aided Design) geometries for their efficient usage in simulations is well documented in literature (e.g., compare [3] for a broad overview), no literature was found regarding the reversed process of automatically enhancing the design fidelity of simple or reduced geometry descriptions to more detailed finite element models. However, Knowledge Based Engineering (KBE) can be seen as a field of research, in which related techniques such as rule based, automated CAD modeling for minimum user input are also addressed. An overview is given in [4].

In order to get good accordance between physical responses of aluminum extrusion profiles and their finite element representations under crash loads, a certain degree of modeling detail is needed [5]. When using solid elements with a fine mesh size, all geometrical details defined in a technical drawing, like radii, can be represented in the model (compare Fig. 1b)). Modeling these details can influence the structural responses since the deformation behavior of the structure (e.g., buckling wavelength of profile walls) might change due to shorter wall lengths in the profile cross section or additional stiffnesses at the profile wall intersection points. Especially when it comes to extrusion profiles under axial compression loads, small variations in the modeling can lead to big differences in the structural responses due to possible bifurcation points [6].

Due to the requirements of full car crash simulations concerning calculation times and costs, instead of solid elements, shell elements are usually preferred and the minimum element length is limited. If a complete 3D CAD model exists as a basis for the extrusion profile model, commercial software such as BETA ANSA or Altair HyperMesh can be used to transform the volumetric CAD model into a mid-surface representation for shell element meshing (compare Fig. 1c)). The radii and fillets from the volumetric model are therefore transformed into rounded corners and thickness increases around profile wall intersection points in the mid-surface representation.

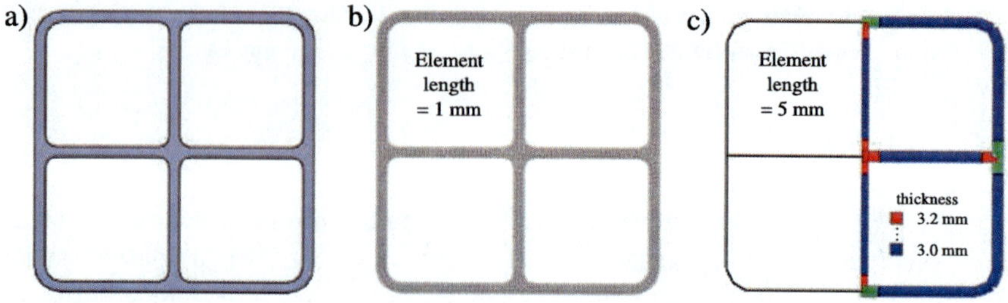

Fig. 1. Different representations of an extrusion profile cross section: a) 3D CAD geometry, b) finite element mesh with volumetric elements and c) mid-surface representation for crash simulation with partly visualized shell element thicknesses derived from the 3D CAD model with Altair HyperMesh

However, incorporating a CAD designer into each design loop in order to create complete 3D CAD models and then translating these into finite element shell meshes would increase development time and costs significantly. Instead, CAE (Computer Aided Engineering) engineers often perform modifications directly on the shell mesh, which is efficient, but could also lead to neglecting details like curvatures and material accumulations. Furthermore, especially for structural optimization methods, a CAD designer cannot be consulted during the optimization process due to the requirement of design process automation. The usage of parametrized 3D CAD models can be beneficial for shape and sizing optimization, but lacks flexibility with arbitrary topologies in a topology optimization.

To encounter these challenges, two simple, yet efficient algorithms are presented in this paper to apply rounded corners and material accumulations on finite element shell representations of extrusion profiles. Instead of a full volumetric CAD geometry description, they only use a simple graph based geometry description of the profile's cross section, which is afterwards translated into a 3D finite element model. The algorithms are described in Chap. 2. They are then applied to profile structures which are investigated in crash load cases under lateral and axial compression (Chap. 3).

2 Algorithms to Enhance Mid-Surface Design Fidelity

In this work, two efficient, yet simple to configure, graph based algorithms to automatically apply rounded corners and to consider material accumulations are described. The first algorithm, described in Sect. 2.2 uses an arc length as a single input parameter in order to apply rounded corners to the profile cross sections as defined by design guidelines for extrusion profiles. In addition, it aims for a homogeneous looking design as a CAD designer would typically do. The second algorithm, which accounts for the consideration of accumulated material at profile wall intersection points, increases the shell element thickness around profile wall intersections and is dependent on a given fillet radius. It is described in Sect. 2.3. Both algorithms are designed to have minimal input while still providing good and robust results for any cross section on hand. These requirements are derived from the usage of these methods within a topology optimization method called the Graph and Heuristic based Topology Optimization (GHT) [7, 8], which has an automated optimization process that allows for no user intervention and which already has many input parameters that need to be considered by the user prior to the optimization.

2.1 Graph Syntax and Translation into Finite Element Model

The algorithms described in this paper use the graph syntax of the GHT, which was modified in comparison to previous works (e.g. [7]) due to considerations not related to the present work. The graph is now simple, planar and directed. The graph represents the profile cross section of the extrusion profile. In an automated process this graph is then translated into a shell element based 3D finite element model of the extrusion profile (see [8] for details). This model can then be used for further mesh modifications and conclusively the evaluation of the design in a (crash) simulation.

An example graph with its syntax is shown in Fig. 2. The LINK vertices in this graph syntax describe the Cartesian positions of edge start and end points in the profile cross section. The edges furthermore inherit information regarding the vertices they connect, the thickness of the individual wall as well as additional information, if they are on the outer boundary of the profile or if they have a curvature. Moreover, a parameter vertex is given, which stores further information regarding the profile.

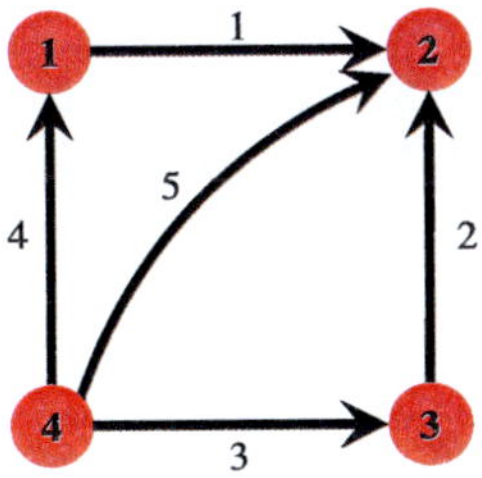


```
VERTEX, LINK, 1, COORDINATES(-40.0, 40.0)
VERTEX, LINK, 2, COORDINATES(40.0, 40.0)
VERTEX, LINK, 3, COORDINATES(40.0, -40.0)
VERTEX, LINK, 4, COORDINATES(-40.0, -40.0)
VERTEX, PARAMETER, 5, NAME(column), LENGTH(400.0), DENSITY(2.7E-6)
EDGE, 1, VERTICES(1, 2), CURVATURE(0.0), THICKNESS(2.5), OUTER
EDGE, 2, VERTICES(3, 2), CURVATURE(0.0), THICKNESS(2.5), OUTER
EDGE, 3, VERTICES(4, 3), CURVATURE(0.0), THICKNESS(2.5), OUTER
EDGE, 4, VERTICES(4, 1), CURVATURE(0.0), THICKNESS(2.5), OUTER
EDGE, 5, VERTICES(4, 2), CURVATURE(0.25), THICKNESS(2.5)
```

Fig. 2. Example graph visualization with GHT graph syntax

Using a graph based geometry description simplifies topology and geometry modification compared to modifying a volumetric CAD geometry. Furthermore, it allows for relatively simple geometric calculations compared to using a finite element or volumetric CAD model.

2.2 Algorithm to Apply Rounded Corners

Due to manufacturing restrictions and design guidelines, extruded aluminum profiles usually come with rounded profile corners [9, 10]. When an aluminum extrusion profile is developed, the CAD designer usually applies a fillet with a different radius to the inner and outer side of a corner in a 3D CAD volume model respectively. The given radius depends on the thickness of the walls, the angle between the walls, the number of connected walls and on their location (inner or outer profile wall). Furthermore, the CAD designer is looking for an homogeneous looking design.

The intuitive attempt of simply applying a user given radius parameter to the profile corners is not suitable since it is leading to bigger rounded corners when the graph edges enclose an acute angle while resulting in smaller corners for obtuse angles (compare Fig. 3a)–b)). This illustrates the importance for the radius to be dependent of the angle between the edges.

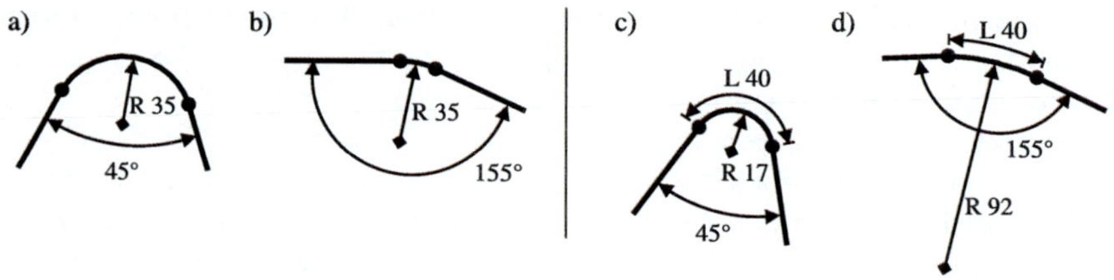

Fig. 3. Same radius of a rounded corner at a) an acute and b) an obtuse angle; Same arc length of a rounded corner at c) an acute and d) an obtuse angle

To still control the rounding of a corner with one geometrical parameter it is proposed to use the arc length of the circle sector of the rounded corner as an input parameter (compare Fig. 3c)–d)). This provides several advantages. First, the calculated radius is now angle dependent which leads to an homogeneous look of the profile section for altered angles. To reflect the fact that rounded corners exclusively between inner or outer walls are often assigned different radii in the design process, an independent perimeter length for the circle sector can be set for each of these cases. Corners between an outer and inner edge are handled as if they were between outer edges. Rounded corners are only applied if edges are not collinear.

For the case that there are more than two edges connected in one vertex, the two predominant (outmost) edges are identified. The algorithm is then applied to these two edges and the remaining edges are linked to the mid-point on the created rounded corner (compare Fig. 4).

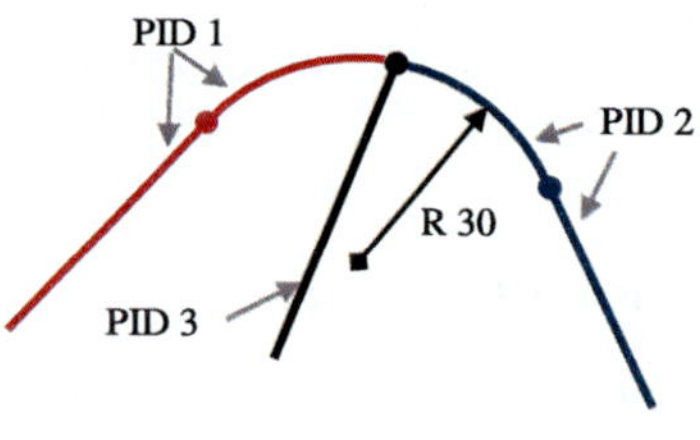

Fig. 4. Rounded corner with multiple edges and Part-ID (PID) assignment for finite element model

The application of the rounded corners is achieved by modifying the graph. First, a new vertex is generated on each of the predominant edges. These can be seen geometrically as a foot of a perpendicular from the center of the intended rounded corner (center of circle sector) on each of the existing predominant edges. Then, another vertex is generated at the middle of the circle sector line. The vertices are then connected by graph edges and the calculated radius is used to calculate the curvature parameter of the graph edge. In the proceeding step the additional edges are connected to the vertex on the middle of the circle sector line. For the two added graph edges the PIDs of the existing edges are assigned individually (compare Fig. 4). The process of applying rounded corners onto an example profile structure is shown in Fig. 5.

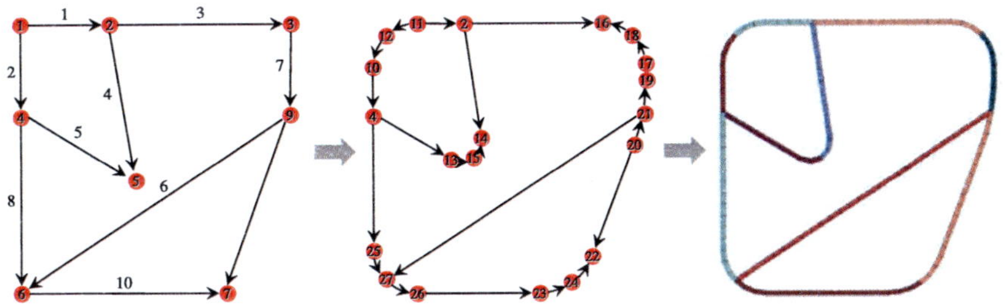

Fig. 5. Unmodified graph, graph with rounded corners and resulting FE model with rounded corners (with visualization of shell element thickness)

Currently this method has some limitations since only the more relevant edge combinations are handled (compare cases 1–3 in Fig. 6). Also, profile wall intersection points (more than two edges) where the inner graph edges enclose angles with a sum of more than 180° are not handled yet, since the algorithm cannot decide which edges are predominant in this case, and since they are subject to the material accumulations presented in the next chapter. Furthermore, edges with a given curvature parameter are also excluded. These cases could be handled by more sophisticated (geometric) considerations in the presented algorithms in the future. The flowchart of the algorithm can be seen in Fig. 6.

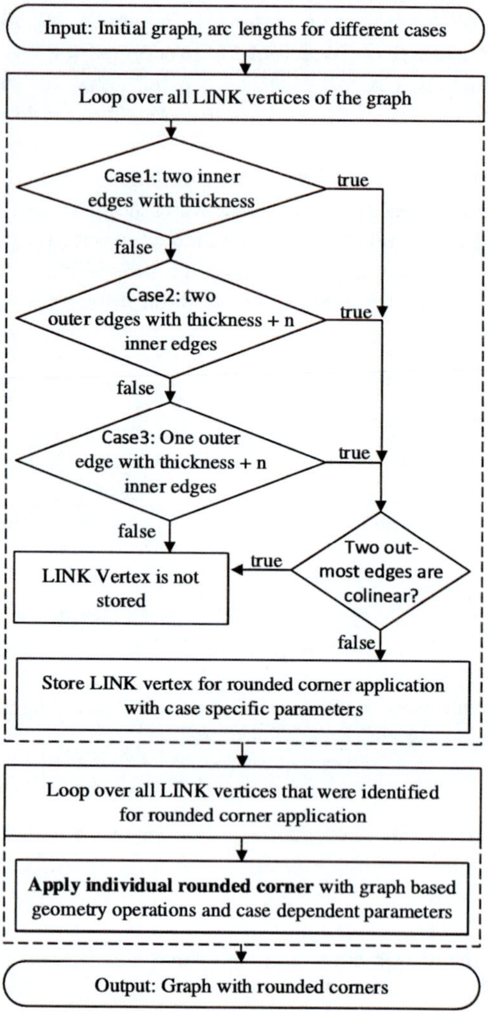

Fig. 6. Flowchart for rounded corner algorithm

2.3 Algorithm to apply material accumulations

In addition to rounded corners, design guidelines and manufacturing constraints require to avoid sharp corners in extruded profiles by adding fillets between wall connections [9, 10]. The presence of these fillets causes an accumulation of material at the profile's walls intersections (compare Fig. 1). It is common in the design process to use the same radius for all fillets in the profile. This is advantageous for the definition of an algorithm, concerning the number of required input parameters, since only a single value for the radius needs to be defined. In order to consider this accumulation of material into the finite element shell model, the elements' thickness near the intersection point is increased without further modification of the mesh's nodal positions.

The first step to determine the required thickness increase of each element is purely geometrical and consists of calculating the area of the additional material caused by the presence of the fillet between a pair of edges. The superficial face of a wall can be obtained from its mid-surface by adding an offset of half the wall thickness to the edge. When a pair of edges connects, the wall intersection is obtained by the crossing of the two offset lines from each of the corresponding edges. The point where these two offset curves intersect ("SP" in Fig. 7) corresponds to the sharp corners of an extruded profile without fillets. The accumulation of material is an inherent consequence of trying to fit a circle that is tangent to both of these surface faces. The accumulation of material corresponds to the area enclosed by the surface curves and the fillet arc (compare Fig. 7). This circular arc can be calculated with the same algorithm used for the creation of the rounded corners (compare Sect. 2.2).

The area of accumulated material can then be calculated from the area of a right triangle formed from the surface intersection point (SP), the arc center point (CP) and its foot point (FP), then removing the area of the circular section with arc's radius and angle θ (compare Fig. 7). This process is repeated for the other edge in the pair. It should be noted that the case shown here is the simplest case where both of the edges have no curvature. Nevertheless, this algorithm has to be able to consider a case where at least one of the edges in the pair has some amount of curvature. This case occurs commonly during the application of rounded corners (compare Sect. 2.2 and Fig. 5). It is geometrically more complicated, since the fitting of a tangent circle to the superficial curves is more complex and multiple cases may occur depending the given fillet radius. To handle this situation, an approximation is done by creating straight curves that are tangent to the curved edge (for calculation purposes only).

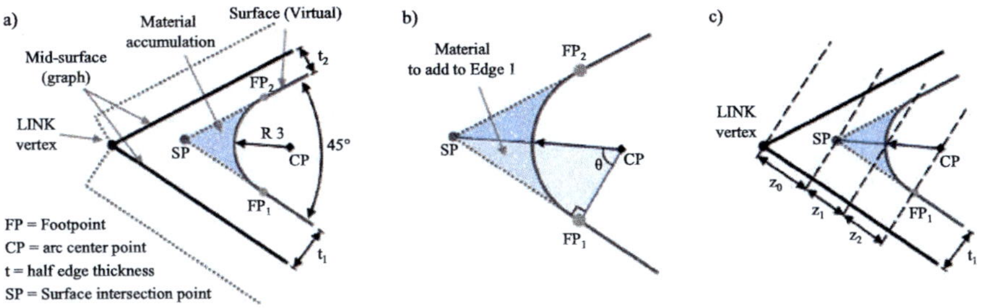

Fig. 7. Material accumulation due to a fillet between a pair of edges: a) Relevant information for calculation of material accumulation, b) detail material accumulation to add to one edge, c) zones on graph edge relevant for material accumulation

The second step, once the area of the material accumulation is known, is to distribute it among the appropriate finite elements. To calculate the thickness that has to be added to each element, three distinct zones (compare Fig. 7c)) have to be defined:

z_0: Zone between LINK vertex and the surface intersection point, where no additional thickness is necessary, because only the wall thickness should be considered.

z_1: Zone between surface intersection and intersection with fillet circle, where the additional material is defined by a right triangle and where the thickness to be added is calculated by the area of a right triangle, the FE length and the element position (compare Fig. 8a)).

z_2: Zone between intersection and fillet circle to foot point, where a segment of the arc has to be considered and where the additional element thickness starts to decrease until the nominal wall thickness is obtained again (compare Fig. 8b)).

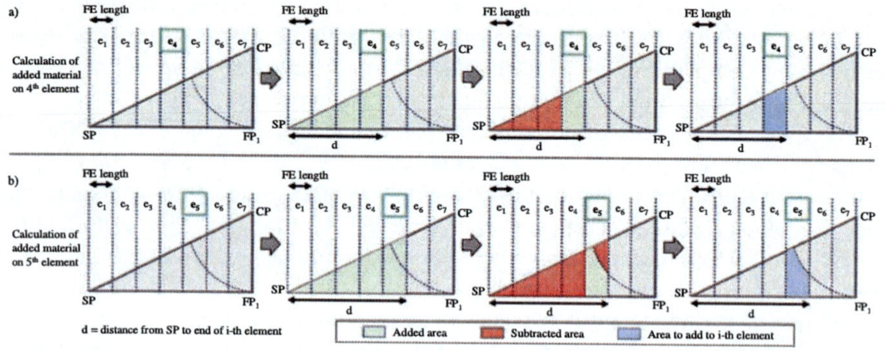

Fig. 8. Material accumulation area calculation for two arbitrary elements at different zones: a) element inside z_1, b) element inside z_2

The next step, now that the required additional thickness of each element is known, is to modify the FE mesh. However, the assembly of the complete model would be difficult with such a local approach, since at this point the additional thickness is stored by element index starting at each LINK vertex (compare Fig. 9). Therefore, the information needs to be transformed to a parameter that could be shared globally. To accomplish this, the required element's additional thickness is stored at certain positions along the edge. These positions are normalized so that the start of each edge has a position value of 0 and the end of an edge, a value of 1. These positions are then called "relative positions of the edge" and the relation between relative position and additional thickness is defined as "relative position map" (compare Fig. 9). In this manner, the superposition of different pairs of edges as well as thickness information at the start and the end of an edge is simple to store and retrieve.

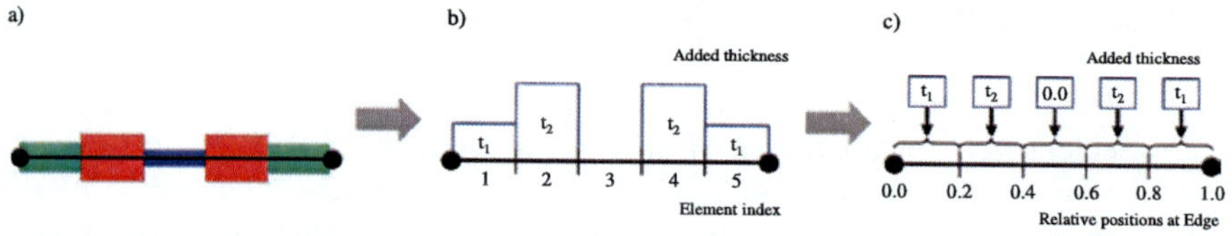

Fig. 9. Abstraction of element's added thickness from FE visualization to relative position map; a) increased element thickness visualization, b) thickness information per element index, c) thickness information at relative positions of edge

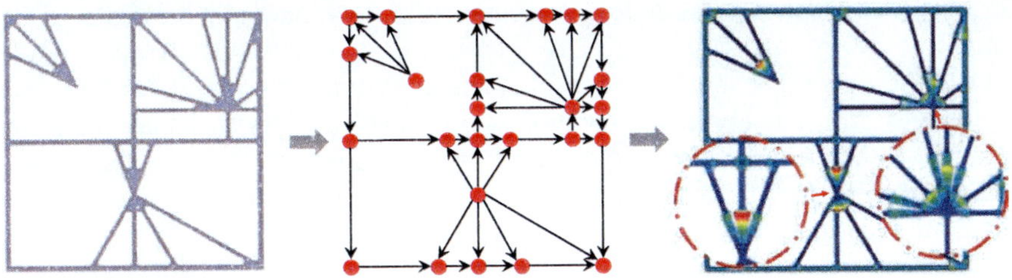

Fig. 10. Artificial 3D CAD geometry, its graph representation, and resulting FE model with accumulation of material (with visualization of shell element thickness and 3D CAD geometry overlay; element lengths are greatly reduced for illustrative purpose)

The appropriateness of the presented algorithm can be shown by applying it on an overly-complex geometry with several wall intersections, several wall angles, and different thicknesses, as shown in Fig. 10. The element length size is greatly decreased for visualization purposes only. Therefore, the thickness to element length ratio is not relevant in this case.

A problem caused by the discretization of additional thicknesses is that if the mesh is not equally distributed (e.g., triangular elements, irregular spline paths), big steps in the thickness between neighboring elements may occur, causing a non-smooth looking distribution of the thicknesses (compare Fig. 11a)). In order to solve this issue, a smoothing option is added to the algorithm, which is simply the interpolation of the thickness information between two positions of the relative position map. Given that only the interpolation between two consecutive points would make sense, a linear interpolation is the most intuitive approach. This interpolation gave good results (compare Fig. 11b)). However, when using this interpolation in a regular mesh, some steps in thickness between neighboring elements were seen due to the sensibility of the linear interpolation. As an alternative, a cubic interpolation was done using cubic Hermite splines, by assuming that the gradient at the interpolation

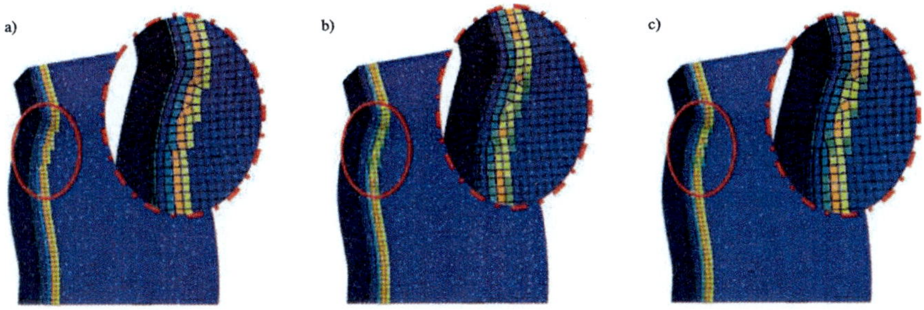

Fig. 11. Smoothing of added element thickness: a) no smoothing, b) linear smoothing, c) cubic smoothing. Element thickness increase is indicated by colors from blue to orange

points is zero (which is an eligible assumption since no variation is desired near these points).

One last aspect that should be considered is that having a single value for all radii may lead to bigger material accumulations for pairs of edges that have very small angles enclosed between them. A CAD designer would try to avoid this, since it would lead to manufacturing problems and a far from optimal mechanical behavior. Unlike the algorithm for rounded corners, as shown in Sect. 2.2, the use of perimeter length for the arc would not be an appropriate solution since the result is not as easy to identify as in the rounded corner case where a perimeter proportional to the FE length would lead to smooth looking geometries. In general, a CAD designer would try to maintain a single radius for all fillets. It should be noted that small material accumulations for obtuse angles is not a problem. Therefore, it is sufficient to include a scaling factor to the given radius in case the angle enclosed between a pair of edges is below a certain threshold. The default threshold value is set to 30°. All mentioned steps are repeated for each continuous pair of edges at all vertices of the graph where at least three edges converge. Once all information is calculated, the center position of each element can be compared against the stored thickness information at each relative position of the edge, and the appropriate element's thickness is assigned to each element. The algorithm's flow chart is shown in Fig. 12.

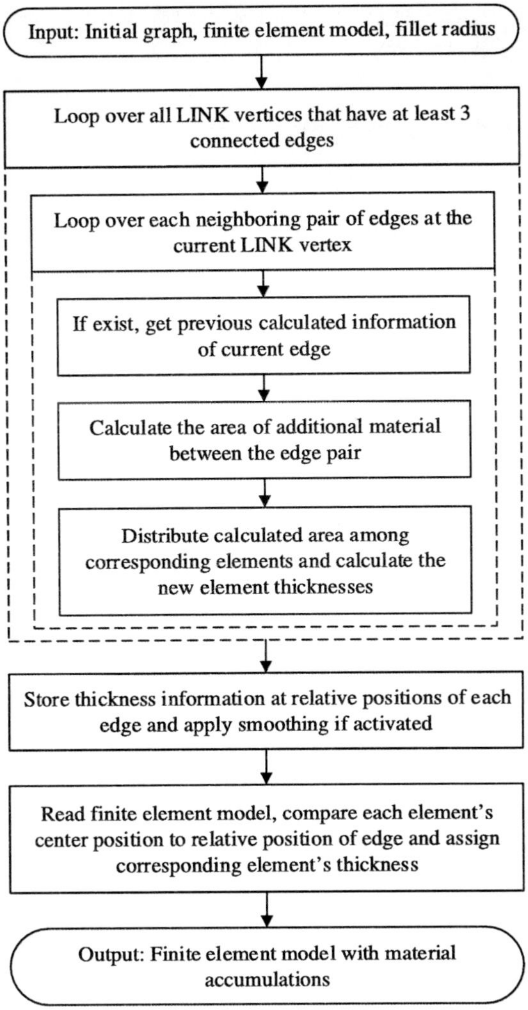

Fig. 12. Flowchart of algorithm for material accumulations

3 Application of Algorithms on Structures in Crash Load Cases

In order to analyze the influence of applying the two algorithms presented in the preceding chapter onto typical vehicle crash structures, two load cases with sub models are investigated.

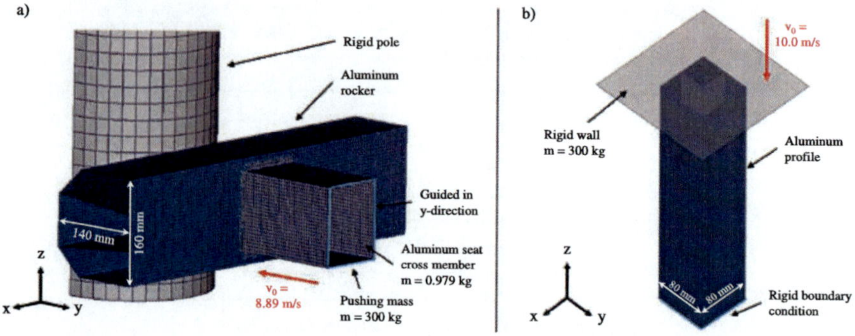

Fig. 13. Investigated crash load cases: a) Lateral load case—pole impact on aluminum rocker structure and b) axial load case—drop hammer impact on aluminum profile structure

The first structure (compare Fig. 13a)) is an aluminum vehicle rocker under a lateral pole impact. The second structure (compare Fig. 13b)) is an aluminum extrusion profile under axial compression load in a drop hammer test. This test can be related to crash structures in the vehicle front and rear end such as crash boxes and longitudinal members. For both load cases the algorithms to modify the structures' geometry are applied separately as well as together and are compared to a reference profile without any modifications. As an explicit finite element solver, PAM-CRASH 2018 is used.

3.1 Rocker Pole Impact (Lateral Load Case)

The structure used in the lateral load case example is an aluminum vehicle rocker under pole impact (compare Fig. 13a)). It is supported by a seat post cross member also made of aluminum with a wall thickness of 5 mm. Both aluminum structures are made of AA6XXX alloy with a yield strength of 280 MPa. At the free end of the seat post, a ridged body with the mass of 300 kg is connected. The vehicle substructure impacts at a speed of 8.89 m/s into the rigid pole. The movement is guided through a boundary condition at the free end of the seat cross member which only allows nodal displacement in y-direction. The substructure is discretized by shell elements with an edge length of 5 mm.

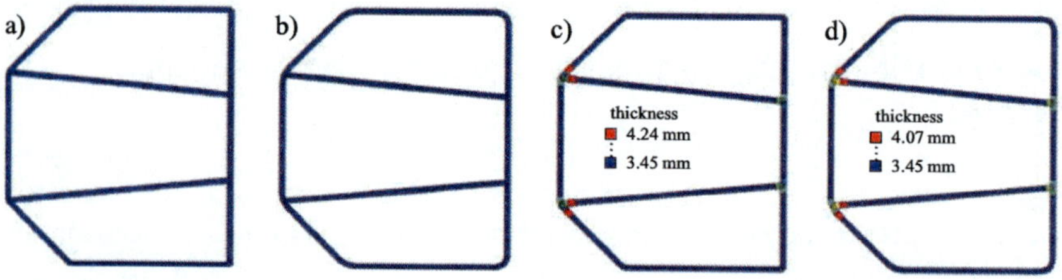

Fig. 14. Rocker profile cross section of finite element model (shell thicknesses are visualized and colored): a) reference profile without modifications, b) profile with rounded corners, c) profile with material accumulations and d) profile with rounded corners and material accumulations

Figure 14 shows the different profile geometries investigated. In this figure the shell element thicknesses are visualized. For a better recognizability they are also colored from blue (3.45 mm) which is the base thickness of the profile to red (4.24 mm for c) and 4.07 mm for d)) which are the biggest shell thicknesses at points of material accumulation. For rounded corners, the arc length is set to 10 mm, while the radius for material accumulations is set to 3 mm with smoothing and acute angle treatment activated.

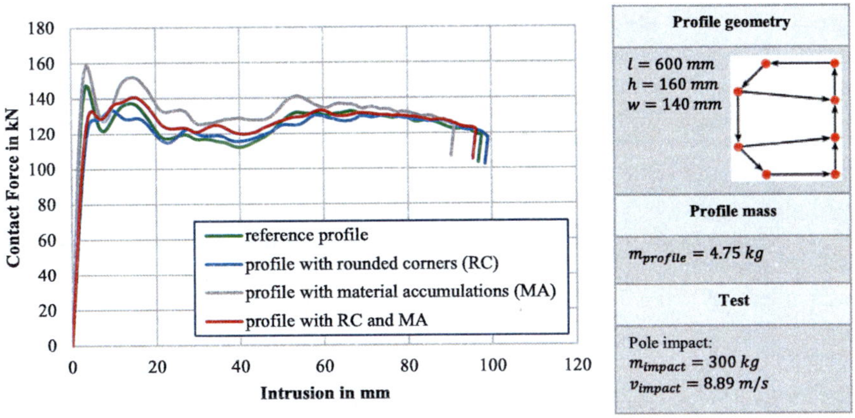

Fig. 15. Contact force—intrusion curve for an aluminum profile under pole impact with consideration of different geometrical details

The resulting contact force—intrusion curves for the different profile cross section geometries under pole impact are shown in Fig. 15. It can be noticed that rounded corners weaken the structure, especially in the first crash phase where the initial force peak is decreased by 13%. In contrast, material accumulations increase the stiffness of the profile significantly (due to the additional stiffness at the intersection points) in the first half of the crash event and therefore decrease the intrusion by 5% compared to the reference profile. When both, rounded corners and material accumulations, are applied, the intrusion only decreases slightly since the weakening effect of the rounded corners and the stiffening of the material accumulations almost compensate each other. Nevertheless, there is still a decrease of 11% for the initial force peak.

3.2 Drop Hammer Impact (Axial Load Case)

For the axial load case a square, four chamber profile made of AA6XXX aluminum alloy with a yield strength of 280 MPa is investigated. The profile is impacted at a speed of 10 m/s by a rigid body plate with a mass of 300 kg. The profile is fixed at the opposite side of the impact. The profile is discretized by shell elements with an edge length of 5 mm. The motion of the impacting rigid plate is guided in z-direction. The contact force is measured between the rigid impact plate and the profile. The load case is shown in Fig. 13b).

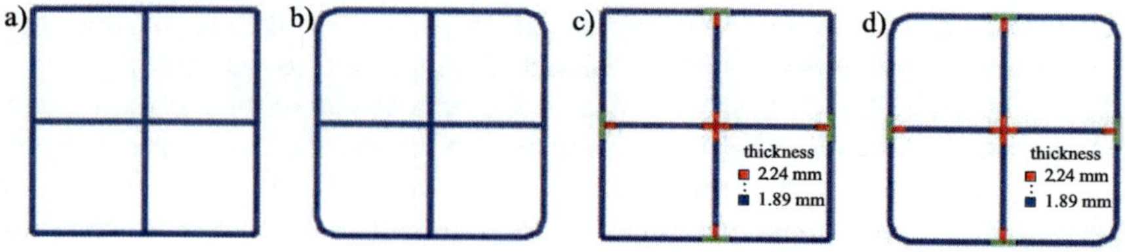

Fig. 16. Profile cross section of finite element model for the axial load case (shell thicknesses are visualized and colored): a) reference profile without modifications, b) profile with rounded corners, c) profile with material accumulations and d) profile with rounded corners and material accumulations

Figure 16 shows the different profile cross sections investigated. As with the rocker profile, the shell element thicknesses are visualized and colored. Blue indicates the base thickness of 1.89 mm, while red indicates the maximum shell thickness of 2.24 mm at the profile wall intersection points with material accumulation. For rounded corners, the arc length is set to 10 mm, while the radius for material accumulations is set to 3 mm with smoothing and acute angle treatment activated.

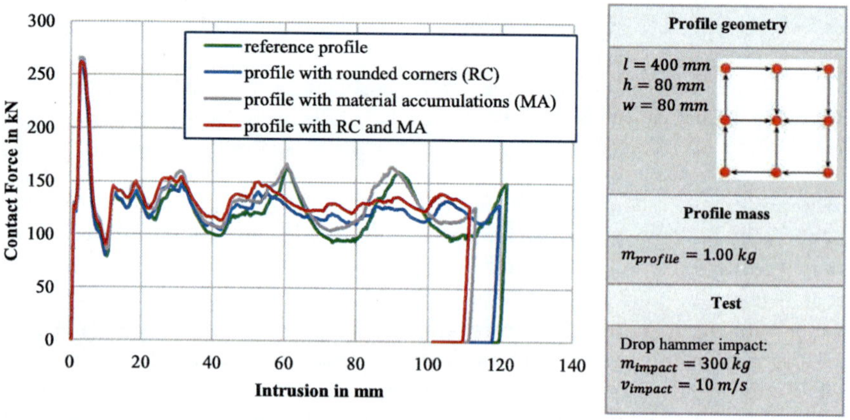

Fig. 17. Contact force—intrusion curve for profile axial impact with consideration of different geometrical details

The resulting contact force—intrusion curves for the different cross section geometries are shown in Fig. 17. The initial force peak is quite similar for all different variations since in this specific case it only depends on the impacted area (compare stress wave propagation, which leads to a critical stress after one reflection at the rigid boundary condition [11]), which is only slightly modified by both algorithms. The application of rounded corners only slightly decreases the intrusion in this load case, although it changes the buckling behavior of the profile which is indicated by decreased force amplitudes for the second half of the contact force—intrusion curve.

When material accumulations at profile wall intersection points are applied, the intrusion is decreased by 7.5% while the buckling mode is similar to the reference profile. These phenomena could be related to investigations presented in [12], where gradual thickness increases at profile intersections and corners lead to higher specific energy absorptions (which can be linked to less intrusion). When both modifications are combined, also their effects are combined, which results in a decreased intrusion with a different buckling mode compared to the reference profile.

4 Conclusion

In this paper, two graph based algorithms to enhance mid-surface design fidelity of finite element models of extrusion profiles are presented. The first algorithm automatically applies rounded corners to extrusion profile's cross section geometries by modifying a graph representation of it, prior to the automated finite element model creation. The second algorithm accounts for material accumulations at profile wall intersection points by modifying the finite element thicknesses close to these points based on graph calculations. The algorithms can be applied in a sequential process, where rounded corners have to be applied before material accumulations. Nevertheless, they can also be used separately. Both algorithms fulfill the requirements regarding a robust and fully automated process as well as having few input parameters. Furthermore, both algorithms have proven their ability to increase the quality of the finite element representation of extrusion profiles and therefore increase the accordance between physical tests and simulation results in terms of geometry and result accuracy. It is shown, that the influence of these modifications is greater for profiles under axial impact load than under lateral impact, where the effects have a tendency to cancel out each other. The algorithms have been implemented in the automatic model building process of the Graph and Heuristic based Topology Optimization where they increase the simulation result's quality.

References

1. Kim, H.: New extruded multi-cell aluminum profile for maximum crash energy absorption and weight efficiency. Thin-Walled Struct. **40**, 311–327 (2002)
2. Friedrich, H.: Leichtbau in der Fahrzeugtechnik. 2. Ausgabe. Springer Vieweg, Wiesbaden (2017)
3. Thakur, A., Banerjee, A., Gupta, S.: A survey of CAD model simplification techniques for physics-based simulation applications. Comput. Aided Des. **41**(2), 65–80 (2009)
4. Stjepandić, J., Wognum, N., Verhagen, W.: Concurrent Engineering in the 21st Century. Springer International Publishing, Switzerland (2015)
5. Fyllingen, Ø., et al.: Brick versus shell elements in simulations of aluminium extrusions subjected to axial crushing. In: 7th European LS-DYNA Conference, Salzburg, Austria (2009)
6. Fyllingen, Ø., Hopperstad, O.S., Langseth, M.: Stochastic simulations of squared aluminium tubes subjected to axial loading. Int. J. Impact Eng. **34**, 1619–1636 (2007)

7. Ortmann, C., Schumacher, A.: Graph and heuristic based topology optimization of crash loaded structures. J. Struct. Multidisc. Optim. **47**, 839–854 (2013)
8. Ortmann, C.: Entwicklung eines graphen- und heuristikbasierten Verfahrens zur Topologieoptimierung von Profilquerschnitten für Crashlastfälle. Dissertation. Shaker Verlag, Aachen (2015)
9. Ostermann, F.: Anwendungstechnologie Aluminium. 2. Ausgabe. Springer Berlin Heidelberg (2014)
10. Mital, A., et al.: Product Development: A Structured Approach to Consumer Product Development. 2nd edn. Elsevier (2015)
11. Schumacher, A.: Optimierung mechanischer Strukturen: Grundlagen und industrielle Anwendung. 2. Ausgabe. Springer Vieweg, Berlin (2013)
12. Sun, G., et al.: Crashing analysis and multiobjective optimization for thin-walled structures with functionally graded thickness. Int. J. Impact Eng. **64**, 62–74 (2014)

Author Index

K. Dröder and T. Vietor (Eds.): *Technologies for economic and functional lightweight design*, Zukunftstechnologien für den multifunktionalen Leichtbau, pp. 383–384, 2021. https://doi.org/10.1007/978-3-662-62924-6